CORROSION

FOR SCIENCE AND ENGINEERING

CORROSION

FOR SCIENCE AND ENGINEERING

Kenneth R Trethewey BSc PhD CChem MRSC FICorr CEng FIMarE

Senior Lecturer in Engineering Materials
University of Southampton
Southampton, UK
[First and Second edition]

John Chamberlain CEng MIM

formerly Principal Scientific Officer
Ministry of Defence, UK
[First edition]

 LONGMAN

Addison Wesley Longman Limited
Edinburgh Gate
Harlow, Essex
CM20 2JE, England
and Associated Companies throughout the world

First edition published 1988
Second edition published 1995
Reprinted 1996

British Library Cataloguing in Publication Data
A catalogue entry for this title is available from the British Library.

ISBN 0-582-238692

Library of Congress Cataloging in Publication Data
A catalog entry for this title is available from the Library of Congress.

Set in 10 on 12pt Monotype Times by 6
Produced by Longman Singapore Publishers (Pte) Ltd.
Printed in Singapore

CONTENTS

Contents

Contents

PREFACE TO THE SECOND EDITION

The first edition of *Corrosion* was received very well. Many readers have provided constructive comments for which we are grateful and upon which we have tried to act, within the usual commercial constraints.

Corrosion continues to create a serious drain on resources, and Chapter 1 has been completely rewritten to provide the latest information. A small addition has been made to Chapter 2 to help readers understand the section on impedance in Chapter 4, Chapter 3 is essentially unchanged.

The theory of corrosion in Chapter 4 has been much revised, whilst continuing to bear in mind that this is a first text. We have tried to retain the mix of practical and theory so that students can relate theory to simple laboratory experiments. Since the last edition, electrochemical impedance spectroscopy has become established as a standard technique. A good description requires much more space than we have allowed, however, we hope the treatment is enough to inform students of what is involved.

Omission of microbiologically induced corrosion from the first edition generated much comment, so we have endeavoured to correct this in Chapter 6.

The localised corrosion forms described in Chapter 7 have been considerably expanded, particularly to update some of the new and very useful techniques being used. Electrochemical noise, whilst still very new and little understood, is here to stay and we felt it an important topic to include.

Chapter 9 is largely unaltered, although an example of fatigue has been added. Similarly Chapter 10, except for a revision of the cases.

Chapter 11 is almost wholly new and is very much a personal statement based upon several years of research. We propose some new ideas about corrosion management on the basis that graduates will enter industry and be faced with an array of corrosion-related problems. We feel that our new perspective on the relationships between people, materials and environments will enhance the modern systems engineering approach. To some extent we are looking into the future, but the way ahead is reasonably clear.

The remaining chapters have all been revised and updated where necessary; it is most surprising how quickly a photograph can look dated. Particular reference has been made to some excellent commercial literature, notably from Sigma Coatings and Haynes International. Grateful acknowledgement is made to these companies for permission to delve deep into their excellent sources of information.

On a personal note, in the ten years since we last began this project, much has changed. One of us (JC) has retired and the other has seen his career wax and then wane with the announced closure of the Royal Naval Engineering College. As we said in the preface to the first edition, the work described in this volume has been very much influenced by activities sponsored by RNEC and we make no apology for the occasional bias to marine topics. We will both retain a soft spot in our hearts for the institution which has treated us so kindly. But life goes on, and KT looks forward immensely to a new start at the University of Southampton which, through its dean of engineering and applied science, Professor Chris Rice, has recognised the value of what has been achieved at RNEC.

As before, many people have helped in the preparation of this new edition. Kathy Hick and her team at Longman have been thoroughly professional, as usual, for which we thank them. Derek Sargeant and David Marsh have been of invaluable assistance in ways too numerous to mention, while Simon Phillips has contributed a great deal to the work on polarisation curves. Our old friends Dick Holt and Geoff Backhouse at Subspection Ltd have again provided useful information for Chapter 16, and Oliver Seibert, who originally provided the stimulus to adopt an approach based on case histories, was kind enough to allow further use of some of his material. A special word of thanks is offered to Professor Pierre Roberge at the Royal Military College, Kingston, Canada, for being such a source of inspiration, a long-suffering sounding board and a great personal friend. His influence on this work has been much greater than he realises.

PREFACE TO THE FIRST EDITION

This book is an attempt to present in simple form a subject which affects every one of us. Without the use of metals, the advance of society would not have occurred, yet we allow our most valuable natural resources to be wasted through corrosion. In many instances, corrosion is inevitable, but it can be controlled. Engineers have a major contribution to make in reducing the unnecessary levels of corrosion but it is the responsibility of everyone to ensure that society uses its metals to the best advantage. An understanding of corrosion and its control is important for all of us, if only for the selfish reason that it saves us money. Studies of the effects of corrosion in society have shown that we seem unable to learn from our mistakes. This problem can be properly addressed only through education and it is therefore the aim of this book to bring to the attention of as many people as possible the reasons for the occurrence of corrosion and the methods of controlling it.

The subject material is a slight expansion of a course of 25 lectures given to engineering undergraduates. Much of the additional material has been added to enable the book to reach a wider readership. In particular, Chapter 2 contains some elementary material drawn from chemistry, physics, metallurgy and electrical science in an attempt to draw together students from various disciplines who may have a weakness in one or other of the subject areas. Many undergraduate students should be confident enough to omit most of this chapter.

Corrosion is a very practical subject and emphasis is placed on experiments, case histories, questions and answers. The experiments in Chapter 3 are simple and can be performed with the minimum of materials and equipment, but are important to the development of an understanding of corrosion. The authors regard them as an essential part of any teaching course. They have been carefully selected and written in such a way that they can be performed with no previous knowledge of the subject. Students should find them both entertaining and instructive. Alternatively, the teacher may prefer to use them as class demonstrations, in which case they may be easily refined with more dedicated equipment. Other, rather more specific but equally important experiments are placed at intervals throughout the early chapters.

The authors make no apology for the simplifications to the corrosion theory. Our experience has shown that the theory given in many other texts defeats all but the brightest students. We have therefore attempted to reduce what is a very complicated subject to a limited number of logical concepts, presented in a more relaxed style with the aid of simplified equations and figures. The question-and-answer approach furthers this aim by trying to anticipate students' questions at various points of the discussion.

The case histories which are reported have been collected from many sources, but a large proportion are from the museum collection at the Royal Naval Engineering College. These, together with the authors' own experience, often relate to marine corrosion, but it is considered that any textual bias towards the marine environment should not affect students' appreciation of corrosion in other environments. A good number of the case histories have been reproduced with the kind permission of Dr Oliver Seibert from his two excellent papers on classic blunders in corrosion protection. The authors found these papers quite inspirational and are sincerely grateful to Dr Seibert for his assistance.

Section 18.1 contains a selection of worked examples, but solutions to the questions in Section 18.2 have not been provided. In most of these cases, the solutions may be lengthy, as well as subjective, and it is expected that these will be discussed with the course teacher.

Some readers may consider that the space allocated to high temperature corrosion is disproportionately small. In the authors' opinion this area of study is more specialised and, for a text which is intended as a first level of study, the treatment given is deemed to be appropriate.

The authors wish to express their sincere gratitude to all those who have helped in the preparation of this book. Indeed, it might never have been written without the first suggestion by Clive Holly to whom the authors are indebted. In particular, acknowledgement is made to the captain, Royal Naval Engineering College, Manadon, for permission to use facilities and materials. Support and encouragement by the dean of RNEC, Captain G C George BSc MSc CEng MIM ndc Royal Navy, are gratefully acknowledged. Dr D J Tighe-Ford BSc PhD CBiol MIBiol kindly allowed us to use the caption to the frontispiece, as well as providing the focus for many lively discussions. Mr D A Sargeant CEng DipEE MIERE AMIEE carefully checked the electrical science section, Dr D L Bartlett BSc MSc PhD assisted with numerous theoretical aspects of the work, and Lieutenant Commander S J Bates BSc PhD Royal Navy advised on the contents of Chapter 9. Professor R N Parkins BSc PhD DSc FIM of the University of Newcastle upon Tyne kindly gave constructive criticism of Chapter 10, while Dr Peter Scott of the Materials Development Division, AERE, Harwell read the section on corrosion fatigue and kindly gave permission for the reproduction of some of his work. Dr Geoff Backhouse of Subspection Ltd provided useful material concerning cathodic protection, while Lieutenant-Commander J M Harrison BSc MPhil Royal Navy offered much help with the section on hot corrosion.

Dr Michael Rodgers was sufficiently daring to accept the original idea and then proceeded to offer every possible assistance in the preparation of the book, including psychological support during moments of authors' desperation. The authors are very grateful to him and his team at Longman. Thanks are due also to an anonymous British referee for checking the whole text so carefully, and to Professor M Marek of the Georgia Institute of Technology who read the first draft and made some very helpful suggestions for improvement. The greatest debt of gratitude is owed to Captain J N McGrath MSc PhD MIM Royal Navy, without whose constructive criticism and constant encouragement the book would have been much the poorer.

Finally, we thank our wives for their patience and support, and our friends and colleagues in the Engineering Materials Section at RNEC for forming such a pleasant social and professional environment over many years.

1 CORROSION AND SOCIETY

Lay not up for yourselves treasures upon earth, where moth and rust doth corrupt and where thieves break through and steal.
(Matthew 6: 19)

1.1 THE LESSONS OF HISTORY

In 1763 a report fell upon a polished oak desktop in an office of their Lordships of the Admiralty. Although the report had only modest significance at the time, its concise and detailed script is now seen as one of the earliest practical solutions to corrosion. But in common with many subsequent studies, its conclusions were frequently disregarded.

Two years before the appearance of the report, the 32-gun frigate, HMS *Alarm* (Fig. 1.1), had had its hull completely covered with a thin copper sheathing. The purpose of the sheathing was twofold. Firstly it was intended to reduce the considerable damage caused by the wood-boring shipworm, and secondly the well-established toxic property of copper was expected to lessen the speed-killing barnacle growth which always occurred on ship hulls. After a two year deployment to the West Indies, HMS *Alarm* was docked in order to examine the effects of the experiment. It was soon discovered that the sheathing had become detached from the hull in many places because the iron nails which had been used to fasten the copper to the timbers had been much 'corroded and eat'. Closer inspection revealed that some nails, which were less corroded, were insulated from the copper by brown paper which was trapped under the nail-head. The copper had been delivered to the yard wrapped in brown paper, which was not removed before the sheets were nailed to the hull. So the obvious conclusion, the one contained in the report of 1763, was that iron should not be allowed direct contact with copper in a sea-water environment if severe corrosion was to be avoided. This type of corrosion by two dissimilar metals in contact was to become known as **galvanic corrosion**, though it is more precisely called **bimetallic** or **dissimilar metal corrosion**.

It would seem that the first occasion in which this advice went unheeded was in 1769 when Commodore the Hon. John Byron began a circumnavigation of the globe in his coppered ship, *Dolphin*. In addition to the frightening prospect of an uncharted Coral Sea reef scything through the hull, 'Foul Weather Jack' lost much sleep because of a repeated thump, thump, thump below the stern

1

Fig. 1.1 HMS Alarm, the Royal Navy frigate which in 1763 was the subject of the first recorded study of bimetallic corrosion after iron nails had been used to secure copper sheathing to the hull.

windows of his cabin. He recorded in his log that he feared the *Dolphin's* very loose rudder would drop off at any time, for its iron pintles, which were in contact with the copper sheathing, were corroded away to needle thinness. With no prospect of repair in such a dire event, this was a quite unnecessary worry in addition to the multitude of others with which he had to contend.

Since then, bimetallic corrosion and many other forms of corrosion have continued to cause service failures, despite their apparently well-publicised effects. In 1962 a report fell upon a plastic desktop in the British Ministry of Defence. This study bore a striking resemblance to its bicentennial antecedent for it revealed that the copper alloy end-plate had fallen off a sea-water evaporator in a submarine because the steel bolts with which it was secured had effectively dissolved through galvanic action (Fig. 1.2). In 1982, the nose wheels failed on two Royal Navy Sea Harriers which had returned from the Falklands War (Fig. 1.3). Studies showed that the same galvanic action which had been so clear to the scientists who had studied HMS *Alarm* 219 years earlier had occurred between the magnesium alloy wheel hub and its stainless steel bearing (Fig. 1.4).

These three cases are from the Royal Navy, but the Royal Navy is by no means unusual in its proportion of corrosion failures. Examples abound throughout the engineering world, and like many examples, they illustrate quite clearly that we do not always learn the lessons of the past.

The term 'corrosion' derives from the Latin, *rodere*, meaning 'to gnaw', in the context of rats; and 'corrodere' means 'to gnaw to pieces'. A very early study of corrosion was made by Robert Boyle. In a book published in London in 1675,

Fig. 1.2 The remains of steel bolts which in 1962 had been used to hold a copper alloy end-plate on to an evaporator aboard a Royal Navy submarine.

Fig. 1.3 A Royal Navy Sea Harrier which suffered nose-wheel collapse because of bimetallic corrosion between the stainless steel bearing and the magnesium alloy wheel.

'*The Works of the Honorouble Robert Boyle, Esq. Epitomized,*' we find that Chapter XVIII of the appendix to Book V, Part IV is entitled, 'Of the Mechanical Origin of Corrosiveness'. Chapter XX of the same sequence is entitled, 'Of the Mechanical Origin of Corrodibility' . [1]

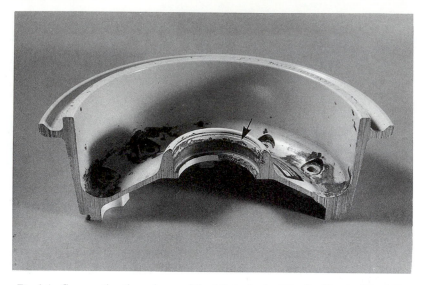

Fig. 1.4 Cross-section through one of the failed wheels of the Sea Harrier (Fig. 1.3). Corrosion caused by the galvanic couple between the magnesium alloy and the stainless steel bearing is evident in the centre of the picture (arrowed).

To the great majority of people, **corrosion** means **rust**, an almost universal object of hatred. Rust has more recently referred specifically to the corrosion of iron. But corrosion is a destructive phenomenon that affects almost all metals. Although iron was not the first metal used by humans, it has certainly been the most used, and must have been one of the first with which serious corrosion problems were obtained. It is not surprising, therefore, that the terms corrosion and rust are almost synonymous.

The great Roman philosopher, Pliny (AD 23–79) wrote at length about *ferrum corrumpitur,* or 'spoiled iron' [2] for by his time the Roman Empire had been established as the world's foremost civilisation, a distinction due partly to the extensive use of iron for weaponry and other artefacts. To the fighters of old, rust was something of a mixed blessing. In the eleventh century, a Norman knight, William de Lacey, lost his way during a hunting expedition into the thickly wooded and swampy Vale of Ewas, in Wales. [3] He came across the ruins of St David's hermitage, whereupon, overcome by an urge to mend his sinful ways, he decided to dedicate the remainder of his days to religious contemplation and rebuilding the chapel. Legend has it that he never for the rest of his life removed his armour. One explanation for this strange behaviour was that it was a self-imposed penance. More likely, however, is that he was prevented from doing so because of corrosion brought about by the dank atmosphere of the valley.

Corrosion of arms and armour has also been advantageous. The techniques of blueing and gilding were frequently used to protect steel objects, for it was found that the application of a variety of heat treatments created highly protective films of oxide (rust). [4] These, with skill, could turn functional weaponry into beautiful works of art. It has always been considered by many that the green patina on copper roofs and other ornamental objects is desirable, rather than unsightly. A well-known lover of rust, sculptor Anthony Caro, spent many years producing abstract steel sculptures. Initially they were painted brown, the colour of rust, but he later allowed them to rust naturally and then preserved it by varnishing. Few people, however, would subscribe to the view that rust is beautiful.

The Romans must have been vexed by the susceptibility of iron to rust, for, in true scientific fashion, Pliny asked himself why iron should corrode more easily than other metals. He did not know how to investigate the problem experimentally, but he arrived at a metaphysical solution: because iron is both the best and the worst of man's servants. Although very useful domestically, it is also the metal of war, slaughter and brigandage. Writing about iron arrows, Pliny said, 'It is a great evil that to enable death to reach human beings more quickly we have taught iron how to fly.' He believed that the nuisance of corrosion compensates for the advantages of the metal because 'the same benevolence of nature has limited the power of iron by inflicting on it the penalty of rust, and the same foresight has made nothing in the world more mortal than that which is most hostile to mortality.'

A remarkable consequence of his philosophy was the technique of corrosion control by religious ceremony. He recorded that this had been used to protect the chains of a suspension bridge built for Alexander the Great, although he was sceptical about its chances of success because the method had failed on previous occasions.

Had Pliny been transported to the twentieth century, he may well have concluded that **Corrosion Costs Society**. It is unfortunate that, in modern usage, the term 'cost' usually implies only financial penalty. In fact, corrosion costs society in three ways:

1. It is extremely expensive financially.
2. It is extremely wasteful of natural resources at a time of increased concern over damage to the environment.
3. It causes considerable inconvenience to human beings and sometimes loss of life.

We shall consider the first item by describing corrosion economics. The next two items will be combined in Section 1.3.

1.2 THE EXPENSE OF CORROSION

Those who work in the field of corrosion frequently try to point out how much it costs as a percentage of a nation's economy. The approximate annual cost of corrosion in the United States was first estimated in 1949 by Uhlig to be $5 billion, 2.1% of GNP. [5] In 1971, a major study of the cost of corrosion in the

5

United Kingdom was carried out by the Government Committee on Corrosion and Protection. [6] The committee concluded that the total cost to the national economy was a staggering £1,365 million (1971 prices), about 3.5% of the GNP. Of this, it was said that about one-quarter could be saved by better and wider use of well-established corrosion protection techniques. The survey did not include the agricultural industry; this was covered in 1981 in a report by the University of Manchester Institute of Science and Technology (UMIST), which reported that corrosion was costing the agricultural industry about £600 million per year and that about half of this could be saved by existing corrosion control technology. [7] By 1975, the estimated cost of corrosion in the United States had risen to $70 billion, 4.2% of GNP. [1, 8]

The enormity of these corrosion costs is at first surprising. How can they be so great? A popular misconception is to reckon the cost of corrosion as the cost of replacement, but there are almost always additional costs. They are known as **indirect costs** and may occur for any of the following six reasons.

Lost production shutdown or failure

Case 1.1:   In 1977 it cost £5,000 per day to take a typical 400 kV transmission line out of service to deal with corrosion damage.

A small corroded pipe in a chemical plant may be replaced for just a few pounds, but the loss of product during the time that the process is stopped can run into thousands of pounds an hour. In the worst possible case the whole plant may be taken out and loss of life may occur.

Case 1.2:   In 1974, at Flixborough, a cyclohexane reactor in a train of six suffered a leak because of corrosion. A temporary assembly was introduced to bypass the leaking reactor but it failed catastrophically before the leak could be repaired. Twenty-nine people died in the explosion.

High maintenance costs

The selection of the material or protection system which optimises corrosion resistance for a given component lifetime greatly reduces the need for costly maintenance during that lifetime. Though the initial cost may be higher, the overall cost is usually much less. The essential factor is the choice of the correct design lifetime. Problems arise when a time scale chosen at the design stage to fit the various cost parameters is extended later in the life of the component. Alternatively, even though maintenance schedules are laid down, they are not adhered to. Both can have disastrous consequences.

Case 1.3:   Four men died and five were seriously injured at the construction site of a UK power station when the single suspension rope of the hoist cage in which they were travelling broke at a point weakened by corrosion and lack of lubricant. The safety gear failed to operate, also because of corrosion, and the cage fell more than 30 m to the bottom of a 60 m shaft. At the inquiry it was stated that the required six-monthly inspection was overdue.

Problems have frequently arisen in the consumer industry when the manufacturer and the consumer have differed in their opinions as to the desirable lifetime of the product. Manufacturers may deliberately select inferior materials for their products using the philosophy that

Manufacturing costs are kept to a minimum.

The product is competitively priced.

The product lifetime is short.

Turnover is maintained at a high level.

Large profits ensue.

In the event, the consumer always pays the price. This applied, in particular, to the transport industry where for several decades corrosion resistance was consistently poor. The Hoar Committee [6] reported that corrosion costs to the transport industry in 1971 were the greatest of all the industries they examined — £350 million per annum. In their defence, car and bus manufacturers blamed the increased use of road salt and the reluctance of the public to pay for increased protection as the major causes of corrosion. However, the committee highlighted the case of exhaust systems as an important example. Here, the replacement of the unprotected mild steel exhaust by, perhaps, an aluminised steel system, it was said, would lead to an increase of life from the then average of two years to an expected six years. For very little extra cost the national annual saving was estimated at £55 million on this item alone.

Such a manufacturing philosophy was shown to be flawed as the power of consumers grew. Those who found the performance of a product unsatisfactory bought the competitor's product. Today, consumer associations have the power to demand quality improvements from offending companies, as well as compensation for the inconvenience of their members. Customer relations has become an important part of all large businesses.

Compliance with environmental and consumer regulations

In Canada, in 1980 the Ontario Rusty Ford Owners Association obtained substantial compensation from the courts for corrosion to their cars. This kind of event heralded a new age of indirect costs arising from consideration of environmental and consumer needs, further discussed in Section 1.3.

Loss of product quality in a plant owing to contamination from corrosion of the materials used to make the production line

In other areas of industry, consumer demands for better quality control have increased overall production costs. In general, the heavy-chemical, oil and petrochemical industries have been found to be more corrosion-conscious than the pharmaceutical industry. It was suggested that this was because of the former industries' experience in the use and storage of highly corrosive substances, and a consequent tendency to overdesign. A corrosion specialist was employed in less than 10% of the pharmaceutical companies questioned, and in the majority of cases problems were dealt with by a maintenance engineer. The industry was found to be very conscious of product quality, though it suffered higher corrosion costs than were necessary.

In the food industry it was reported that companies were suffering from a wide variety of corrosion problems, which had led to high maintenance costs, but there was a lack of experience to tackle them. The food industry was said to be conservative and reluctant to change processes and equipment that had proven satisfactory products.

High fuel and energy costs as a result of steam, fuel, water or compressed air leakage from corroded pipes

Serious problems arose in the United States where in 1981 the report of the Nuclear Regulatory Commission to Congress in Washington stated that the 'vast majority' of steam-generating pressurised water reactors were suffering failures of stainless steel cooling tubes. While stressing that there was no danger to the public it was estimated that the maintenance bill would top $6 billion.

Extra working capital because of increased labour requirements and larger stocks of spare parts

Case 1.4: A medium-sized UK engineering company, believing that it had no problems with corrosion, decided to look deeper into the question. It discovered that corrosion was in fact costing £43,000 a year. By improving materials handling and paying greater attention to stock control and records, the company was able to save over £10,000 of this sum.

The above factors illustrate quite clearly that, to maximise profit, **no manager can afford to ignore corrosion.** Meanwhile, in the immediate future managers will be concerned with other matters than just economics.

1.3   THE SOCIAL IMPLICATIONS OF CORROSION

In the history of the use of materials, the past 150 years have been closely associated with alloys of metals such as iron, aluminium and copper. Our highly developed civilisation could not exist without them. Yet corrosion is their Achilles heel. The degradation of the metallic materials with which we go about our business is very largely due to corrosion, and, in a society which focuses more and more on dollar-cost, we have seen how expensive corrosion is. The reader might be forgiven for asking why the treatment of corrosion has not been pursued as earnestly as a cure for AIDS. This very difficult question is further discussed in Chapter 11. Part of the answer concerns the way society is structured. The scientific and engineering communities are strongly structured and rely for their financing upon long-standing practices. Corrosion falls between the traditional disciplines, so it is frequently considered to be out of the mainstream and somehow less important. Virtually all metals suffer corrosion, so its effects permeate nearly every aspect of human endeavour, and this fact alone makes the study of corrosion and its control *more* important, not less. Thus we find that corrosion is mostly ignored by chemists and electrochemists who, not being familiar with metallurgy, do not wish to use systems involving real alloys. Although studied as a minor part of mechanical engineering courses and an even smaller part of electronic/electrical engineering courses, corrosion is the main reason why engineered systems will ultimately fail. The following sections provide evidence in support of this.

The nuclear industry

The nuclear industry has most of the corrosion problems of other industries and some that are all of its own. Right from the start, the potential for disaster was recognised and tackled head-on by adopting high grade materials in many parts of the systems. Zirconium alloys, among others, were needed and brought their own corrosion problems and solutions. [9,10] Many would argue that the very extent of the nuclear industry, for both peaceful and military uses, is evidence of its success, but the growing worldwide demands for acceptable environmental performance has alienated others to the cause of nuclear power, particularly after events at Three Mile Island and Chernobyl.

The industry has certainly had its share of corrosion costs. For boiling water reactors (BWRs) capacity factor losses due to corrosion problems averaged over 6% between 1980 and 1991, reaching a peak value of 18% in 1982. The corresponding numbers for pressurized water reactors (PWRs) are an average capacity loss of 5% and a peak of 8% in 1982. It is estimated that corrosion problems have cost the nuclear utility industry more than $5 billion since 1980. In addition, repairs and mitigating measures are thought to have cost the average US light water reactor (LWR) more than $1 for every megawatt-hour of electricity produced since 1980, i.e. >$0.5 billion throughout the industry; this has also resulted in radiation exposures of about 100 rem per year. [11]

The unique problem of long-term storage of high-level nuclear waste has occupied many human-years. Corrosion behaviour has long been known to be aggravated under high-level irradiation [12] and the task of designing suitable containers which will maintain high-level waste in a safe condition for ten thousand years has stretched designers to the limits. Projection of behaviour even ten years into the future has been difficult enough, but significant progress has now been made in the resolution of the problem. [13]

Fossil fuel energy systems

The problem of acid rain attracted considerable attention long before a greenhouse effect was identified and shown to involve increased levels of oxides of carbon, nitrogen and sulphur. Acid rain, exacerbated by acidic sulphur gas emissions from fossil fuel power stations, was shown to have a devastating effect on the ecology of regions hundreds of miles downwind. In the United States, the first clean air legislation was enacted in 1967 and, after some further amendments to the law in the 1970s, the Environmental Protection Agency (EPA) was set up to regulate air pollution and specify a rigorous enforcement timetable. Industrial (pulp and paper, municipal incinerators and chemical plants) and coal-fired electric power plants were identified as the primary sources of sulphur oxides and particulates. Since the major source of SO_2 was identified as the coal-fired power plants, the electric power industry was targeted by the EPA for major reductions in these air pollutants. The first flue gas desulphurisation (FGD) or scrubber system at a coal-fired plant in the United States was installed and operational in 1968. By 1995, 110 sites with 261 units had been required to comply with the legislation. [14]

The aerospace industry

The particular demands of the aerospace industry have created a manufacturing environment with a very high and proper regard to optimising corrosion performance, as a result of which the safety record has been exemplary and the efficiencies of operation have been outstanding. In little more than 50 years, passenger travel has progressed from the 50-seater Comet airliner to the point at which 800- and even 1,000-seater aircraft are close to manufacture. In the early days of the world's first jet-engined metal fuselage, there were several fatal crashes from fatigue failure caused by cracks emanating from the window openings. Today, very few examples of fatal accidents are caused by corrosion failure, despite total reliance on high performance light alloys of the fundamentally reactive element, aluminium. It is not that corrosion is unimportant or has in some way been banished from the industry; on the contrary, aerospace applications offer much scope for many and serious corrosion failures. It is simply that the correct materials management practices have been adopted from drawing-board to aircraft maintainer. Many facets are involved: careful materials performance evaluation prior to use; adoption of good design principles; application of effective barrier coatings; regular and efficient inspection and maintenance schedules. All these have created a success story from which other industries could learn.

The electrical and electronics industry

Many might believe the electronics industry would be unaffected by corrosion. Not so. Aluminium is used for tracks on most devices, and aluminium corrodes. There is also a high possibility of a strong galvanic interaction with gold, used for connectors. Many critical systems must operate in quite hostile environments — temperatures ranging from well below zero to 40 °C or humidities up to 100% — and cope with airborne particles or insects. This kind of versatility in corrosion performance is a tall order by any standards.

Corrosion of power cables has been a considerable problem. It was often attributed to 'sheath damage' or 'water in the cable'. A variety of causes of damage to the protective sheath were found, ranging from gunshot damage and pinholes, caused by lightning strikes on overhead lines, to rodent damage — attack by gophers and squirrels — and damage by termites or other insects. Problems with direct buried cable were often solved by burying another to replace the failed cable and the reasons for the failure were not found.

Fifty years ago, the main material for cable sheathing was lead, protected from corrosion by asphalt-impregnated jute coating. Even then, there was an extensive history of corrosion problems. By the end of the 1940s, extruded or drawn aluminium sheaths were in use, soon to compete, in the 1950s, with polythene, although lead continued to dominate the market until the 1960s. In the 1960s an ever increasing number of failures caused by water in the cable led to development of sophisticated methods of sheathing involving coated aluminium and clad metals of copper adjacent to a number of steels. [15]

The marine and offshore industry

The marine environment is probably the most aggressive common environment in which metals are expected to operate. The corrosion performance record is mixed. Startling achievements have been made offshore, particularly since the development of the new generation of deep-water platforms in the North Sea oilfields during the 1970s. This technology has now progressed to the even deeper waters of the Gulf of Mexico and continues to develop. Much more will be said in Chapter 16, but the idea of creating over 200 of the world's largest steel structures, immersing them in violent seas and expecting them to operate with efficiency and safety for over 25 years is both astonishing and true. Furthermore, the record has been one of great success, save only for the Piper Alpha disaster in 1988 in which 167 men died as a result of a fire following a mechanical (not corrosion) failure.

The same cannot be said of the shipping industry. The battle against corrosion was first waged with hulls made of puddled iron, commencing with the construction in the Scottish Clyde shipyard of the fast passage barge, *Vulcan* in 1819. In 1822 the *Manby* became the first iron steamship and the first to make an international voyage under power. Besides corrosion, marine fouling was a serious and continuing problem with which iron hulls had to contend. [16] Despite the difficulties, many of the iron hulls gave long years of service and survived remarkably well, even when lost or abandoned, as the example of the submarine, *Holland I*, showed (Case 13.9).

Despite the recent advances made in materials [17] and coatings technology, [18] civilian merchant fleets have suffered badly from poor investment by owners and builders and there is a long and sorry catalogue of major disasters, many attributable to corrosion in the first instance, caused by poor management at every level.

Case 1.5: The loss of the 170,000 tonne *Derbyshire* in 1980 with all 44 crew members resulted in much speculation that the ship had suffered a major structural failure as a result of corrosion fatigue brought about by a combination of poor initial design and heavy corrosion of the structure. Although proof was never obtained, several sister ships were found to be suffering from identical problems. The ship was only four years old at the time of the loss.

During 1990–91 over 30 large bulk carriers were lost or damaged and more than 300 crew died. A principal surveyor from Lloyd's of London reported that he was commonly seeing corrosion from an original steel thickness of 12 mm down to 3 mm in 10–15 years of life. For decades, unscrupulous owners have consistently cut purchase, operating and maintenance costs, taking advantage of the impotent international regulation of merchant shipping. Thus, in contrast to the success of identical technology in the offshore industry, shipping has suffered from every possible aspect of corrosion and management failure.

Case 1.6: In 1993, the oil-tanker *Braer* went aground in the Shetland Islands to the north of Scotland, discharging the entire contents of her 100,000 tonne cargo onto the fragile ecosystems of the islands. Only the unexpected beneficial influence of a severe storm in breaking up and dispersing the oil saved the islands from economic and biological disaster. The accident resulted from the simple single failure of a marine propulsion system after sea-water was ingested in the storm. Such an event is easy to predict given the poor management practised onboard, but the total lack of backup systems is directly attributable to the cost-cutting practices endemic in the industry.

Navies too have had their problems, although often they have been more attributable to the extreme complexity of modern military systems than to cost-cutting. [19] In the modern climate of reduced defence expenditure, however, this latter aspect of marine corrosion may develop a greater importance than we have seen hitherto.

The automotive industry

Much has been written about cars and their corrosion. [20–22] It has been said that no other consumer product suffers so much from environmental factors. A mild steel car body, for example, suffers from high and low temperatures, from precipitation, and from increasing concentrations of reactive components in the atmosphere and along the road. The car body should survive the grinding effect of splashing gravel, should offer comfortable protection to its passengers, should be easily and economically repairable on damage, should maintain a pleasant appearance throughout the years and, in future, should be easily disassembled for recycling purposes. Meanwhile, its lifetime should be guaranteed. This is a tall order!

In the 1980s, increasing the lifetime of car bodies became more important for manufacturers than control of exhaust emissions or fuel economy. A code for Canada in 1981 specified that a body should last a minimum 1.5 years (60,000 km) before suffering cosmetic (outside in) corrosion, 5 years (200,000 km) for perforation (inside out) corrosion and 6 years (240,000 km) for structural corrosion. The code projected for North America in 1990 was to have no cosmetic corrosion in 5 years and no perforation corrosion in 10 years. This resulted in a rise in the use of precoated steel, especially galvanised steel products which, in 1993, seemed to have satisfied the requirements. [23]

In 1981, passenger cars and commercial vehicles were being scrapped in Europe at the rate of 3 million per year, attributable to one of three causes: accidents, obsolescence and corrosion. Since the amount of corrosion often determines whether a car can be economically repaired, the number scrapped as a result of accidents is affected by the amount of corrosion. McArthur [24] published evidence that there might be a correlation between the number of serious injuries suffered in road accidents and the age (amount of corrosion) of the vehicle. He argued that it was reasonable to expect that a corroded vehicle was less able to absorb the energy of an impact than was a new vehicle.

McArthur also reported the results of a survey which found that the average life of a motor car was increasing, both in total miles run and lifetime (50 per cent survival of initial registrations). The lifetime of the average car was 9.9 years, while for Volvo the figure was 17.9 years. In the United States, Chandler [25] reported that in 1974 the 'median life' of a car was 4.9 years, which in 1984 had risen to 6.0 years, evidence in support of what many people already believed: the motor industry had at last reacted to the public demand for more corrosion protection.

Meanwhile, many car owners, in their rightful attempts to own vehicles which remained rust-free, had resorted to one of the many rustproofing treatments which could be applied to new vehicles. In 1982 a report by the Office of the Attorney General in New York State claimed that consumers were being defrauded of at least $11 million annually because of the poor quality of treatments. A remarkable 83% of all cars inspected failed to meet acceptable standards. Most of the manufacturers' rustproofing warranties were said to contain limitations or conditions that rendered them essentially worthless. [26]

The construction industry

Despite its well-known propensity to rust, iron, whether alone or as a reinforcement in concrete, has been the basis of civil engineering projects since the world's first iron bridge was constructed over the River Severn at Coalbrookdale in England by Abraham Darby in 1780. Many such examples of wrought iron bridges continue to perform perfectly after modest, but regular maintenance.

Fig. 1.5 The Royal Albert Bridge at Saltash in Cornwall. Built largely of wrought iron, by Brunel in 1859, it has spent its lifetime in a salt-laden marine atmosphere, yet still provides the main rail link into Cornwall.

Figure 1.5 shows the Royal Albert Bridge at Saltash, completed in 1859 by Brunel to carry a main railway line across the River Tamar between Devon and Cornwall. The 450 m bridge is made of two 155 m spans of wrought iron, the upper portions of which are comprised of two ellipsoidal tubes, 4 m high and 5.5 m broad. The bridge has remained in good condition with regular maintenance and has been given a clean bill of health until 2035. Brunel's other famous bridge at Bristol, the Clifton Suspension Bridge, has been maintained in safe condition by regular attention, and Telford's Menai Straits Bridge, a suspension bridge in North Wales, has also stood the test of time. First opened in 1826, it was reconstructed from 1938 to 1940 to meet the traffic requirements current at the time, and was the first large bridge to be metal sprayed in the United Kingdom. It provides an excellent example of the good use of protective coatings in a hostile environment, for after four decades the original coating remained undamaged.

In cases such as these, the engineers and designers of the past applied commonsensical principles born out of sound engineering practices. The results were often long-lasting structures which showed good corrosion resistance and were comfortably within the margins of safety to allow normal operation if something unforeseen should occur. In today's society, the burden of difficult financial restrictions and highly competitive tenders has enforced the use of the most advanced and often untried engineering techniques in order to save money. Operating margins have been trimmed to such a fine extent that when a problem arises, as it must in any novel situation, severe penalties ensue. Thus, the supposed advantages of modern engineering design seem to have created a plethora of problems which underpin the conclusions of the Hoar Report. The following examples illustrate this.

Case 1.7: In the United Kingdom, the innovative design of the suspension bridge over the River Severn was heralded with great optimism when the bridge was opened in 1966. Problems were first noticed in 1978 when broken wires within the raked steel hangers were revealed by cracking in the paintwork. The steel deck embodied an innovative design in which aerodynamic closed box sections intended to eliminate the marine atmosphere from inside the boxes. Yet, within a short time, it was found that cracking of the steel deck box-welds had occurred, and by 1983 grave concern was expressed over the future of the main arterial link between England and South Wales. The Department of Transport insisted that, while there may have been some fatigue failure of wires in the suspension/hanger system, the problem had been caused by unexpectedly high traffic loading, aggravated by corrosion, as a result of inadequate protection against the salt-laden atmosphere of the Severn Estuary. Estimates of the cost of repair spiralled upwards and in 1983 reached £30 million. A new bridge is presently under construction (see also Case 13.3).

Case 1.8: Problems were experienced with the 350 m long Pelham Bridge in Lincoln. A dual carriageway supported on 23 rows of columns, it had been constructed in 1957, at a time when it was impossible to foresee the great increase in the use of de-icing salt on roads which was to occur during the 1960s. Penetration of salt had caused severe corrosion of the steel reinforcement in the concrete road bridge decks. The engineer reported that the cost of repairing the faults had proved very high, even though they amounted to little more than poor design and construction details. The same report concluded there were many similar cases in England.

In the snow belt of the United States, use of road de-icing salts rose from 0.6 million tonnes in 1950 to 10.5 million tonnes in 1988. Besides the corrosion damage to the bridges, in 1974, motor car corrosion damage was estimated at $5 billion and at this time manufacture of electrocoated bodies was not common. A report from the New York Department of Transport stated that, by 2010, 95% of all New York bridges would be deficient if maintenance remained at the same level as it was in 1981. The obvious deterioration had been seriously accelerated since the mid 1960s when the extensive use of road salt as a de-icer began. Rehabilitation of such bridges has become an important engineering practice about which much has been written. [27]

Case 1.9: In Paris, the award-winning cultural showpiece, the Centre Pompidou, showed signs of severe corrosion in 1981, less than five years after its completion in late 1976. Monitoring of the corrosion began in 1979 when the first effects were found on the 'inside out' building of steel-framed construction with external steel Warren trusses supported between corner columns. Fire protection was achieved by mineral fibre blankets encased in stainless steel sheeting on the trusses.

The nodes were protected with cement, plastic coated to prevent water penetration. It appears that somehow water must have penetrated the node protection, for during the winter of 1978–79 freezing of the trapped water caused spalling. As a result of danger to the public from falling insulation the coatings were removed leaving the steelwork protected only by the 7.5 μm thick metal-sprayed zinc coating.

While bridges are by no means the only structures susceptible to corrosion, they provide good examples of the catastrophic effects of the most trivial corrosion problem.

Case 1.10:   On 15 December 1967 Point Pleasant Bridge in Ohio collapsed killing 46 people. It was found that the cause of the disaster was a stress corrosion crack 2.5 mm deep in the head end of an eyebar. Here, the metal had a low resistance to fracture once a notch had been initiated. Failure resulted, and the bridge collapsed.

Sometimes, corrosion problems can be caused by the most unexpected agencies.

Case 1.11:   In December 1979 the City of Westminster, London, reported that it had a problem with falling lampposts. It was suggested that the main culprits for the corrosion that had occurred at the base of the posts were dogs. The city's pets were daily depositing about 2,000 litres of urine, mostly at the base of the city lampposts and this caused a great increase in the rate of corrosion. One of many problems experienced with the Charing Cross railway bridge in central London was also attributable to this surprising cause. Repeated visits from dogs had caused severe crevice corrosion in a part of the structure that was impossible to maintain.

The chemicals industry

Corrosion is an obvious and well-recognised problem for industrial chemical systems handling, say, hydrofluoric acid or hot concentrated sodium hydroxide solutions, and much care is taken to use the most efficient materials. However, the technology is both difficult and dangerous. There are many examples within these pages of how even a slight change in operating conditions, the presence of an unexpected impurity or change in the concentration of process chemical can result in a sudden dramatic rise in corrosion rate. As shown by the tragedies at Flixborough (Case 1.2) and Bhopal, the results can be terrible, yet the success rate is generally good. One of the biggest single causes of corrosion is stress-corrosion cracking, said in 1975 to be responsible for one-third of all failures.

Biomedical engineering

The single life-limiting factor for a prosthesis or other medical implant is its corrosion performance. Fatigue and corrosion fatigue of artificial heart valves or hip joints is a serious problem. A heart valve suffers stress cycling at 60–100 cycles per minute, 24 hours a day for many years. This represents an engineering challenge as serious as any, and there have been instances in the United States of sudden fatigue failure of a heart valve, causing the patient to die. A ball-and-socket hip joint made from a combination of titanium and stainless steels experiences severe mechanical loads during jogging or even simply descending stairs. And its working environment of warm saline serum solution would stretch any alloy to its limits. Manufacturers have had to face serious underperformance of current designs. Implants supposed to last twenty years have needed replacement after only two or three. These are but two examples to indicate the seriousness with which corrosion impacts upon late twentieth-century medicine and how the effects are far more devastating to people than simply those of dollar-cost.

The defence industry

Failure of military systems in action due to materials or corrosion failure is a much feared phenomenon and is dealt with by reliability, a wider topic that arose from the needs of the defence industry. The old adage about the war being lost for the sake of a nail in a horseshoe almost came about in the build-up to the Gulf War.

Case 1.12: In 1992, the *Washington Post* carried a report about the failure of a water pipe in the Pentagon at a time when troops were being sent to Saudi Arabia. The corroded pipe released considerable amounts of water into the basement within inches of the electrical closets which, had they failed, would have shut down the Pentagon.

Many defence systems are carried around passively for years in the expectation that, when the time comes, they will operate instantly and with complete efficiency at the touch of a button. It may be almost impossible to check out the system completely, as in the case of a torpedo or missile, for example, without actually firing it, yet they are systems of extreme complexity manufactured with lightweight corrodible alloys operating in corrosive and widely varying atmospheres. It is frequently the unlikely combination of circumstances which results in a failure at the critical moment, as happened to the ammunition of the British Army in India (Chapter 10).

1.4   CORROSION AND THE ENVIRONMENT

Acid rain, the greenhouse effect and the depletion of the ozone layer are just some of the observed changes to our fragile planet that have led to demands for control over processes likely to affect the environment. This change in social perspective has already led to considerable amounts of international agreement and legislation and will only increase in the foreseeable future. Control of corrosion is an integral part of this pressure, for, properly implemented, engineering systems can perform more efficiently for longer periods, with less wastage of material and energy resources and greatly reduced pollution. Scientists and engineers must now take fully into account the environmental effects of their activities.

Fact: It has been calculated that in the United Kingdom, 1 tonne of steel is converted completely into rust every 90 seconds. Apart from the waste of metal, the energy required to produce 1 tonne of steel from iron ore is sufficient to provide an average family home with energy for 3 months.

The avoidance of catastrophe is obvious, yet there is a marked difference between the level of importance attached by international opinion to the avoidance of a major passenger aircraft accident and that of the incident involving the oil-tanker, *Braer* (Case 1.6). Even the terrible aftermath of the *Exxon Valdez* spillage in Alaska, broadcast worldwide, did not prevent the *Braer* incident and many others besides. Pollution is still allowed to happen because of cavalier attitudes and lack of international action.

In almost every aspect of industry, the technology now exists to use alloys in ways that are efficient and environmentally sound. In the water treatment industry, the need to control effluent from waste water and sewage systems is being successfully addressed. The corrosion environments in which materials must operate are some of the toughest, but the combination of correct materials selection (Chapter 15) with efficient protective barrier coatings (Chapter 14) has provided excellent solutions. Solid waste disposal has been seen as a good alternative to landfill, which can also generate welcome energy as a by-product, providing the gaseous emissions are controlled. The high temperature corrosion performance for the materials used in the incinerators are well within the scope of modern materials [28–31].

Rail transport has for many years been unfashionable, but the need to reduce the volume of cars on the roads and the resulting pollution from exhaust gases has given new impetus to the development of electric-powered mass rail transit systems. The corrosion problems associated with the use of high current land-based systems in which current leaks from the system into nearby metallic structures and causes very rapid corrosion are well understood and well within the scope of existing technology. [32]

The 'polluter pays' policy, now established in the United States but not internationally, spells danger for the old management technique of saving inital costs without caring about those which follow. The corrosion failure of a chemical reactor which results in a leak of pollutants into the environment is a very serious financial liability. Much more favourable is the alternative of investing more in the cost of the initial design to achieve long life and low maintenance.

Sometimes, the need for change to more environmentally acceptable processes introduces new problems. For example, it has been suggested that the chlorination of water, a process frequently adopted where biocides are necessary, should be replaced by a process involving dissolved ozone, but there is insufficient performance data. [33] Corrosion control techniques have had both good and bad effects. Organotin formulations used as antifouling coatings on ships were so effective that they badly polluted the environment and have now been largely banned. The use of cadmium as a sacrificial material is severely restricted because of its toxicity, whereas the environmental effects of using hundreds of thousands of tonnes of zinc to protect steel platforms in the sheltered and shallow waters of the North Sea remain to be evaluated. These, however, are just some of the disadvantages to be set against a very large number of advantages. On balance, much is now being achieved.

1.5 CONCLUDING REMARKS

The place of corrosion and its control in society parallels that of engineering at large. In his excellent book, *To Engineer is Human*, Petroski [34] suggests that for all their years of laboratory and drawing-board effort, scientists and engineers have advanced our society by *doing* and then checking performance in action. Petroski believes that large engineered systems are simply too complicated for accurate performance predictions. And in reality, no laboratory test programme can give reliable mathematical predictions of the outcome.

In the descriptions of various industries given in Section 1.3, the story is mostly one of success in control of corrosion: our bridges and buildings *do* stay up, aircraft safety *is* extraordinarily high, and cars can survive extremely well the rigours of our world yet we *learn* more from the failures. In order to build upon this success into the new millenium, we must continually mull over the reasons for past failures. Learning by practical example is therefore a central theme of this book. The well-established corrosion science of materials forms the foundations of understanding, but the practical aspect is a pivotal component that needs to be integrated to achieve effective corrosion control. Finally, it is critical to appreciate the integrative nature of all aspects of corrosion because of the need to actively *manage* materials in systems, discussed in Chapters 11 and 12. Ill-considered cost-cutting in times of financial stringency is almost certain to increase the number of serious corrosion failures.

1.6 REFERENCES

1.   Hackerman N 1993 A view of the history of corrosion and its control. In *Corrosion 93 plenary and keynote lectures*, edited by R D Gundry, NACE International, Houston TX, pp. 1–5.
2.   Pliny 1938 *Natural history of the world,* Heinemann, London.
3.   Fancourt L D 1970 Lanthony priory — history and guide, published privately.

4. Williams A R 1986 The knight and the blast furnace, *Metals and Materials* **2**:485–9.
5. Uhlig H 1949 *Chemical and Engineering News* **97**:2764.
6. Hoar T P 1971 *Report of the committee on corrosion and protection, HMSO, London.*
7. Editorial 1982 *Corrosion Prevention and Control* **29**(1):1.
8. Editorial 1980 *Corrosion Prevention and Control* **27**(3):1.
9. Cox B 1994 Modelling the corrosion of zirconium alloys in nuclear reactors cooled by high temperature water. In *Modelling aqueous corrosion*, edited by K R Trethewey and P R Roberge, Kluwer Academic, Dordrecht, The Netherlands, pp. 183–200.
10. Gordon B M 1993 BWR structural materials corrosion — a NACE 50th anniversary perspective. Paper 175 in *Corrosion 93, New Orleans LA*, NACE International, Houston TX.
11. Jones R L 1993 Corrosion experience in US light water reactors — a NACE 50th anniversary perspective. Paper 168 in *Corrosion 93, New Orleans LA*, NACE International, Houston TX.
12. Baroux B and Gorse D 1994 The respective effects of passive films and non-metallic inclusions on the pitting resistance of stainless steels — consequences on the pre-pitting noise and the anodic current transients. In *Modelling aqueous corrosion*, edited by K R Trethewey and P R Roberge, Kluwer Academic, Dordrecht, The Netherlands, pp. 161–82.
13. Shoesmith D, Ikeda B and King F 1994 Modelling procedures for predicting lifetimes of nuclear waste containers. In *Modelling aqueous corrosion*, edited by K R Trethewey and P R Roberge, Kluwer Academic, Dordrecht, The Netherlands, pp. 201–38.
14. Ross R W 1993 The evolution of FGD materials technology. Paper 414 in *Corrosion 93 plenary and keynote lectures*, edited by R D Gundry, NACE International, Houston TX.
15. Bow K (ed) 1993 *History of corrosion in power and communication cables*, NACE International, Houston TX
16. Walker F M 1990 Iron's contribution to modern shipbuilding, *Marine Technology* **Dec**: 778–82.
17. Sedriks J 1993 Advanced materials in marine environments. Paper 505 in *Corrosion 93 plenary and keynote lectures*, edited by R D Gundry, NACE International, Houston TX.
18. Munger C G 1993 Marine and offshore corrosion control, past present and future. Paper 549 in *Corrosion 93 plenary and keynote lectures*, edited by R D Gundry, NACE International, Houston TX.
19. Trethewey K R 1992 An overview of current naval marine corrosion problems and their solutions. In *Proceedings of INEC92, Plymouth UK*, Institute of Marine Engineers, London.
20. Baboian R (ed) 1981 *Automotive corrosion by de-icing salts*, NACE International, Houston TX.
21. Baboian R (ed) 1992 *Automotive corrosion and protection*, NACE International, Houston TX.
22. McArthur H 1988 *Corrosion prediction and protection in motor vehicles*, Ellis Horwood, Chichester.
23. Soepenberg E N 1993 History of materials technology for the automotive industry — a continuing environment–materials–energy debate. Paper 546 in *Corrosion 93, New Orleans LA*, NACE International, Houston TX.

24. McArthur H 1981 Motor vehicle corrosion — safe at any age? *Corrosion prevention and control* **28**(3): 5–8.
25. Chandler H E 1984 Update on corrosion control coatings for autos, *Metal Progress* **Feb**: 39–43.
26. Editorial 1982 *Corrosion Prevention and Control* **29**(2): 1.
27. Broomfield J P 1994 Five years research on corrosion of steel in concrete: a summary of the strategic highway research program structures research. Paper 318 in *Corrosion 93, New Orleans LA*, NACE International, Houston TX.
28. Lai G Y and Sorrell G (eds) 1992 *Materials performance in waste incineration systems*, NACE International, Houston TX.
29. Krause H H 1993 Historical perspective of fireside corrosion problems in refuse-fired boilers. Paper 200 in *Corrosion 93 plenary and keynote lectures*, edited by R D Gundry, NACE International, Houston TX.
30. Stringer J 1993 Materials for fossil energy systems — past, present and future. Paper 225 in Corrosion 93 plenary and keynote lectures, edited by R D Gundry, NACE International, Houston TX.
31. Antony C M, Lai G Y and Kannair M D 1993 Alloy performance in municipal solid waste incinerators. Paper 214 in *Corrosion 93, New Orleans LA*, NACE International, Houston TX.
32. Szeliga M J 1994 Stray current corrosion: the past, present, and future of rail transit systems. Paper 300 in *Corrosion 94, Baltimore MD*, NACE International, Houston, TX.
33. Brown B E and Duquette D J 1994 A review of the effects of dissolved ozone on the corrosion behaviour of metals and alloys. Paper 486 in *Corrosion 94, Baltimore MD*, NACE International, Houston TX.
34. Petroski H 1992 *To engineer is human: the role of failure in successful design*, Vintage Books, New York.

1.7 BIBLIOGRAPHY

Abbott W H 1993 *Atmospheric corrosion of control equipment*, NACE International, Houston TX.
Anon 1983 *Prevention and control of water-caused problems in building potable water systems*, NACE International, Houston TX.
Anon 1987 *Metals handbook, volume 13: corrosion*, ASM International, Ohio.
Anon 1992 *Seventh international symposium on corrosion in the pulp and paper industry proceedings*, TAPPI.
Anon 1992 *Solving corrosion in air pollution control equipment*, NACE International, Houston TX.
Anon 1993 *Engineering solutions to industrial corrosion problems*, NACE International, Houston TX.
Asperger R G, Teevens P and Ho-Chung-Qui D 1989 *Corrosion control and monitoring in gas pipelines and well systems*, NACE International, Houston TX.
Atkinson J T N and Droffelaar H V 1982 *Corrosion and its control — an introduction to the subject*, NACE International, Houston TX.

Brown B F, Burnett H C, Chase W T, Goodway M, Kruger J and Pourbaix M 1991 *Corrosion and metal artifacts — a dialogue between conservators and archaeologists and corrosion scientists*, NACE International, Houston TX.

Burke P A, Asphahani A I and Wright B S 1986 *Advances in CO_2 corrosion*, vol. II, NACE International, Houston TX.

Cohen A 1993 *Historical perspective of corrosion by potable waters in building systems*, NACE International, Houston TX.

Craig B D 1989 *Handbook of corrosion data*, ASM, Ohio.

Cubicciotti D 1991 *Environmental degradation of materials in nuclear power systems — water reactors*, NACE International, Houston TX.

Delinder L S V 1984 *Corrosion basics — an introduction*, NACE International, Houston TX.

Dillon C P 1982 *Forms of corrosion — recognition and prevention*, NACE International, Houston TX.

Drayman-Weisser T 1992 *Dialogue 89 — the conservation of bronze sculpture in the outdoor environment: a dialogue among conservators, curators, environmental scientists and corrosion engineers*, NACE International, Houston TX.

During E D D 1991 *Corrosion atlas*, Elsevier, Amsterdam.

Fontana M G 1986 *Corrosion engineering*, 3rd edn. McGraw-Hill, New York.

Gellings P J 1985 *Introduction to corrosion prevention and control*, 2nd edn. Delft University Press, Delft, The Netherlands.

Gordon B M 1993 BWR structural materials corrosion — a NACE 50th anniversary perspective. Paper 175 in *Corrosion 93, New Orleans LA*, NACE International, Houston TX.

Gundry R 1993 *Corrosion 93 plenary and keynote lectures: a 50 year history of corrosion prevention and control*, NACE International, Houston TX.

Koch G H and Thompson N G 1984 *Corrosion in flue gas desulfurization systems*, NACE International, Houston TX.

Lochmann W J and Indig M 1980 *Materials and corrosion problems in energy systems*, NACE International, Houston TX.

MacDonald D D, Begley J A, Bockris J O, Kruger J, Mansfeld F B, Rhodes P R and Staehle R W 1981 *Materials Science and Engineering* **50**: 19–42.

Manning M I 1988 *Metals and Materials* **4**: 213–17.

Moniz B J and Pollock W I 1986 *Process industries corrosion — the theory and practice*, NACE International, Houston TX.

Parkins R N 1991 *Engineering solutions for corrosion in oil and gas applications*, NACE International, Houston TX.

Payne G 1993 Material failures in North Sea water injection systems. Paper 66 in *Corrosion 93, New Orleans LA*, NACE International, Houston TX.

Treseder R S, Baboian R and Munger C G 1991 *NACE corrosion engineer's reference book*, 2nd edn. NACE International, Houston, TX.

Tuttle R N and Kane R D 1981 *H_2S corrosion in oil and gas production — a compilation of classic papers*, NACE International, Houston TX.

Vik E A and Hedberg T 1991 *Corrosion and corrosion control in drinking water systems*, NACE International, Houston TX.

Widley J F 1990 Aging aircraft, *Materials Performance* **29**(3): 80–5.

Wranglen G 1985 *An introduction to corrosion and protection of metals,* Chapman and Hall, London, p. 24.

2 ENABLING THEORY FOR AQUEOUS CORROSION

Science arises from the discovery of identity amidst diversity.

(W J Jevons, 1874)

Corrosion is an interdisciplinary subject, in other words, it combines elements of physics, chemistry, metallurgy, electronics and engineering. Many of us who work in the field of corrosion often have a background in one or other of the mainstream subjects but not in all. Thus an electrochemist does not always appreciate the metallurgical or engineering aspects of corrosion, while metallurgists and mechanical or structural engineers have no reason to understand completely the electrical principles behind corrosion testing. Similarly, a student often finds it difficult to understand corrosion through lack of knowledge in one or more of these subjects. The purpose of this is to provide the necessary theory from all relevant subjects to enable you to understand corrosion better. You may find that some of the subject matter which follows is elementary and thus you should study only those sections with which you are not familiar.

2.1 ENERGY: THE RULE OF LAW

It is hard to think of a natural process which is not governed by energy changes. Corrosion is a problem caused by nature; it affects material substance and is governed by energy changes. This chapter is therefore concerned with the fundamentals of energy and substance.

The study of energy changes is called **thermodynamics**, a very precise subject with many definitions, variable quantities (called parameters) and equations. The discussion presented here is restricted to some basic concepts.

We shall often refer to **systems**, which can be defined as any specified part of the universe in which we are interested. Around the system is an imaginary boundary that separates it from the rest of the universe, the **surroundings**. In the universe:

Energy can neither be created nor destroyed.

All spontaneous changes occur with a release of free energy from a system to its surroundings at constant temperature and pressure.

The first statement is the First Law of Thermodynamics and is extremely important when considering the changes which occur when metals corrode. The second statement is one form of the Second Law of Thermodynamics. When corrosion occurs in nature it is a spontaneous process and therefore occurs with a release of free energy. Left to its own devices, nature is extremely concerned always to minimise energy. Water runs downhill, cups of coffee go cold and spinning tops slow down and stop. All are consequences of the Second Law of Thermodynamics, which is most simply stated in the form

Heat will not flow of its own accord from a cold place to a hot place.

Human interference can keep the water at a high level, the coffee hot or the top spinning, but it is necessary to supply energy to the system. We know that many metals corrode given sufficient time, so it would appear that nature minimises the energy of metals through corrosion.

There are many forms of energy but the driving force for corrosion comes from **chemical energy**. This is partly derived from energy stored within the chemical bonds of substances, called the **internal energy** of the system. Only a proportion of this internal energy is available as the useful energy which powers our engines or the destructive agent which causes a corrosion reaction. This available energy is called the **free energy**. We examine free energy in this chapter and in Chapter 4.

An important concept which helps to explain the rates of corrosion reactions is **transition state theory**. Consider the following equation:

$$A + B \rightarrow C + D \tag{2.1}$$

The equation is a shorthand which can be taken to mean: two species, A and B, known as the reactants, interact to form two new species, C and D, known as the products. To produce the new species it is essential that A and B do not just come into contact with each other, but physically join together, forming an intermediate species, AB. In reality, this may happen for only the briefest instant, and then only when the reactants possess sufficient energy and the correct orientation for the joining to occur. AB is called the **transition state**, and it is the reorganisation of the transition state which leads directly to the products, C and D.

We can now use a diagram called an energy profile to describe the free energy changes which occur during the reaction (Fig. 2.1). The y-axis in the diagram is free energy, G; energy changes will appear as ΔG, in which Δ is used conventionally to mean 'a change in'. The x-axis is called the **reaction coordinate** and can be thought of as the extent to which the process has progressed. It is not necessarily a time-scale, but can be thought of as such in the first instance. Theory tells us that the transition state must be of higher free energy than the sum of the free energies of the separate species, A and B. Traditionally, this amount is given the symbol $\Delta G^{\ddagger}$. We are considering a spontaneous reaction, so

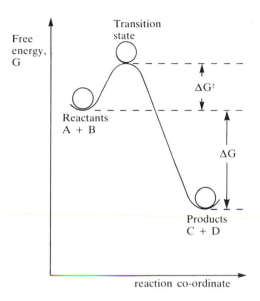

Fig. 2.1 Free energy profile for the reaction which converts reactants, A + B, into products, C + D, via the transition state.

the energies of the products, C and D, must be less than the energies of the reactants, A and B, by an amount ΔG. Once formed, the transition state can either revert to reactants or can progress to the products. By changing into C and D, the transition state is able to reach a lower free energy, so this is the favoured of the two possibilities.

In simplest form, the rate of a corrosion reaction can be expressed as:

$$\text{rate} = \text{rate constant} \times [\text{reactants}] \tag{2.2}$$

The quantity in square brackets is a measure of the amount of substance. This will be explained in the next section. The rate constant can be shown to be related to the size of the free energy barrier:

$$\text{rate constant} = C \exp(-\Delta G/RT) \tag{2.3}$$

where C and R are constants, and T is the absolute temperature. Inspection of the equation shows that as T increases, so also does the rate constant (and hence the rate) but when the size of the barrier (ΔG) is increased, the rate constant decreases. Equation (2.3) is a modified **Arrhenius equation**, named after the scientist who first described this important theory.

Now consider the reverse process. Overall, it cannot be spontaneous, because there is an increase in free energy upon conversion of C and D into A and B. Furthermore, the new reactants, C and D, have to cross a bigger free energy barrier. Transition state theory says that the reverse process is possible, but occurs at a much reduced rate, represented by an equation similar to eqn (2.3) in which the activation free energy has been increased from $\Delta G^{\ddagger}$ to $(\Delta G^{\ddagger} + G)$, as

25

in Fig. 2.1. The reverse process is possible only on the molecular scale, where the energies of an individual C and D pair may be such as to allow the formation of the transition state. Remember that our rule applies to an overall free energy change for a bulk system. The reverse process occurs at a rate far less than the rate of the forward process, so the net reaction observed on the large scale will always appear to be a steady conversion of A and B into C and D. For the reverse process to occur in the bulk system, energy must be supplied to the system (e.g. in electrolysis).

The use of energy profiles is of considerable assistance in the understanding of corrosion processes and will be frequently used throughout our discussions of corrosion theory.

2.2   MATTERS OF SUBSTANCE

All substances are ultimately composed of **atoms**, **molecules** or **ions**. Currently, 104 different kinds of atoms, or elements, have been discovered. Many of these are extremely rare or can be made only in highly specialised nuclear reactions. Several surprising statements can be made about the elements.

Fact: Just five of the 104 elements, oxygen, silicon, aluminium, iron and calcium, make up 91% of the matter in the Earth's crust.

Fact: The amounts of oxygen and silicon together make up 75% of the mass.

Fact: Of the 104 elements, the great majority are metals.

Since corrosion is a problem that afflicts metals, the third fact provides a further incentive to understand corrosion.

Corrosion is about substance and energy. We began by discussing energy and now we continue with an examination of the properties of substances, starting with some definitions which allow us to state the amount of substance.

Most people have agreed to measure the amount of substance in kilograms, incorrectly called a weight, but correctly described as **mass**. Today, the phrase **amount of substance** is used specifically to refer to the number of atoms, molecules or ions in a given quantity of material, **not** a mass measured in kilograms. Amount of substance is one of the seven fundamental units of the SI system of measurement, accepted worldwide and used throughout this book. The unit is called the **mole** and abbreviated **mol**.

> **One mole of a substance is the molecular weight of that substance expressed in grams.**

The reason for the need to distinguish a number of atoms from a mass is obvious to anyone who has been fooled by the child's conundrum, which weighs more: a kilo of lead or a kilo of feathers? Lead is intrinsically more dense than a feather but the term 'weight' has a different implication and the answer to the question, is that they both weigh the same. At the root of the answer lies the fact that the basic particle of lead, the lead atom, has more mass than the basic particle of a feather, let's say a carbon atom, and for a given number of atoms the lead will weigh much more. Thus we see the need to define the number of atoms in a given amount of substance, as well as its mass. Unfortunately the rephrasing of the conundrum, Which weighs more: a mole of lead or a mole of feathers? makes more sense and less fun. The major significance of this unit is that

One mole of material A and one mole of material B have equal numbers of molecules but different masses.

It is unusual to find isolated atoms in nature. The oxygen we breathe is better described as molecular oxygen, for it consists of **molecules** in which two oxygen atoms have combined together. The water we drink, and which aids many corrosion reactions, is composed of molecules in which two atoms of hydrogen have combined with one atom of oxygen. Thus discrete atoms are often of less importance in corrosion than the molecules derived from them. By undergoing such combinations, the atoms are able to attain lower energy states.

Question: *How do I calculate the number of moles of oxygen in a given mass of gas?*

First take the base unit of oxygen gas, the oxygen molecule, which consists of two atoms of elemental oxygen. Next find the atomic weights of all the constituents of the base unit (in this case, oxygen) and sum them. A table of atomic weights can be found in many science texts. Rounding off the weights to the nearest whole number, oxygen has an atomic weight of 16 and a molecular weight of 32. This number, expressed in grams, gives the mass of 1 mol of oxygen, i.e. 32 g. The same method is used for liquids and gases. A water molecule consists of $2H + O$, so the molecular weight is $2 + 16 = 18$. Hence one mol of water has a mass of 18 g.

An atom can be thought of as a diffuse cloud of **electrons**, tiny particles with negligible mass and a single unit of negative charge, surrounding a positively charged nucleus. The nucleus contains virtually all the mass and is made up of **neutrons** and **protons**. The number of protons determines the element and is called the **atomic number**. The number of electrons always balances the number of protons because atoms are always electrically neutral; although composed of charged particles, atoms have no net charge.

The electrons in atoms are arranged into different levels of energy, called **orbitals**. The electrons in the orbital of highest energy determine the reactive properties of the material. When reactions occur between atoms or molecules it is changes in the distribution of electrons between the participating atoms that determine the net thermodynamic energy changes. There are three ways in which atoms combine:

By **sharing** electrons to form **covalent** bonds.

By **exchanging** electrons to form **ionic** bonds.

By **sharing** electrons to form **metallic** bonds.

Although covalent bonds are extremely important in the study of materials, they are much less important in corrosion, so we need not discuss them.

Ionic bonding occurs when electrons are exchanged. One atom may lose an electron to another, in which case, the first has become an ion with a net positive charge and the second has changed into an ion with a net negative charge. The positive ion is called a **cation**, the negative ion, an **anion**. If the process occurs spontaneously, both ions are lower energy states than the atoms from which they were made. In the solid state the ions are held together by the electrostatic attraction between positive and negative charges. They form highly ordered arrangements called **crystal lattices**.

We shall refer frequently to sodium chloride. When dissolved in water, sodium chloride produces an extremely aggressive corrosion medium similar to sea-water. It is formed when metallic sodium reacts with toxic chlorine gas to yield a harmless white non-metallic solid, better known as common salt. The sodium and chlorine atoms have become **ions** and the reaction illustrates an important fact. All ions are electrically charged, so their properties are different from those of the neutral atoms that produce them.

Using the chemical shorthand in which sodium is represented by Na, chlorine by Cl, and e^- is an electron, we write the reactions which describe the formation of salt as

$$Na \rightarrow Na^+ + e^- \tag{2.4}$$

$$Cl + e^- \rightarrow Cl^- \tag{2.5}$$

Equation (2.4) means that a sodium atom gives up one electron to form a positively charged sodium ion, while eqn (2.5) means that a chlorine atom accepts an electron to form a negatively charged chloride ion. Reactions such as eqn (2.4) generate electrons and are known as **oxidation** reactions. Reactions such as eqn (2.5) consume electrons and are called **reduction** reactions. According to these definitions, all corrosion reactions are oxidation reactions.

For clarity, we have expressed the event as two separate equations, (2.4) and (2.5), but it is common to add them together and write them as

$$Na + Cl \rightarrow Na^+ + Cl^- \tag{2.6}$$

In this form there is no direct indication of the electron changes which have occurred, though they are easy to deduce. As in all chemical equations, both sides should balance for mass and electrical charge. In a sense, it is rather meaningless to refer to sodium chloride as NaCl. The large ordered array of ions in the crystal lattice requires the formula, $(NaCl)_n$, but in calculations it helps simply to use NaCl.

When ionic materials are dissolved in water the ions separate and are dispersed at random within the molecules of water. However much salt we use to make the solution, the number of sodium ions is always equal to the number of chloride ions. This statement is more important than it might at first appear; it is the **principle of electroneutrality**. Whenever a positive ion is created, a negative ion must also be created.

Question: *How are the proportions of ions to water molecules calculated?*

This is a most important question for it introduces our method of measuring the amounts of substance in solution. Let us take sodium chloride again. To make a 3.5% solution (in order to crudely mimic sea-water) 35 g of crystals are weighed out and dissolved in water, say 250 ml. Further water is then added to make the total volume 1 litre. One way of expressing the solution composition is to say that the concentration of sodium chloride is 35 grams per litre. However, as we saw earlier, this is not necessarily as useful as expressing the number of moles which have been dissolved in the water. For this we need the molecular weight of sodium chloride, obtained from the relevant atomic weights. As $Na = 23$ and $Cl = 35.5$, we obtain $NaCl = 58.5$. A mole of sodium chloride is 58.5 g, but the weight dissolved was $35 g = 35/58.5 \text{ mol} = 0.6 \text{ mol}$. When 1 mol has been dissolved in a litre of water we say that the solution is **molar**, thus our 3.5% sodium chloride solution is 0.6 molar, abbreviated 0.6 M.

We have still not answered the original question, however. To do this we need to calculate the number of moles in a litre of water. Since a litre of water weighs 1000 g and 1 mol is 18 g then there are $1,000/18 = 55.55$ mol of water in a litre. In the 3.5% solution there is 0.6 mol of salt and we make the approximation that the salt will have taken the place of 0.6 mol of water molecules in order to maintain a constant volume of 1 litre. There is one final point to make before the answer is obtained. A solid assembly of ions was dissolved assuming a formula NaCl. It is now that the usefulness of the mole becomes even more apparent, for 1 mol of sodium chloride dissolves to give 1 mol of sodium ions and 1 mol of chloride ions. The ratio of sodium ions to chloride ions is exactly 1. Thus, the answer to the original question is that the ratio of either ion to water molecules is

$$0.6/(55.55 - 0.6) = 0.011$$

More precise texts use a parameter called an **activity** rather than a concentration; coupled with **molality**, a slightly different way of measuring the concentration of solutions and not to be confused with molarity. It is considered that in an introductory text such as this, molalities and activities are an unnecessary complication and they will not be discussed further. The excellent work by Atkins, *Physical Chemistry*, describes all the terms precisely and is recommended for a full study of this discipline.

An understanding of the bonding of metals is rather more difficult to achieve. In practice this is not necessary at our level of discussion and we shall think of metals as an assembly of atoms in an ordered crystal structure.

Question: *How can a metal be crystalline? It does not look at all like a diamond or a grain of sugar, neither does it have similar properties.*

The property which determines whether a material is crystalline is internal ordering of atoms, usually over a large number of atomic distances. The atoms within metals are ordered in very regular patterns and hence metals are crystalline in nature, despite their apparent dissimilarity to more obvious crystalline materials. It is the special type of bonding in metals which gives them their characteristic properties and distinguishes them from other crystalline, non-metallic substances.

When a metal atom undergoes a corrosion reaction, it is converted into an ion by a reaction with a species present in the environment. Using the symbol M for a metal atom contained within its solid structure, we can represent corrosion by the following equation:

$$M \rightarrow M^{z+} + ze^- \tag{2.7}$$

The integer, z, usually has the values 1, 2 or 3. Higher values of z are possible, though rare. Equation (2.7) is the most general form of a corrosion reaction and states that metal atoms can lose more than one electron. Of the possible values of z, 2 is most common. The value of z is called the **valency** and it is not uncommon for metals to have more than one valency. For example, iron, the most important engineering material of all when used in steel, has 2 and 3 as common valencies, a factor which greatly complicates study of the rusting process.

Equation (2.7), though correct, does not represent a complete process, according to the principle of electroneutrality. A positive ion has been generated from a neutral species, but a negative ion has not. An electron is not an ion and cannot move about freely in solution. Electrons are considered insoluble, i.e. they remain in the solid. If the solid is an electrical conductor, they can be relatively free-ranging in the material. Equation (2.7) is called a **half-reaction** because of its incompleteness but is nevertheless very useful. A further consequence of the principle of electroneutrality is that the 'free electrons' apparently generated in eqn (2.7) must be 'consumed' elsewhere. This is achieved by generating a negative ion to balance the positive ion. A corrosion process is thus more completely described by an equation in which negative ions as well as positive ions are formed. For example, eqns (2.4) and (2.5) are half-reactions, eqn (2.6) describes the complete process and free electrons are not generated in this equation. The formation of anions in corrosion processes need not involve the metal at all and may take place at a location quite remote from the corrosion of the metal. We shall discuss the nature of the second half-reaction of a corrosion process in Chapter 4.

A fundamental definition of corrosion:

> **Corrosion is the degradation of a metal by an electrochemical reaction with its environment.**

Four keywords occur in this definition and require discussion. Corrosion relates to **metals**. This means that only a half-reaction such as eqn (2.7) can be a true corrosion reaction. The second half-reaction, though it describes a process essential for corrosion, is not itself a corrosion reaction.

By using **degradation** we assume that corrosion is an undesirable process. There are circumstances in which this is not true, in which case the process is not usually referred to as corrosion.

The degradation of the metal involves not just a chemical but an **electrochemical** reaction, electron transfer occurs between the participants. Electrons are negatively charged species, and their transport constitutes an electrical current, so electrical reactions are influenced by electrical potential (see Section 2.7).

The **environment** is a convenient name to describe all species adjacent to the corroding metal at the time of the reaction, eqn (2.7). The topic of environment and its control is the subject of Chapter 13.

Environments that cause corrosion are called **corrosive**. A metal that suffers corrosion is called **corrodible**.

Today, there are those who widen the definition of corrosion to include all processes in which degradation of **materials** occurs, for example, the attack of a rubber hose by a chemical. This extension has occurred because it is useful for practising engineers who are concerned on a day-to-day basis with failure of materials. At this stage it is best to consider corrosion as the degradation of metals alone.

The most important material in aqueous corrosion processes, apart from metals themselves, is water. We shall devote the next section to a more detailed look at the properties of water and ions in solution.

2.3   PROCESSES IN SOLUTION

Water is a neutral molecule in which two atoms of hydrogen are combined with one of oxygen. It does, however, undergo limited **dissociation** into a **hydrogen ion** and a **hydroxyl ion**:

$$H_2O \rightleftharpoons H^+ + OH^- \tag{2.8}$$

Equation (2.8) is an example of what is generally referred to as an **equilibrium**, denoted by $\rightleftharpoons$. At any given temperature and pressure, the degree to which this process occurs is constant. If it were possible to remove some of the ions from the right-hand side of the equation, more water molecules would dissociate to restore the balance of the equilibrium. Such an equilibrium process can be expressed mathematically using the Law of Mass Action which is stated as follows:

For the equilibrium process

$$A + B \rightleftharpoons C + D \tag{2.9}$$

then

$$\frac{[C][D]}{[A][B]} = \text{constant} \tag{2.10}$$

Square brackets express the concentration of the species they enclose. When equilibrium exists, the constant is represented by K, whilst the substitution of non-equilibrium values for [A], [B], [C] and [D] in eqn (2.10) requires the use of symbol J. Applying this law to equilibrium (2.8) we have

$$\frac{[H^+][OH^-]}{[H_2O]} = \text{constant} \qquad (2.11)$$

It can usually be assumed that the concentration of water in a dilute solution is constant. In the last section, the effect of adding 0.6 mol of salt to 55.55 mol of water made only a small change. In more dilute solutions, the effect becomes even smaller. Thus, for dilute solutions, we write $[H_2O] = 55.55$ and obtain a new constant in eqn (2.12):

$$[H^+][OH^-] = \text{constant} \qquad (2.12)$$

We define standard conditions of temperature and pressure as $25\,^\circ$C and 1 atmosphere. The constant in eqn (2.12) has been measured experimentally, as 10^{-14} under standard conditions. This value and relationship (2.12) are extremely important, for they form the basis of a scale of **acidity**. Many metals are dissolved by acids and acidity is thus closely related to corrosion. The common factor in the properties of acids is the presence in aqueous solution of the hydrogen ion. The 'opposite' of an acid is an **alkali**, by which is meant that acids are neutralised by alkalis, and that alkalinity is associated with hydroxyl ions. As eqn (2.8) shows, water is a source of both acid and alkali in equal quantities and must therefore represent a **neutral** substance.

The accepted method of defining acidity is by means of a term called **pH**. It is measured on a scale from 0 to 14 and is defined as follows:

$$pH = -\lg[H^+] \qquad (2.13)$$

Throughout this book we shall use 'lg' to mean logarithm to the base 10.) At first sight eqn (2.13) may appear a little complicated but it soon becomes very easy to understand and use successfully. The term in square brackets is the concentration of hydrogen ions expressed as a molarity. Thus a concentration of 10^{-8} M when substituted into eqn (2.13) gives a pH of 8, while one of 10^{-2} M leads to a pH of 2. A concentration of 2×10^{-4} is slightly more difficult to convert:

$$\begin{aligned} pH &= -\lg(2 \times 10^{-4}) \\ &= -(\lg(2) + \lg(10^{-4})) \\ &= -(0.3 - 4) \\ &= 3.7 \end{aligned}$$

Returning to eqn (2.8), it is apparent that every water molecule produces one hydrogen ion and one hydroxyl ion. Therefore, in pure water

$$[H^+] = [OH^-] \qquad (2.14)$$

Substitution into eqn (2.12) yields

$$[H^+]^2 = 10^{-14} \tag{2.15}$$

so

$$[H^+] = 10^{-7} \tag{2.16}$$

Thus the pH of pure water is 7. On the scale of pH, if the concentration of hydrogen ions is greater than 10^{-7} M then the pH is in the region 0–7, which is acid, while a concentration of hydrogen ions of less than 10^{-7} gives a pH in the region 7–14 which is alkaline.

It is possible to define the concentration of alkali (hydroxyl ion) in the same way, i.e. pOH = $-\lg[OH^-]$. However, it is not often necessary to use this quantity because, by taking negative logarithms to the base 10 of eqn (2.12) we get

$$pH + pOH = 14 \tag{2.17}$$

and pOH is always readily calculated from pH. You should try the questions at the end of the chapter to gain experience in these quantities. Figure 2.2 summarises the different pH and pOH values in a memorable way. A point worth emphasising is that pH 7 is neutral only under the standard conditions described. As the temperature goes up, so more dissociation occurs and the pH of neutral water falls. This can be extremely important for corrosion reactions in a chemical plant operating at high temperature and pressure.

It is important to appreciate that an ionic material such as sodium chloride in solution possesses very different properties from that of the solid. No longer are the ions bound together tightly by strong electrostatic forces. The dissolution process completely destroys the crystal lattice of the solid state. If the solution is dilute the sodium ions may be separated from the chloride ions by relatively large distances. The ionic material is said to be **dissociated** when the ions have been separated in solution.

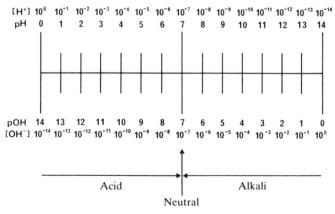

Fig. 2.2 The scales of pH and pOH.

The addition of a second soluble ionic material to the solution merely adds new types of ions which become randomly dispersed among all the other species present. It is quite wrong to think of a solution merely in terms of the individual solids that were added to the water. Undissociated solid may be present in the solution, though the substance is not truly in solution because the distribution of ions in the solid and in the solution is not homogeneous: we say that the solid is a different **phase** to that of the solution. Although these points may seem obvious to some readers, it is a common error to consider a solution as a mere mixture of one or more solids in water, rather than a homogeneous dispersion of a number of different ions. There is still a force of attraction between the positive and negative ions while in solution, but this force is greatly reduced by the presence of the water molecules.

2.4   METALS: MIGHT AND BLIGHT

Engineers are rightfully keen to exploit the might of metals but it is usually the task of the corrosion engineer to patch up the blight of corrosion. We have discovered so far that corrosion is the destruction of metals and that as many as 80% of all known elements are metals. With so many metals to choose from, engineers exploit surprisingly few. Even then, they frequently prefer alloys to pure elements.

The first elements to be used in a practical sense were gold and copper, two metals which exist uncombined in nature. Gold was too soft to be of use for tools or weapons, but copper was hammered into crude tools about 8000 BC. By 4000 BC the Egyptians found that copper could be obtained by heating a suitable ore in a kiln, while about 3000 BC, it was discovered that by alloying copper with tin much more durable weapons could be made than if either of the pure metals was used. Since then, it has become common knowledge that alloys have a wider range of useful properties than pure metals.

Metals are exceedingly useful for a number of reasons. Apart from being typically opaque and lustrous, they have such characteristic properties as ductility, high strength/weight ratio, and the ability to conduct heat and electricity. It is obvious that metals are extremely important in modern engineering, yet many can be badly affected by corrosion. What makes them so useful yet so vulnerable? Unfortunately the answer to such a crucial question cannot be provided in a few paragraphs.

We have already noted that the bonding of metals in solid structures is different from that in non-metals: this is largely why metals have such characteristic properties. The most significant difference in the bonding mechanism is that, while the electrons in covalent and ionic substances are usually strongly localised around their parent atoms or ions, in metals a proportion of the electrons are free to move throughout the material. This gives metals the property of **electrical conduction**.

But what of the solid structure of metals? Many books have been written on this subject alone and it is impossible to give more than the briefest coverage here. However, some aspects are vital to the understanding of the corrosion mechanisms and their control. These will be discussed next.

2.5   METALS IN THE MELTING POT

The solid state is one of order and permanence. The paper you are reading does not disintegrate before your eyes because its constituent atoms are held firmly within their molecules, and the molecules too are fixed in space by other bonding forces which maintain short, interatomic distances. The same is not true for liquids; shape is not retained because the energies of the atoms, molecules or ions are sufficient to overcome the forces which would have held them together in the solid state. Although they remain closely packed, there is little resistance to free movement within the bulk. All liquids can be made to form solid, ordered structures if their energies are reduced sufficiently by lowering the temperature. Solids melt when their constituent species are given sufficient energy to overcome the bonding forces; liquids freeze when that energy is taken away.

Metals are not different from other materials in this respect. Since the extraction, purification and fabrication of metals usually involves solidification from a **melt** — a quantity of molten metal — it is important to consider the events which take place during solidification. Imagine a pot (crucible) containing a sample of molten, pure metal. If the temperature of the sample is plotted as time proceeds, a graph rather like that shown in Fig. 2.3 is obtained. Initially, at A in the diagram, the crucible of molten metal contains the metal atoms jumbled together in a loosely packed assembly. At the temperature represented by the freezing point, B, the metal atoms begin to organise themselves into highly ordered arrays. At a given temperature for a given metal, the same array is always formed, although a variety of array patterns are found among the many different metals. The solidification process occurs with a release of energy, which is commonly called the **latent heat of fusion**. Under ideal conditions, the release of this energy maintains pure metals at a constant temperature for the duration of the solidification process (B to C in the figure), compensating for the natural tendency of the system to cool down to the temperature of its surroundings.

The number of crystals which nucleate depends upon the rate of cooling. Rapid cooling causes many nucleation sites in the liquid; conversely, slow cooling yields only a few crystals that continue to grow relatively slowly. The rate of cooling metals during the casting (solidification) process is very important in determining the mechanical properties of the metal. As will later become apparent, it also significantly affects corrosion properties.

Figure 2.4 illustrates schematically the solidification of a pure metal. Once a crystal has been formed (Fig. 2.4(a)), even though it may be only a tiny cluster

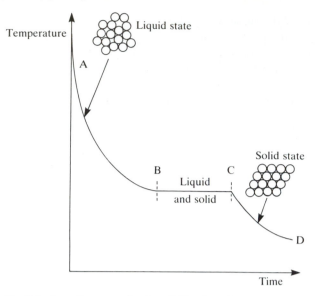

Fig. 2.3 A cooling curve for the solidification of a pure metal. Alloys show similar behaviour with discontinuities at B and C (known as 'arrests') but which are not at the same temperature.

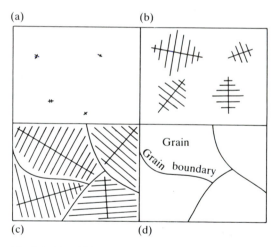

Fig. 2.4 Dendritic growth and solidification: (a) nucleation of crystals in the melt; (b) growth of crystals into dendrites; (c) complete solidification; and (d) final grain structure.

of atoms, it is more natural for further solidification to occur on the established solid array rather than creating new crystals. Let us suppose the crystal is a tiny cube. The rate of cooling is faster from the corners than from the edges. Likewise, cooling from an edge is faster than from a face, so the growth of the crystal is fastest at the corners which generate branches. These branches 'fill in' with solid material as the growth on the edges and faces tries to catch up with the growth at the corners. When a crystal has become branched (Fig. 2.4(b)) it is known as a **dendrite** and the process of solidification of metals is often referred to as **dendritic growth**. The orientation of each of the dendrites is different. Eventually, the dendrites grow so large that their outermost atoms touch and movement becomes restricted (Fig. 2.4(c)); the dendrites become fixed in their random orientations and the liquid which remains between the branches of the dendritic arms solidifies to give the final solid sample (Fig. 2.4(d)). At this point, represented by the end of the horizontal portion of the cooling curve, C in Fig. 2.3, cooling continues in the normal manner, in the solid state.

Once completely formed in the solid state, the individual crystals are known as **grains**. In the zone between any two grains where the crystal pattern changes orientation, the atoms are mismatched to the lattices in the grains. These zones are called the **grain boundaries**. Crystal lattices are rarely perfect and the next section discusses possible imperfections, together with their significance in corrosion.

2.6 DEFECTS IN METAL STRUCTURES

We have tended to assume that, when metals solidify to form highly ordered crystal lattices, there are no faults in the stacking arrangements. This is not the case. Metal lattices always contain imperfections, known as **defects**, that often have a considerable effect upon the corrosion properties of the metal.

In fact, we have already discussed one way in which the crystals are imperfect. Grain boundaries are regions of mismatch between adjacent lattices, each with a different orientation. The grain structure of metals arises from solidification processes during casting. It is also substantially affected by the mechanical treatment received during working and fabrication. The malleability of metals means that mechanical processes can cause substantial deformation of the grains and can fracture otherwise perfect parts of the lattice.

Question: *What types of defects occur within the grains themselves?*

Figure 2.5 shows three sorts of defects which can occur with individual atoms.

Vacancies: where there is an atom missing from a lattice site.

Substitutional defects: where a foreign atom occupies a lattice site which would have been occupied by a host atom.

Interstitial defects: where an atom occupies a site which is not a normal lattice site and is squeezed in between atoms of the host lattice. The interstitial atom may be a host atom or a foreign atom.

Single-atom defects, or **point defects** as they are more correctly known, are very significant in the theory of alloying. Although they are defects in a perfect

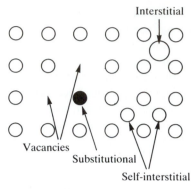

Fig. 2.5 Point defects in a crystal lattice.

lattice, they are used to great advantage in improving the mechanical properties of metals. They also play a role in some corrosion mechanisms, notably hydrogen embrittlement, selective attack, oxidation and hot corrosion, all of which rely upon the diffusion of species through a metal lattice.

A second type of defect occurs within the grain structure when planes of atoms, rather than individual atoms, are not perfectly fitted into the lattice. These are known as **line defects**. One example of a line defect is a **dislocation**. The mobility of dislocations and point defects is intimately concerned with such properties as strength, hardness and toughness. Two important types of dislocation are

Edge dislocations: where an 'unfinished' plane of atoms is present between two other planes (Fig. 2.6(a)).

Screw dislocations: where a plane is skewed to give it a different alignment to its immediate neighbour (Fig. 2.6(b)).

Dislocations and their role in corrosion mechanisms are dealt with in Chapters 9 and 10. **Volume defects** show up on a macroscopic scale, where relatively large parts of the volume of a metal do not conform to its perfect structure. Volume defects are created during manufacture.

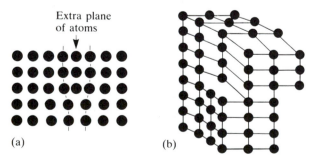

Fig. 2.6 Two types of dislocation in crystal lattices: (a) edge dislocation and (b) screw dislocation.

Voids are holes within the material. They may be caused by entrapment of air, by release of gas into its mould or by trace amounts of moisture turning to steam on contact with the hot liquid metal. Voids can also be caused by interdendritic shrinkage during solidification.

Cracks are much more severe than the mere mismatch of grain boundaries for they allow ingress of corrodents on a macroscopic scale. They are created during casting, usually as a result of uneven rates of cooling and the setting up of stresses within the cast. Cracks can also be created during forging and are very common in and adjacent to weldments.

Inclusions are particles of foreign matter embedded in the solid and therefore not part of the crystal lattice structure of the metal itself. The elements which make up the inclusions may solidify first and then become entrapped as individual particles within the dendrites of the metal as these in turn solidify. Alternatively, they may be the last to solidify after the host metal dendrites have been formed and the inclusions are again trapped in the grain boundaries.

In the ideal situation, the manufacturer's product should be free from all volume defects, but in practice this is very difficult to achieve in large-scale production. The variation in the amount of such defects in samples of metal of supposedly the same specification can lead to a wide variation in corrosion behaviour. Obviously, if very small amounts of material are required then it is possible to obtain a material relatively free from such defects, but in the engineering world at large, volume defects are common.

2.7 ELECTRICAL SCIENCE

Electricity is the passage of charged particles between two defined points, normally connected by a metal wire. The charged particles are electrons, the flow of electricity is a **current**. Electricity can also pass through aqueous solutions, but the electrical charge in this case is carried by positive and negative ions.

The amount of charge carried by an electron is known accurately and is measured in **coulombs** (C). When an amount of charge is passed at a constant rate we talk about the flow of current, I, measured in **amperes**, (A). Thus, when a charge of 1 C passes a given point in 1 s an electrical current of 1 A is flowing.

The driving force that causes the charged particles to move is analogous to the force that causes water to flow downhill, except that we do not imagine the current physically flowing downhill. We do, however, say that when there is a difference of **potential** between two points then the current will flow from the high potential to the low potential. We talk of a potential gradient and of the current *flowing down a potential gradient*.

Electrical potential is measured in **volts**, (V). Often, a driving force from a cell, or other source is called an **electromotive force**, abbreviated emf.

The flow of electric charges is impeded by a quantity called **resistance**, and between any two points there is always some resistance to the passage of the current (except in the special case of superconductors, not described here). In metal wires, where the electron flow is extremely efficient, the resistances involved are low, though not always negligible. In solution, the resistance is much bigger and may need to be taken into account. The passage of an *ion* current is quite different from the passage of an *electron* current. Although the ions are charged to either one, two or sometimes three units of electron charge, the mass transported is greater by a factor of about 2,000; the resistance to the motion of such massive particles is therefore much greater than for electrons. Electrical resistance is measured in **ohm** (Ω) .

We use a relationship known as **Ohm's Law** to equate potential, current and resistance:

$$\text{volts} = \text{amperes} \times \text{ohms} \tag{2.18}$$

or, using the commonly accepted symbols:

$$V = IR \tag{2.19}$$

Charge, J, is related to current by

$$\text{coulombs} = \text{amperes} \times \text{seconds} \tag{2.20}$$

or

$$J = It \tag{2.21}$$

It should be obvious to the reader that the resistance to electron flow offered by a conductor will increase with the length, l, of the conductor, but decrease with the area of cross-section, A. The proportionality constant is called the **resistivity** or **specific resistance** and is written ρ. Thus:

$$R = \frac{\rho l}{A} \tag{2.22}$$

By inspection, the units of resistivity are seen to be $\Omega\,\mathrm{m}$.

Conductance is defined as the inverse of resistance. The unit of conductance used to be the **mho** (Ω^{-1}) but recently it became the **siemens** (S). Likewise, **conductivity**, κ, is the inverse of resistivity, i.e. $\kappa = 1/\rho$. The units of conductivity are $\Omega^{-1}\,\mathrm{m}^{-1}$, or $\mathrm{S}\,\mathrm{m}^{-1}$. Conductivities of aqueous solutions are an important consideration in the understanding of corrosion processes.

The ideas we have put forward so far have implied that we can remove charge from one point and 'pile it up' at another. By using a circuit, or loop, around which charge is transported, charge is never accumulated at one point. In physics, the study of electrostatics often involves the storage of charge, and devices which operate by storing charge are very important in all areas of electronics. However, the passage of continuous **direct current, dc**, is only achievable when there is a circuit.

It is now time to make a historical point. When a decision was made about the sign of the charge which constitutes electrical current in wires, it was decided that it should be positive. With the benefit of hindsight we know that the wrong choice was made: **electron current is negative**. Unfortunately, by the time this was realised, it had become conventional in electrical science to designate current flow as a movement of *positive* charges around a circuit. Many circuit diagrams in textbooks still indicate conventional current in this way.

In this text we shall refer almost exclusively to current as *electron flow* because most of the discussions will consider how electrons are produced in corrosion reactions. Discussions concerning ion currents will involve the passage of both positive and negative ions. When the term **conventional current** is used it will refer to the electrical engineering concept of current as a flow of positive charges around a circuit. Consistent with this is the concept of a positive current being driven from a high potential to a low potential, i.e. a positive potential difference.

Consequently electrons flow from a higher negative potential to a lower negative potential.

It is unfortunate that the legacy of that incorrect choice remains, for it is possible for students setting up experiments to confuse the role of the terminals of electrical instruments, or to find the application of positive and negative signs somewhat muddled. In some areas of corrosion it is vital to know the role of each terminal, and to distinguish positive from negative. One Royal Navy ship suffered severe corrosion damage to the hull because the positive and negative terminals of electrical equipment had been wired wrongly (Case 16.3). Our discussions will try to clarify these aspects as they arise, but some comment is appropriate at this stage, particularly since there seems to be a discrepancy between an electrical engineer's terminology and that of a corrosion engineer.

Most laboratory corrosion measurements can be made using simple dc electrical circuits, (Fig. 2.7(a)). A source of emf is represented by the symbol (1)

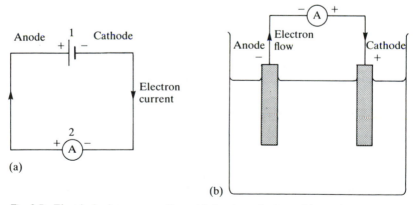

Fig. 2.7 Electrical science conventions. (a) Anode, cathode, positive and negative definitions in a battery (1), and the direction of electron current flow which results in a positive current reading on an ammeter (2). (b) Anode, cathode, positive and negative definitions in a corrosion cell and the direction of electron current flow in the circuit. With an ammeter connected as shown, a positive current reading will be indicated.

while symbol (2) is used for an ammeter. It is conventional to use the shorter line of the battery symbol for the negative terminal and the longer line for the positive. The negative pole, the **cathode**, provides the electron current which flows around the circuit towards the attraction of the positive pole, the **anode**. In batteries, the electrons are generated at this cathode by anodic electrode processes such as the oxidation reactions discussed in Section 2.2. When the negative (low or black) terminal of the ammeter is connected to the negative terminal of the battery, as in Fig. 2.7(a), the ammeter will register a positive current because

> **Conventional current flows from the positive (high or red) terminal to the negative (low or black) terminal, while electrons flow in the reverse direction**.

An aqueous corrosion cell (Fig. 2.7(b)) is explained in Section 4.2. Electrons for the external circuit are still produced at the negative electrode, but this negative electrode is called the **anode** because convention leads us to think of the anode as supplying the positive current (positive ions) through the solution. This change of terminology often causes confusion. If you have not yet read Chapter 4 you will not understand why this change of terminology is necessary. You may find it useful to reread this section after having studied Sections 4.2 to 4.5. By then you should be familiar with the processes occurring at the electrodes of corrosion cells and batteries. Meanwhile, here is a summary.

Electrons flow from negative to positive through the metallic circuit.

Conventional (positive) current flows from the anode of a dc power supply into the metallic circuit.

Conventional (positive) current flows from the anode of an electrolytic or corrosion cell into the electrolyte.

Laboratory corrosion studies often involve measurements of current and potential. Current measurement is achieved with an **ammeter**, inserted into the circuit in series so that the entire current flow in the circuit passes through it. Remember that ammeters are constructed to give positive deflections or positive digital readings when electrons pass through them from the negative (low or black) terminal to the positive (high or red) terminal.

The use of ammeters in corrosion current measurements should always be treated with caution because their finite resistance adds to the resistance within the corrosion cell and can affect the reactions. Instruments known as **zero resistance ammeters** (ZRAs) are used for accurate measurements. The experiments described in this book will be adequately carried out using a modern electronic multimeter used in the current mode.

The measurement of potential difference between two points is performed with a **voltmeter**, connected across the points of interest *in parallel*. The aim of such a measurement is to allow as little current as possible to be diverted from the circuit. This is achieved by having an extremely high resistance within the voltmeter. All traditional voltmeters draw some current from the circuit and will inevitably lead to a small error in the measurement, but the greater the internal resistance, the smaller the error. Many modern electronic instruments have resistances of the order of gigohms and are quite suitable for most work. Older instruments should be treated with more caution. Errors in the measurement of potential differences have been overcome by the use of **potentiometers**, which are able to measure potential yet draw no current. These are not as simple to use as voltmeters, but should be used for accurate work. However, the modern electronic multimeter used in the volts mode is adequate for the experiments of this text. Remember that positive potential differences are indicated from the high (positive or red) terminal to the low (negative or black) terminal.

The discussion so far has centred on direct current in which the emf is not deliberately varied. However, when it *is* deliberately varied, usually with sinusoidal frequency, ω, the current I is also observed to vary sinusoidally and at the same frequency. When the current in a circuit changes, the circuit opposes the change, a property known as **inductance**. Thus, when the current increases, inductance tries to hold it down and *vice versa*. And circuits show similar opposition to changes in voltage, a property known as **capacitance**. When parts of a circuit are able to store charges, capacitance is present. Corrosion involves solid metals in contact with heavy ions, so capacitive effects must be involved. And the theories applied to electrical circuits must also be applied to corrosion systems.

When either inductance or capacitance is present, there is a **time lag** or **phase shift** between the sinusoidal variation of the voltage and potential. In dc circuits the voltage seldom varies except at switch-on and switch-off; only then does capacitance produce any effects. In ac circuits the voltage is continuously changing, so the effect of capacitance is continuous. The amount of capacitance present in a circuit depends upon its construction and the electrical devices present. The capacitance may be so small that its effect on circuit voltage is negligible. If a circuit contains only resistive elements, the inductance is zero and the phase shift is also zero. A pure inductance produces a phase shift in which the voltage leads the current by 90°, whereas a pure capacitance produces a phase shift in which the current leads the voltage by 90°.

Figure 2.8 shows these effects for voltage and current in a circuit containing a pure inductance. The voltage is represented by a rotating vector, V, and the

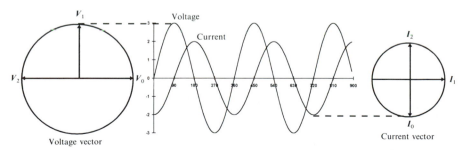

Fig. 2.8 Relationship between voltage and current vectors in an alternating current circuit containing only a pure inductance.

current by the rotating vector, I. At the starting point, represented by O, the voltage is V_0, and the current is I_0, lagging behind by 90°. As the voltage vector rotates (anticlockwise) to V_1, the current vector also rotates to I_1, but remains 90° behind. Note that, because of the continuously changing nature of voltage and current in ac circuits, it is necessary always to use vector addition and subtraction when computing circuit parameter values.

In alternating current/voltage situations, Ohm's Law, eqn (2.19), is amended such that resistance is replaced by a new parameter, **impedance**, Z, a term which correctly implies that current flow is being hindered by various elements in the circuit, such as resistors, capacitors or inductors. But unlike dc, the resistance to current flow is not constant; impedance is a vector quantity that must be computed accordingly. The standard way is to separate parameters into real and imaginary components. Thus, for current, the real component is

$$I_x = |I| \cos(\omega t) \tag{2.23}$$

and the imaginary component is

$$I_y = |I| \sin(\omega t) \tag{2.24}$$

where

$$|I|^2 = |I_x|^2 + |I_y|^2 \tag{2.25}$$

Similar equations can be written for voltage. Using standard notation, in which j is used to denote $\sqrt{-1}$, $I = I_x + I_y = I' + jI''$ and $V = V_x + V_y = V' + jV''$ the impedance is given by

$$Z = Z' + jZ'' = \frac{V' + jV''}{I' + jI''}$$

Works listed in the bibliography provide further discussion, but all readers should now be equipped to progress to a study of the theory of corrosion.

2.8 GLOSSARY

Acid: a solution in which the concentration of hydrogen ions exceeds the concentration of hydroxyl ions.
Alkali: a solution in which the concentration of hydroxyl ions exceeds the concentration of hydrogen ions.
Anion: ion with negative charge.
Capacitance: the property of an ac circuit that opposes any change in circuit voltage.
Cation: an ion with positive charge.
Corrosion: degradation of metals by an electrochemical reaction with the environment.

Defect (in solid metal): an arrangement of atoms that does not match the ideal lattice structure.

Dendrite: a characteristically ordered crystal grown from a melt.

Dissociation: breakup of an ionic solid into discrete ions.

Electrical potential: a source of electricity.

Electromotive force: electrical potential; a source of electricity.

Environment (in corrosion): an aggregation of chemical species immediately adjacent to a metal surface; at a given temperature and pressure the environment influences corrosion reactions.

Free energy: the portion of a substance's internal energy available to perform work.

Impedance: the total opposition to current flow in an ac circuit; the vector sum of resistance, inductance and capacitance.

Inductance: the property of an ac circuit that opposes any change in the current flow.

Ion: an atom with electrical charge.

Mole: the SI unit of substance; molecular weight in grams.

Molecular weight: each atom in a molecule has an atomic weight; the molecular weight of the molecule is the sum of these atomic weights.

Neutral (solution): a solution having equal concentrations of hydrogen and hydroxyl ions.

Oxidation: a reaction that generates electrons.

Reduction: a reaction that consumes electrons.

Surroundings: everything outside the system under consideration.

System: specified amount of substance under consideration.

Transition state: an arrangement of atoms intermediate between reactants and products.

Valency: the characteristic number of electrons lost when a metal atom becomes an ion.

2.9   WORKED EXAMPLES

E2.1   What is the molecular weight of hydrated copper (II) sulphate, $CuSO_4 \cdot 5H_2O$?

The atomic weights are:

$Cu = 63.5$ $S = 32$ $O = 16$ $H = 1$.

Therefore the molecular weight is:

$63.5 + 32 + (4 \times 16) + (5 \times 18) = 249.5$

E2.2   How do you make a solution of copper (II) sulphate that is 0.4 M?

A solution that is 1 M contains 249.5 g per litre of solution.
A solution which is 0.4 M contains $0.4 \times 249.5 = 99.8g$ per litre.

E2.3   The elements nickel and oxygen both have a valency of 2. When nickel is heated in air it reacts with oxygen to form the ionic solid material, nickel oxide. Write equations to describe the processes involved.

The nickel undergoes an oxidation reaction:

$$Ni \rightarrow Ni^{2+} + 2e^-$$

The oxygen undergoes a reduction reaction:

$$O + 2e^- \rightarrow O^{2-}$$

The two processes together lead to:

$$Ni + O \rightarrow Ni^{2+} + O^{2-}$$

According to our earlier definition, the oxidation of nickel is a corrosion reaction. Try to be clear about the two definitions of oxidation: a reaction of a metal with oxygen or a reaction in which electrons are generated. In this example, the nickel is oxidised in both senses.

E2.4   Write a corrosion reaction for iron.

Iron has two valencies, 2 and 3. You can therefore write two possible reactions:

$$Fe \rightarrow Fe^{2+} + 2e^-$$

$$Fe \rightarrow Fe^{3+} + 3e^-$$

E2.5   Under standard conditions, what is the concentration of (a) hydrogen ions and (b) hydroxyl ions in a solution of pH 8.4?

By definition,

$$8.4 = -lg[H^+]$$

so

$$0.6 - 9 = +lg[H^+]$$

Thus

$$lg[H^+] = lg(10^{0.6} \times 10^{-9})$$

And

$$[H^+] = 4 \times 10^{-9} M$$

Since

$$[H][OH] = 10^{-14}$$

$$[OH^-] = 10^{-14}/(4 \times 10^{-9})$$
$$= 2.5 \times 10^{-6} M$$

E2.6 An emf of 0.2 V drives an electrical current of 2 mA through a solution of sodium chloride for 1 hour.

(a) How much charge has been passed?
(b) Assuming there is no resistance elsewhere in the circuit, what is the resistance represented by the solution?
(c) By what means is the charge carried through the solution?

(a) The charge (C) passed is the current (A) × time (s):

current time $= 0.002 \times 60 \times 60$
$$= 7.2C$$

(b) From Ohm's Law,

$$V = IR$$

$$R = V/I$$

$$R = 0.2/0.002$$
$$= 100\Omega$$

(c) The charge is carried by the ions in the solution:
Na^+, Cl^-, H^+, OH^-.

2.10 PROBLEMS

P2.1 Potassium chloride is an ionic substance. An aqueous solution of potassium chloride is commonly used in corrosion experiments.

(a) What is the molecular weight of potassium chloride?
(b) How do you make a solution which is 0.1 M?
(c) What is meant by a saturated solution?

P2.2   What is the pH of a solution containing 10^{-3} M hydrogen ions?

P2.3   Nitric acid is a liquid with the formula HNO_3. When added to water it dissociates into a hydrogen ion and a nitrate ion, NO_3^-.

(a)   What is the molecular weight of nitric acid?
(b)   If 6.3 g is contained in a litre of water, what is the concentration of hydrogen ions?
(c)   What is the pH of this solution?

P2.4   When iron rusts it forms a chemical compound with atoms of oxygen. Iron can have a valency of 2 or 3; oxygen has a valency of 2.

(a)   Give two possible ratios of iron/oxygen atoms in rust.
(b)   Suggest two possible formulae for rust.

P2.5   You are given three liquids:

(a)   A solution containing 0.1 M sodium chloride
(b)   A solution containing 0.5 M hydrochloric acid
(c)   Tap-water.

Which solution will conduct electricity the best?

2.11 BIBLIOGRAPHY

Askeland D R 1988 *The Science and Engineering of Materials*, SI edn. Van Nostrand Reinhold, London, UK.
Atkins P W 1994 *Physical Chemistry*, 5th edn., Oxford University Press, Oxford.
Chapman J M (ed) 1962 *Basic Electricity*, 2nd edn., The Technical Press, Oxford (The Brolet Press, New York).
Higgins R A 1987 *Materials for the Engineering Technician*, 2nd edn, Hodder and Stoughton, London, UK.

3   PRACTICAL AQUEOUS CORROSION

It is a capital mistake to theorize before one has data.
(Sir Arthur Conan Doyle; *The Memoirs of Sherlock Holmes*)

Most of the experiments described in this chapter are suitable for use in college or school science laboratories. They are in their simplest form, and when students or teachers are familiar with them, may be refined using dedicated apparatus. In some cases, basic metallurgical equipment is advantageous, for example, grinding and polishing facilities. The experiments can be performed either by students or as demonstrations for classes. As demonstrations they form excellent visual aids to accompany lesson material. The authors have used the experiments described for many years and are fully convinced of the value of even the simplest experiment in understanding corrosion. They are described early in the book on the basis that doing is learning. At this stage it is not necessary to understand the theory behind the practice. Comments have been added where necessary, and a discussion occurs in Section 3.3.

Corrosion was defined in Section 2.2 as the degradation of a metal by an electrochemical reaction with the environment. From a practical point of view there are four keywords in the definition: metal, degradation, electrochemical and environment. Each of the experiments illustrates the importance of one or more of these keywords and an understanding of all four is necessary for an appreciation of the mechanisms of corrosion and their control.

3.1   COMMON APPARATUS

Power supply

A source of direct current is required. Commercial devices are available which are both potential and current controllable. Otherwise, an ordinary 12 V car battery charger is quite suitable. Dry cells may also be used if they can be mounted for easy connections.

Inert electrodes

Electrodes are conducting media that provide an interface between the electron current in wired electrical circuits and the ionic current of aqueous solutions. Inert electrodes are best made of platinum foil or gauze, approximately 100 mm^2, which should have platinum wires attached, leading away from the solution. Electrical connections can be made to these wires but should not be allowed to get wet. A cheaper but satisfactory alternative is to use carbon rods to which suitable connections can be made, either by crocodile clips or a more permanent method. Some carbon rods disintegrate slowly and contaminate the solutions, but they are quite adequate for quick simple experiments.

Working electrodes

Copperplate is preferred in a number of experiments designed to explore comparatively small differences in the system. The electrodes should be as similar to each other as possible or false observations may be made. It is desirable to prepare a number of copper plate electrodes, say 150 mm × 50 mm × 3 mm, which should be kept scrupulously clean between experiments.

Multimeters

Many experiments require measurements of current and potential. Modern electronic multimeters are versatile and easy to use. They can be used in current, voltage and resistance modes. More traditional ammeters are also suitable. However, errors in potential readings may arise when voltmeters are used in parallel. This is less likely with modern voltmeters, which have much greater internal resistance.

Use of solutions

A common solution used in corrosion experiments is 3.5% sodium chloride. This solution is best made up in bulk and is obtained by dissolving 35 g of laboratory grade reagent in a litre of water. When preparing solutions, use deionised or distilled water whenever possible. These forms of water will be denoted by the term 'pure' in the descriptions to distinguish them from tap-water. Add the appropriate chemical and stir with a glass rod until dissolved.

Use of acids

When reference is made to acids, e.g. nitric acid, it is the commonly supplied concentrated reagent which should be used. If dilution is necessary, details will be given. Dilution of acids should always be performed by addition of acid to water, **NOT** water to acid. In the case of sulphuric acid, this rule is essential. In Expt 3.12 water is added to nitric acid in contravention of this general rule, but if done carefully, is quite safe. *Safety glasses, rubber gloves and protective clothing should always be worn when handling acids.*

Laboratory glassware

Common laboratory glassware such as beakers, test-tubes, Petri dishes, glass rods and tongs should be available. A large quantity of sodium bicarbonate and a bucket of tap-water should be on hand in case of acid spillages. Paper towels and rags are useful for spillages of aqueous solutions and for general drying purposes.

3.2   THE EXPERIMENTS

Experiment 3.1

Aim: To demonstrate the different electrical conductivities of liquids.

Apparatus
   Power supply
   3 × 250 ml beakers
   2 inert electrodes, such as carbon rods
   ammeter, switch and connections
   liquids: pure water, white spirit, stock aqueous sodium chloride
The apparatus is connected as shown in Fig. 3.1. The power supply should be capable of providing a voltage of 2–5 V.

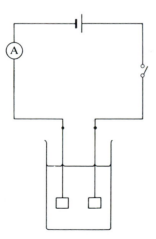

Fig. 3.1   Apparatus for Expt 3.1.

Procedure
1. Pour 200 ml white spirit into a beaker.
2. Immerse the electrodes.
3. Close the switch.
4. Record the current.
5. Open the switch.
6. Remove the electrodes from the beaker and test the liquid. Clean and dry the electrodes.
7. Repeat steps 1 to 6 using pure water in the second beaker.
8. Repeat steps 1 to 6 using stock aqueous sodium chloride in the third beaker.

Note: You should have measured zero current using white spirit or pure water.

Question:
Why does the aqueous sodium chloride conduct electricity whereas the other liquids do not?

Experiment 3.2

Aim: To demonstrate the importance of the environment in the process of rusting.

Apparatus
> 6 test tubes and stand
> 6 uncoated steel nails ~ 50 mm long
> sealant
> 1 cork bung
> chemicals: sodium nitrite, sodium chromate, acetone

Procedure
1. Degrease the nails by washing with acetone; dry thoroughly.
2. Place each test-tube in the stand and add a nail to each one.
3. Leave the nail in test-tube 1 exposed to the air.
4. Half cover the nail in test-tube 2 with tap-water.
5. Completely cover the nail in test-tube 3 with tap-water.
6. Fill test-tube 4 with boiled tap-water. Allow to cool from boiling, but while still hot, add the bung and seal so that no air can get into the tube. **CAUTION:** *Take care with scalding water.*
7. Prepare aqueous solutions of sodium nitrite and sodium chromate, each 10 g in 100 ml tap-water.

8. Completely cover the nail in test-tube 5 with aqueous sodium nitrite.
9. Completely cover the nail in test-tube 6 with aqueous sodium chromate.
10. Leave at least overnight and preferably for a week. Observe the effects of the different environments upon the nails.

Note: Sodium nitrite and sodium chromate in solution are called inhibitors. Inhibitors will be discussed in Chapter 13.

Exercise: Correlate any corrosion you have observed with the different environments you have created.

Experiment 3.3

Aim: To illustrate the relative energy levels of different metals by measurements of potential difference.

Apparatus
voltmeter
1× 250 ml beaker
electrodes (in any convenient geometry or size): copper, iron, zinc, tin, lead, magnesium, silver, platinum, graphite rod
electrical connections: crocodile clips for quick and easy attachment to metals
stock aqueous sodium chloride
The arrangement of the apparatus is shown in Fig. 3.2.

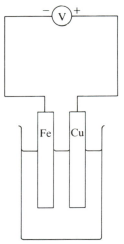

Fig. 3.2 Apparatus for Expt 3.3.

Procedure

1. Pour 200 ml stock sodium chloride solution into the beaker.
2. Use copper and iron as the two electrodes. Connect the iron to the black (negative) terminal of the voltmeter, copper to the red terminal and immerse the copper and iron in the solution in the beaker. Take care that the electrodes do not touch, and that the electrical connectors do not make contact with the solution.
3. Note the magnitude and sign of the potential when it has become reasonably steady. The potential is not constant.
4. Keep the copper in position, but exchange the iron for each of the other available metals in turn.
5. Compile a table of potentials, placing them in numerical order from the most positive to the most negative.

Note: You have shown that different materials have different energy levels. These levels are measured as a series of potentials which develop spontaneously when dissimilar metals are in electrical contact in an aqueous environment. The potentials are often called galvanic potentials and the series, a galvanic series.

Questions

What is the significance of the potentials you have found?
If you measured a different sign for some of the potentials, what does it tell you?
Draw a circuit diagram and examine the flow of electrons in the various cases. What is happening at the metal surface in solution?

Experiment 3.4

Aim: To illustrate the difference in energy levels between a metal and its oxide by measurements of potential difference.

Apparatus

1 × 250 ml beaker
2 copper electrodes
Bunsen burner
tongs
stock solution of aqueous sodium chloride

Procedure

1. Using the tongs, hold one copper electrode in the Bunsen flame until heavily oxidised (blackened).
2. Carefully cool the copper. Try to keep the black solid intact; the black oxide is brittle and may fall off. Use the oxidised copper as one of the electrodes in the apparatus of Fig. 3.2.

3. Measure the magnitude and sign of the potential between the electrodes.

Note: The potential you have measured is a difference in energy levels between the copper and its oxide.

Questions
Draw a circuit and indicate the direction of the current flow. Which of the two electrodes supplies electrons into the wires of the circuit? Which of the two electrodes is of higher energy?

Experiment 3.5

Aim: To illustrate the effect of temperature upon energy levels by measurements of potential difference.

Apparatus
2 × 250 ml beakers
2 copper electrodes
large filter paper
hotplate
stock solution of aqueous sodium chloride
The apparatus in Fig. 3.3 is for use in Expts 3.5 to 3.7.

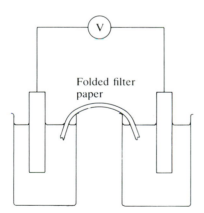

Fig. 3.3   Apparatus for Expts 3.5 to 3.7.

Procedure
1. Set up the apparatus of Fig. 3.3.
2. Immerse the copper electrodes in the solutions.
3. Heat the solution in one beaker using the hotplate.

4. When the solution is almost boiling, switch off the heat.
5. Connect the voltmeter and measure the magnitude and sign of any potential difference between the two electrodes.

Notes

The filter paper is folded into a strip and bridges the two beakers. It must be saturated with the solution from the beakers before any attempt is made to measure potential.

A Bunsen burner may be used to heat the solution but care should be taken not to burn the filter paper. It may be necessary to raise the non-heated solution on a small stand.

Question: Which of the two electrodes is in a higher energy state, the hot or the cold copper?

Experiment 3.6

Aim: To illustrate, by measurements of potential difference, the generation of energy differences because of changes in the concentration of oxygen in the environment.

Apparatus
2 × 250 ml beakers
2 copper electrodes
large filter paper
supply of nitrogen or argon
stock solution of aqueous sodium chloride

Procedure
1. Set up the apparatus as in Fig. 3.3 using copper electrodes.
2. Bubble nitrogen or argon through *one* of the solutions for 5 minutes.
3. Measure the magnitude and sign of any potential difference.

Note: The aim of bubbling gas through the solution is to reduce greatly the level of dissolved oxygen in that solution. Argon is much more efficient than nitrogen and much more expensive, so it may not be available.

Question: Which electrode is at a higher energy, the one with more oxygen in the solution or the one with less oxygen in solution?

Experiment 3.7

Aim: To show, by measurements of potential difference, how a change in concentration of the ions in solution affects the energy levels of metals.

Apparatus

> 3 × 250 ml beakers
> 2 copper electrodes
> large filter paper
> chemical: copper (II) sulphate

Procedure
1. Prepare an aqueous solution of copper (II) sulphate, 20 g in 200 ml.
2. Add 100 ml copper sulphate solution to each beaker.
3. Set up the apparatus as in Fig. 3.3, using the copper sulphate solution instead of sodium chloride.
4. Immerse the copper electrodes in the solutions.
5. Connect the voltmeter and monitor any potential differences between the two electrodes.
6. Fill the third beaker with tap-water and pour quickly into one of the solutions of copper sulphate.
7. Observe any changes of potential.

Question: Which electrode is at a higher energy, the one in the more dilute solution or the one in the less dilute solution?

Experiment 3.8

Aim: To illustrate the effect of mechanical work on the energy of a metal and how this creates localised corrosion effects.

Apparatus

> 1 piece of mild steel 30 mm × 30 mm × 10 mm
> metal punch
> chemicals: copper (II) ammonium chloride, hydrochloric acid

Procedure
1. Grind the steel specimen to remove coarse scratches.
2. Using the punch, mark the specimen so that an easily identifiable symbol is stamped into the surface of the metal to a depth of about 1 mm.

3. Grind the specimen until the identification mark is just invisible. Polish to achieve a fair mirror finish.
4. Prepare a solution of copper (II) ammonium chloride, 12 g in 100 ml.
5. Prepare an acidic solution of copper (II) ammonium chloride, 12 g in 100 ml. To this solution add 5 ml hydrochloric acid.
6. Slowly pour the non-acidified solution over the surface of the specimen, allowing the solution to run to waste.
7. Immerse the specimen in the acidified solution and leave for 1 hour.
8. Remove the specimen. The copper coating should be easily rubbed off under a running tap. The specimen should show the original identification mark.

Notes

Step 6 is essential because it provides an underlying loosely adherent layer and facilitates easy removal of the copper coating in step 8.

This experiment is the basis of a forensic technique to help identify stolen property.

Experiment 3.9

Aim: To illustrate regions of different energy within metals and how these lead to localised corrosion effects.

Apparatus

1 convenient cast brass object
commercial brass polishing fluid and cloth
cotton wool
chemicals: iron (III) chloride, hydrochloric acid

Procedure

1. Polish the specimen until a mirror finish is obtained.
2. Prepare an acidified aqueous solution of iron (III) chloride, 5 g in 100 ml. To this solution add 2 ml hydrochloric acid.
3. Dip a cotton wool ball into the solution and rub gently over the polished surface for about 15 seconds. **CAUTION:** *Avoid skin contact with the etchant; use rubber gloves.*
4. Wash under a running tap then dry.

Notes

The procedure is known as etching and is commonly used in metallurgical laboratories for examination of the grain structure of metals.

Sand cast brass objects work well in this experiment because they usually have a grain size large enough to be seen with the eye: many grain structures are visible only under a microscope.

Question: Which part of the metal is of higher energy, the area within a grain or the grain boundary?

Experiment 3.10

Aim: To use colour changes as an indicator of chemical reactions in corrosion.

Apparatus

2 Petri dishes
2 uncoated steel nails ~ 50 mm long
chemicals: agar-agar, sodium chloride, phenolphthalein, potassium ferricyanide, copper (II) ammonium chloride, hydrochloric acid, ethanol or clean methylated spirits

Procedure
1. Prepare an aqueous solution of potassium ferricyanide, 1 g in 100 ml.
2. Prepare a solution of phenolphthalein, 1 g in 100 ml ethanol (or clean methylated spirits), diluted with 100 ml water.
3. Heat 2.5 g agar-agar and 7.5 g sodium chloride in 250 ml deionised water until a uniform suspension is just obtained. Add 2 ml of the solution of potassium ferricyanide and 2 ml of the phenolphthalein solution.
4. Place an uncoated nail in one of the Petri dishes then pour over some of the liquid agar-agar mixture. Allow to cool.
5. Prepare a solution of aqueous copper (II) ammonium chloride, 12 g in 100 ml. To this solution add 5 ml hydrochloric acid.
6. Immerse a nail in the acidified solution for about 5 minutes.
7. Remove the nail and carefully dry.
8. Abrade one-half of the nail to remove the copper coat.
9. Repeat step 4 using the partially plated nail.
10. Leave both dishes for 12 hours. Effects may be observed within a couple of hours.

Notes

Phenolphthalein solutions turn pink in the presence of hydroxyl ions.

Potassium ferricyanide solutions turn blue in the presence of iron (II) ions, Fe^{2+}.

Exercise: Describe the colours you observe and explain any changes over time.

Experiment 3.11

Aim: To use colour changes as an indicator of chemical reactions in corrosion.

Apparatus

1 piece of mild steel plate in any convenient size or geometry

dropping pipette

Agar-agar indicator solution from Expt 3.10.

Procedure

1. Pipette about 1–2 ml agar-agar indicator solution used in step 4 of Expt 3.10 on to the cleaned surface of the steel plate so that a well-rounded droplet is obtained.
2. Allow to stand for an hour and observe the colour changes inside the droplet.

Note: If the stock solution of agar-agar has gelled, simply reheat and stir to obtain a uniform fluid suspension. Fresh solutions should always be used. After a few days, the efficiency of the agar-agar is greatly reduced. Once the experiments have been completed, properly dispose of the agar gels. Do not allow them to stand around in the laboratory for numbers of days.

Exercise: Correlate any colour changes with possible differences in the oxygen levels within the solution.

Experiment 3.12

Aim: To illustrate the formation of surface films which protect metals against corrosion.

Apparatus
- 1 × 250 ml beaker
- 1 × 2,000 ml beaker or plastic bucket
- 1 piece of clean mild steel sheet about 100 mm × 50 mm × 2 mm
- glass rod
- tongs
- chemical: nitric acid

Procedure
1. Carefully pour 100 ml nitric acid into one 250 ml beaker and fill the 2,000 ml beaker or bucket with tap-water.
2. Gently stand the steel plate in the acid.
3. Allow the specimen to stand undisturbed for a minute.
4. **VERY** carefully add 100 ml of water to the 250 ml beaker, pouring down its side so as not to disturb the acid.
5. Use the glass rod to tap or, if necessary, scratch the surface of the steel as it stands in the solution. When the reaction starts, use the tongs to remove the specimen and place it in the 2,000 ml beaker or bucket of water.

CAUTION: *Beware of spillage when the reaction starts. Sodium bicarbonate should be on hand to spread on any acid spillage. Safety glasses should be worn. The experiment should be carried out in a fume cupboard. Carefully dispose of solutions with copious quantities of tap water. Any item contaminated with acid should be dropped into the 2,000 ml beaker or bucket of water.*

Note: Nitric acid is an oxidising agent; it is capable of transforming metals into oxides.

Questions

Why does the metal not corrode very vigorously in the concentrated acid?

Why does it corrode when it is knocked or scratched?

Experiment 3.13

Aim: To illustrate the formation of surface films which protect metals against corrosion.

Apparatus
 1 piece of clean, polished mild steel about 100 mm × 50 mm × 2 mm
 2 × 250 ml beakers
 glass rod
 chemicals: nitric acid, copper (II) sulphate

Procedure
1. In one beaker dissolve 10 g copper (II) sulphate in 100 ml water.
2. Pour 100 ml nitric acid into the second beaker.
3. Ensure the steel is scrupulously clean and polished. Without touching the sides dip the steel vertically into the beaker of copper sulphate solution, so the solution half covers it. Leave for 1 minute.
4. Remove the steel slowly from the solution and allow excess solution to drain off.
5. Dip the steel vertically into the nitric acid until the acid level is just above the level reached by the copper sulphate solution. Leave for 2 seconds; the pink film dissolves.
6. Carefully remove from the acid, allow excess acid to drain off then immediately dip the steel into the copper sulphate solution so the level of the electrolyte does not rise above the level reached by the acid. Do not impact the steel with the bottom of the beaker.
7. Remove the steel at once from the copper sulphate solution; the wetted surface should not be coloured pink. Immediately strike the wetted surface with the glass rod. The metal should turn pink, as if by magic. If the surface turns pink before it has been hit, repeat steps 5 to 7.

Notes:
This experiment takes a little practice, but is worth the effort and is probably better demonstrated to a class. The specimen quality is very important. The first immersion causes the steel to turn pink. The surface should be uniformly coloured, otherwise the specimen is of doubtful quality and the experiment may well be unsuccessful. The immersion in the nitric acid causes the colour to disappear. The second immersion in the copper sulphate is crucial. It must be done carefully and quickly, and neither the steel nor the solution must be agitated in any way. Contact of the specimen with the

beaker must be avoided. The copper sulphate must not reach a higher level than did the nitric acid. If the steel turns pink or brown before it has been struck then the experiment has failed. Either repeat steps 5 to 7 or try again with a new specimen.

Questions

What is the pink coloration?

Why does it disappear in nitric acid?

Why does it not reappear during the second dip into the copper sulphate?

Why did the colour reappear only after the impact of the glass rod?

Experiment 3.14

Aim: To illustrate the formation of surface films which protect metals against corrosion.

Apparatus

1 piece of aluminium sheet about 100 mm × 50 mm × 2 mm

screwdriver

dropping pipette

chemicals: mercury (I) nitrate, nitric acid

Procedure

1. Prepare an aqueous solution of mercury (I) nitrate, 1 g in 100 ml. To this solution add 1 ml of nitric acid. **CAUTION:** *Toxic solution; use rubber gloves.*
2. Pipette a drop of solution on to the clean surface of the aluminium to give a drop of about 20 mm diameter.
3. Using the pointed corner of the screwdriver, score the sheet, commencing on one side of the drop and passing through it to the other side.
4. Dry the metal with an absorbent paper towel and observe it for 10–15 minutes.

Note: Keep the specimen in a draught-free environment because the corrosion product is extremely feathery and will be destroyed by the slightest breeze. Even heavy breathing near the corrosion product will spoil the result.

Experiment 3.15

Aim: To illustrate the crevice corrosion of stainless steels.

Apparatus
    1 × 250 ml beaker
    1 piece stainless steel about 100 mm × 50 mm × 2 mm
    1 rubber band
    chemicals: iron (III) chloride, hydrochloric acid
    stock sodium chloride solution

Procedure
1. Tautly wrap the rubber band around the length of the specimen.
2. Prepare an acidified solution of iron (III) chloride, 5 g in 100 ml. To this solution add 2 ml of hydrochloric acid.
3. Pour 100 ml stock sodium chloride solution into the beaker and add 5 ml acidified iron (III) chloride solution.
4. Stand the specimen in the beaker and leave for 24 hours.
5. Remove the specimen, wash under a running tap then dry.
6. Remove the rubber band.

Note: The best material is a 430 type stainless steel, or other member of the 400 series, but a sample of type 304 will probably give the desired result. A simple identification test is to measure the magnetic property of the stainless steel. Alloys in the 400 series are magnetic, but those in the 300 series are not.

Exercise: Comment upon the performance of stainless steel in the given electrolyte, and give a reason for the behaviour you have observed.

Experiment 3.16

Aim: To illustrate the effect of the combination of stress and a specific environment upon 70/30 brass.

Apparatus
    1 × 250 ml beaker
    tongs
    2 identical pieces of hard drawn 70/30 brass tube 20 mm length, 20 mm diameter, 2 mm wall thickness.
    chemicals: mercury (I) nitrate, nitric acid

Procedure

1. Heat treat one piece of tube for 1 hour at 500°C. Alternatively, heat in a Bunsen flame for 5 minutes and cool under a running tap.
2. Prepare an acidified aqueous solution of mercury (I) nitrate, 1 g in 100 ml. To this solution add 1 ml nitric acid.
3. Make sure the two specimens are distinguishable.
4. Immerse both specimens in the solution and leave for 10–15 minutes.
5. Using tongs remove the specimens and inspect for cracks.

CAUTION: *The solution is very toxic and should be handled carefully; use rubber gloves throughout.*

Notes

The heat treatment is to relieve stresses residual from the manufacturing process.

If the residual stresses are sufficiently high the tube may disintegrate. Gentle pressure in the tongs may be enough to achieve this. If, on the other hand, the tube thickness is too great or the residual stresses are too small, the material may not crack at all.

Question: What is the silvery coating on the tubes after removal from the solution?

3.3   DISCUSSION

Corrosion is a chemical reaction of a metal with its environment as a result of which electric current flows. All the environments which have been considered in the experiments have been aqueous, but this does not mean that corrosion is absent when water is absent. Many corrosion reactions occur in what can be considered as dry environments. And remember that corrosion can occur in the atmosphere where the presence of water vapour condensing on a cold surface, together with an ionic material from any available source, is enough to cause corrosion for the same reasons as if the metal were immersed. The presence of both water and an ionic material is complementary; current can only be carried through water by free ions; the water causes dissociation of the ionic solid to produce the necessary free ions. **Experiment 3.1** was designed to show that electric current flows in solutions only if they possess ions; current flows in aqueous sodium chloride. If there are no ions, as in white spirit, or very low quantities of ions, as in pure water, there is no current flow and no electrical circuit.

Experiment 3.2 is an introduction to corrosion reactions. It is designed to show how variations in the environment cause very different degrees of corrosion. The requirement for ions, oxygen and moisture to be present is clearly demonstrated. An interesting lack of corrosion is obtained in the presence of inhibitors, discussed in Chapter 13.

In the first experiment a current was driven around the circuit using a source of potential. In the other experiments this was absent, or was it? Of course not, it was merely disguised. In **Experiment 3.3** a source of potential was created by connecting two different metals together. *Remember in Chapter 2 it was shown that electrical potential can be considered to be the same as energy, creating a 'force' which drives electric current from a high energy state to a low energy state.* If it occurs spontaneously, this is in keeping with the rule that nature is striving to achieve low energy states. Corrosion is a spontaneous process and we stimulated it by connecting dissimilar metals in Expt 3.3. The potential difference measured was different for each combination, and by examining the direction in which the current flowed in the circuit you should have been able to tell which electrode was supplying electrons into the external circuit. Remember the corrosion reaction illustrated by eqn (2.7) which shows how the metal generates electrons for the external circuit. In any of the experiments, the metal which does so must be corroding.

Some of the most common corrosion mistakes made by engineers relate to galvanic effects and dissimilar metal couples. This is the subject of Chapter 5 and will be discussed again in the theory of Chapter 4.

It should be clear that it is not necessary to have two different metals connected together in order to obtain corrosion. The object of Expts 3.4 to 3.9 is to show how corrosion may occur on the surface of a single piece of metal.

Question: *Then why do the experiments use two electrodes?*

The two identical electrodes make it easier to measure potentials; they are linked as one piece of metal via the external circuit. By using identical electrodes we can explore the effects produced by small changes in their surface condition or by small changes in the environment. Once this is achieved, the energy differences lead to potential differences, which in turn cause corrosion.

Experiment 3.4 showed the potential difference between a metal and one of its chemical compounds, in this case the oxide. Remember, a metal behaves like an assembly of atoms, but in its oxide it exists as ions. You should have discovered that the metal is in a higher energy state than its oxide and provides the electrons for the circuit, i.e. it corrodes. Thus, in any piece of metal that is partially coated with oxide, the metal will form a local corrosion cell with the oxide; it will corrode. If there are no pores in the film and the oxide layer completely covers the metal, this cannot occur; it is protected from corrosion and we say that it is **passivated**.

An obvious way of giving a material energy is to heat it. In **Experiment 3.5** a metal was heated and compared with another piece of the same metal at room temperature. The sign of the potential you measured should have told you that the hot electrode had more energy than the cold electrode. The implication of this is that in a metal structure where there are temperature variations, local corrosion cells can be established in which the hot areas begin to corrode in preference to the cold areas.

In **Experiment 3.6** a similar effect was created by the expulsion of oxygen from one of the solutions. The result obtained should have been that the electrode in the solution with less oxygen was supplying the electrons and therefore corroding. Again, we can say that any system where there is a variation in the concentration of oxygen is a possible corrosion cell. This is a vital conclusion with respect to what happens inside crevices or in other areas which have restricted access to air. Thus we see again that water is not the only important ingredient for corrosion; oxygen also plays a part. Crevice corrosion is discussed in Chapter 7.

Experiment 3.7 showed how changes in the concentrations of ions in the environment also generate potentials. A change in the concentration of liquid passing down a metal pipe, for example, could produce a corrosion problem for a process engineer.

Experiment 3.8 is interesting as an act of magic. An identification mark, apparently removed, has been made visible again. The reason why this can be done is that when the mark is punched into the metal, the material is mechanically worked and some of the energy is transmitted to the volume of metal immediately below the mark. Even though the mark is removed the energy remains. An etching process causes preferential corrosion in those areas where the newly exposed surface has more energy, and magically the mark reappears.

Experiment 3.9 is not an obvious corrosion experiment but the method is fundamental in metallographic analysis. It works because grain boundaries are of higher energy than areas within grains. Atoms at grain boundaries have higher energy because they are mismatched to the regular crystal lattice; nature would prefer them to be neatly located in ideal sites. Contact with a corrosive solution, the etchant, causes the grain boundary areas to corrode more than the areas within the grains. This transforms a mirror-smooth surface into a surface with visible topography. The experiment should have shown the large grain structure common in cast brasses. Grain boundary corrosion is discussed again in Section 6.1.

The use of indicators in **Experiment 3.10** makes a colourful demonstration. The blue colour which develops shows that iron is corroding (anodic) and the regions which turn pink correspond to the lower energy areas (cathodic). Nails are made by stamping wire and shaping the ends. You should observe that the head and the tip are anodes (blue coloration) because they have been subjected to more mechanical work. When half the nail has been coated with copper, a small galvanic cell exists and the steel end corrodes, turning blue, while the coppered half goes pink. Sometimes, if the coated copper layer is porous, the opposite will occur.

An interesting variation is described in **Experiment 3.11** in which the colour changes beneath the droplet show that the area at the centre of the drop is the zone of corrosion. This is a result of the geometry of the droplet; the diffusion path of oxygen through the environment is longer at the centre of the drop than at its edge. For a detailed discussion see Section 7.2. The colour changes are therefore indicative of accessibility by oxygen and reinforce the conclusion of Expt 3.6: areas with less oxygen are likely to corrode.

In **Experiment 3.12** the lack of corrosion of steel in nitric acid is examined. Because nitric acid is an oxidising agent, it causes a very thin layer of unreactive oxide to form on the surface. This effectively excludes the environment and there is no appreciable corrosion, even when the acid is diluted. (The film cannot be generated in dilute acid alone.) However, the film is very fragile; a small tap is usually enough to break it and initiate rapid corrosion.

The most visually dramatic of all the experiments and one which is worth practising is **Experiment 3.13**, an extension of the previous one. Immersion of the steel in the copper sulphate solution produces a thin plating of pink copper on the surface. A detailed discussion is provided in Section 4.1. When the plate is then immersed in nitric acid, the copper dissolves and the steel is given a very thin and fragile protective oxide film. The second immersion in copper sulphate does not plate copper because the protective coating acts as a barrier. When the steel is removed from the solution, mechanical damage to the inert film allows copper ions access to the steel beneath so the pink copper plates out immediately.

A similar effect is illustrated by **Experiment 3.14** in which the surface oxide film on aluminium is damaged while immersed in a corrosive medium. In air, aluminium rapidly oxidises to repair any damage to the adherent pore-free corrosion-resistant layer. If the film is disrupted while the metal is immersed in a corrosive solution, then the resulting reaction is rapid, and in this case spectacular.

Returning to crevice corrosion, **Experiment 3.15** shows clearly the fallibility of believing that stainless steel is the answer to all corrosion problems. Stainless steel is particularly susceptible to corrosion in areas where there is restricted access by oxygen. The mechanisms are discussed in Section 7.1.

Another form of corrosion, the subject of Chapter 10, is illustrated in **Experiment 3.16**. Here, the combination of a specific environment and a specific material in which there is residual stress can lead to catastrophic failure. The example chosen is one of the most rapid for ease of demonstration, but in other guises the problem is widespread throughout the engineering world. The stresses which remain in a tube after fabrication are sufficient to cause the rapid failure observed in the experiment. Elimination of the stress by the heat treatment renders the material much less susceptible.

Now that the scope of corrosion has been examined in the laboratory and you have begun to see some of the reasons why metals corrode, we shall look in detail at the theory of aqueous corrosion.

4   THE THEORY OF AQUEOUS CORROSION

The whole of science is nothing more than a refinement of everyday thinking.

(Albert Einstein, *Out of my Later Years*)

In nature, most metals are found in a chemically combined state known as an ore. Ores may be oxides, sulphides, carbonates or other more complex compounds, and because many have been present in the Earth's crust since it was formed, we may presume that their chemical condition is somehow preferred by nature. This preference is a result of fundamental laws which make up the study of thermodynamics, and using its terminology we say that ores and other such compounds are in low energy states. In order to separate a metal such as iron from one of its ores, e.g. iron oxide, it is necessary to supply a large amount of energy. This is usually done by heating (with charcoal added to react with the oxygen in the ore) in a blast furnace to about 1,600°C. Therefore metals in their uncombined condition are usually high energy states. This is illustrated in Fig. 4.1 where an energy profile has been used to chart the thermodynamic progress of a typical metal atom from being combined in an ore, separated as a metal atom and recombined as a corrosion product. In the figure our rather vague term, energy, has been more specifically called **free energy**, following Section 2.1.

Thermodynamic laws tell us that there is a strong tendency for high energy states to transform into low energy states. It is this tendency of metals to recombine with components of the environment that leads to the phenomenon known as corrosion. A good definition of corrosion is

> **Corrosion is the degradation of a metal by an electrochemical reaction with its environment.**

You might like to recap this definition by referring back to Section 2.2.

The low energy corrosion product generated when a metal degrades is obviously not the same as an ore, although it may be similar, as in the case of iron oxide and rust, and the energies of the ore and the corrosion product may be comparable.

Question: *Why does an uncombined metal exist at all?*

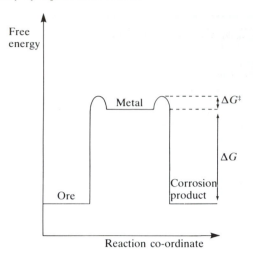

Fig. 4.1 A thermodynamic energy profile for metals and their compounds.

The free energy difference between a metal and its corrosion product, ΔG, in Fig. 4.1 represents only the *tendency* of the metal to corrode; it tells us nothing about the corrosion *rate*. The reason it tells us nothing is because there is an energy barrier between the metal and the corrosion product. Metal atoms must surmount this barrier before they can corrode and more energy must be supplied for this to occur. In our example, the energy barrier is called the **free energy of activation**, represented by the symbol $\Delta G^{\ddagger}$. The size of the free energy of activation determines the rate of a corrosion reaction; its rate constant we shall denote by k_{corr}. The rate of a corrosion reaction, v, can be expressed as

$$v = k_{\text{corr}}[\text{reactants}] \tag{4.1}$$

where

$$k_{\text{corr}} = A\exp(-\Delta G^{\ddagger}/RT) \tag{4.2}$$

A is an undefined constant, R is the universal gas constant and T is the absolute temperature.

It is well known that hydrogen and oxygen have a strong tendency to form water. The relatively high energy states of the gases can be converted into the low energy state of water with an explosive release of energy, but no measurable reaction occurs in a mixture of the gases until a spark is introduced. The necessary energy has then been provided to initiate what becomes a chain reaction, and without it the rate of reaction is extremely slow, even though its thermodynamic tendency is large. The next section examines in more detail the effect of thermodynamics on corrosion and the benefits obtained from its study.

4.1 THERMODYNAMICS OF CORROSION REACTIONS

We have seen that free energy is the single factor which determines whether or not corrosion will take place spontaneously. This is not surprising:

All interactions between elements and compounds are governed by the free energy changes available to them.

In Section 2.1 it was stated that, for a reaction to occur spontaneously, there must be a *net release* of free energy. In this treatment, individual free energies of species are expressed as G and the net change of energy in a reaction is given by ΔG. In line with our notion that natural (spontaneous) changes involve a transition from a high to a low free energy state, it is conventional to denote energy *given out* by a *negative sign,* whereas a net *absorption of energy* by the system is given a *positive sign.*

For a spontaneous reaction to occur, ΔG must be negative.

In Fig. 4.1 you can see that for the metal to corrode spontaneously, the change in G must be negative. At room temperature most chemical compounds of metals have lower (more negative) values of G than the uncombined metals, hence our original premise:

Most metals have an inherent tendency to corrode.

In reality, most metals we use in everyday life have already 'corroded', that is, they have invisible layers of oxides on their surface which render them less corrodible than they might otherwise be. This property is called **passivation**. Aluminium is so reactive towards oxygen that it immediately forms a passive surface layer of aluminium oxide upon exposure to air. This layer is very thin, so we are not ordinarily aware of it, but it is remarkably protective, separating the reactive metal atoms beneath the surface from any corrosive environments. Only when a passive oxide film does not completely cover the surface, or is itself attacked by a species in the environment, does the metal suffer some form of corrosion. **Anodised** aluminium (Section 15.4) has undergone a treatment in which the passive film has been 'reinforced' and rendered even more protective. As a result of the presence of passive films, aluminium is not as corrodible as its chemical reactivity suggests. There are many other examples.

When hot-rolled steel is made, the high temperatures at which it is worked cause oxidation of the surfaces, but the oxide film varies considerably in the degree to which it protects the steel, depending upon the proportions of the various iron oxides in the film. However, because the rate of reaction between iron atoms and oxygen is very slow at ambient temperatures, cold-rolled steel does not have the benefit of an oxide film and its bright finish is much more susceptible to rusting. Steel is made stainless, i.e. more corrosion-resistant, by alloying with chromium, which can form a more protective layer of chromium oxide over the steel surface. As we shall see in Chapter 7, the corrosion properties of stainless steels are strongly governed by chromium oxide films.

Titanium is a remarkably non-corrodible metal because of the presence of an even more protective surface film of titanium oxide.

In summary, the actual nature of the solid surface exposed to an environment is complex. The electrical resistivities of some surface films may be quite high when measured in bulk, but in the thicknesses of films present on metals they do not prevent the passage of electricity between solid and liquid.

Question: *Why don't gold, platinum and other precious metals corrode?*

The theory we have developed so far gives two explanations. First, the energetics may not be favourable. Consider the following reactions and the respective free energy change per mole:

$$Mg + H_2O + \frac{1}{2}O_2 \rightarrow Mg(OH)_2; \ \Delta G° = -596 \, kJ \, mol^{-1} \tag{4.3}$$

$$Cu + H_2O + \frac{1}{2}O_2 \rightarrow Cu(OH)_2; \ \Delta G° = -119 \, kJ \, mol^{-1} \tag{4.4}$$

$$Au + \frac{3}{2}H_2O + \frac{3}{4}O_2 \rightarrow Au(OH)_3; \ \Delta G° = +66 \, kJ \, mol^{-1} \tag{4.5}$$

We shall use the symbol ($\circ$) to represent parameters at 298 K and 1 atmosphere pressure, i.e. **standard state parameters**. The free energy data show clearly that the corrosion reactions of copper and magnesium have negative values of $\Delta G°$, while that for gold is positive. Thus, thermodynamics alone tell us that copper and magnesium are expected to corrode naturally in wet aerated atmospheres, whereas gold is not.

A second answer to the question is that the free energy of activation may be too great, so the rate is very slow. A detailed discussion of corrosion rates will be reserved for later sections of this chapter.

It is not just the precious metals which can remain uncorroded in nature. Iron objects have been found remarkably preserved after centuries of immersion at the bottom of peat bogs. This is almost certainly because of the total exclusion of oxygen which, as we found in Expt 3.2, is often necessary for corrosion to occur. The example is a good one because it reminds us that the environment is just as important as the metal.

There may always be special circumstances why corrosion does not occur when expected, but the general principles outlined above form a useful guide.

Experiment 4.1

Place a piece of iron or mild steel about 100 mm × 50 mm × 4 mm in a 250 ml beaker containing a solution of 10 g copper (II) sulphate in 100 ml water. After about 10 minutes you will see that the iron becomes coated with copper, an effect we shall call **replating**. If the solution were to be analysed it would be found to contain iron ions.

The example in Expt 4.1 has been chosen because it illustrates a number of points and aids our development of a theoretical approach. Although there are countless combinations of metal and environment in which corrosion may be observed, it is still possible to reduce most of them to a set of common features. In Section 2.2, a corrosion reaction was defined by the general equation (2.7), but it was pointed out that this was only one half of an overall process in which the electrons produced in eqn. (2.7) are consumed by another reaction. In aqueous corrosion, these two reactions occur at the solid/liquid interface; the electron-producing reaction is an **anodic** reaction, also called an **oxidation** process. Note that the term used in this sense is not concerned with a reaction involving oxygen. The electron-consuming reaction is one of **reduction** and is termed **cathodic**.

Both anodic and cathodic reactions may occur on the same metal surface.

Question: *What is happening in the beaker in Expt 4.1?*

We can observe that the iron is corroding; in other words, the metal is being degraded into ions according to the simplified equation:

$$Fe \rightarrow Fe^{2+} + 2e^- \tag{4.6}$$

In contrast, copper ions are being replated from the solution:

$$Cu^{2+} + 2e^- \rightarrow Cu \tag{4.7}$$

This cathodic reaction is the reverse of a corrosion reaction, something not true of all cathodic processes. Other cathodic reactions will be discussed below. Equations (4.6) and (4.7) are called **half-reactions** because they show how electrons are being produced or consumed, but we as observers are unaware of the electrons because they flow around a circuit and do not accumulate in bulk at any given point. The complete effect, which we do observe, is represented by the addition of the two reactions:

$$Fe + Cu^{2+} \rightarrow Fe^{2+} + Cu \tag{4.8}$$

Notice that although eqn (4.7) would be expected to have a positive value of ΔG, the reaction is driven by an overall reduction in free energy, the free energy change for eqn (4.8). Since we know that this reaction is occurring spontaneously, we know that $\Delta G < 0$.

All corrosion is dependent upon temperature because the free energy states of the species depend upon temperature. To calculate the value of ΔG at any given temperature, use the following important thermodynamic equation:

$$\Delta G = \Delta G^\circ + RT \ln J \tag{4.9}$$

where, for a reaction

$$jA + kB \rightarrow lC + mD$$

J is defined as

$$J = \frac{[C]^l[D]^m}{[A]^j[B]^k} \tag{4.10}$$

The expression for *J* contains any arbitrary (non-equilibrium) values corresponding to the non-equilibrium free energy change, ΔG, in eqn (4.9). If the system reaches a point where there is no net change of free energy, we say the system is in equilibrium and $\Delta G = 0$. Then, $J = K$, where *K* is the equilibrium constant. This leads to

$$\Delta G^\circ = -RT\ln K \tag{4.11}$$

According to Section 2.3, terms in the square brackets in eqn (4.10) represent the concentrations of the species in the solution. More precise work requires the use of a quantity called an activity but this will not be used here. Concentrations are expressed in molarities (moles per litre), see Section 2.2.

In the case of the copper/iron system we can use eqn (4.8) to substitute for *J* in eqn (4.9), thus

$$\Delta G = \Delta G^\circ + RT\ln\frac{[Fe^{2+}][Cu]}{[Cu^{2+}][Fe]} \tag{4.12}$$

It can be shown that the terms involving the concentration of a pure solid can be replaced by unity.

Remember that we are considering the corrosion of iron in a solution of copper sulphate, i.e. a *spontaneous* reaction, and it is the free energy change which is driving the reaction.

Question: *How can I measure these energy changes and hence study the corrosion in more detail?*

Experiments 3.3 to 3.7 showed that these free energy differences are measurable as electrical potentials and flow of current. Thus, electrical measurements are one way of studying corrosion in more detail. The exact methods involved will be discussed later, but at this stage we can be content to define the expression which relates electrical potential with free energy. This equation is due to the great scientist, Michael Faraday, who expressed the work done (the free energy change of the corrosion process) in terms of the potential difference and the charge transported:

$$\Delta G = (-zF)E \tag{4.13}$$

Equation (4.13) is known as **Faraday's Law**. The symbol *F* represents the charge transported by one mole of electrons and has the value of 96,494 coulombs per mole. The potential, *E*, is measured in volts, and *z* is the number of electrons transferred in the corrosion reaction; $z = 2$ in the case of the iron reaction above. A negative sign is necessary to indicate the conventional assignment of negative charge to electrons, and since *z* and *F* are positive constants, this leads to a positive measured potential when the reaction is spontaneous. The equation tells us that free energy change is directly measurable with electrochemical potential.

Again, using the superscript (∘) to represent standard conditions, we can rewrite eqn (4.13) as

$$\Delta G^\circ = -zFE^\circ \tag{4.14}$$

The concentration of pure, solid substances is always treated as unity, thus we can substitute into eqn (4.12) to obtain

$$-zFE = -zFE^\circ + RT \ln \frac{[Fe^{2+}]}{[Cu^{2+}]} \tag{4.15}$$

Dividing each term of eqn (4.15) by $-zF$,

$$E = E^\circ - \frac{RT}{zF} \ln \frac{[Fe^{2+}]}{[Cu^{2+}]} \tag{4.16}$$

This equation can be written more generally as

$$E = E^\circ - \frac{RT}{zF} \ln \frac{[products]}{[reactants]} \tag{4.17}$$

Equation (4.17) is known as the **Nernst equation** and it has great theoretical and practical significance. As a consequence, it is common to introduce the numerical values into the equation. Using the standard temperature, 298 K, and $R = 8.3143 \text{ J mol}^{-1} \text{ K}^{-1}$, together with conversion into logarithms to the base 10, eqn (4.17) becomes

$$E = E^\circ - \frac{0.059}{z} \lg \frac{[products]}{[reactants]} \tag{4.18}$$

where E is the non-equilibrium potential generated by the reaction, [reactants] is the reactant concentration and [products] is the product concentration. If these concentrations were chosen to be the equilibrium values, there would be no driving force ($\Delta G = 0$) and E would be zero. Notice that this condition would allow the evaluation of the material constant, E°.

4.2 THE BASIC WET CORROSION CELL

Corrosion science involves a study of **electrodics**, electrochemical processes which take place at **electrodes**. An electrode is essentially the boundary between a solid phase (metal) and a liquid phase (aqueous environment) and these processes take place across the phase boundary. They involve effects of both mass and charge; mass is usually transferred (in both directions) between the solid metal and the liquid environment, whilst charge is exchanged between atoms and ions. We tend to think of *atoms* in solid metals, but we tend to think of *ions* in the electrolyte, in solid films of surface oxide, or in corrosion products.

Note also that electrons as isolated particles never cross the interface; all charge is carried in electrolytes in the form of ions.

It is usually possible to identify different regions of a corroding metal/electrolyte interface at which the electrodic processes occur. If the reactions are net anodic (electron-producing), we call that part of the interface an **anode**; if they are net cathodic (electron-consuming) we call that part of the interface a **cathode**. There are many reasons for different regions of an interface to be anodes and cathodes. Some of them were investigated in the earlier experiments, and the next few chapters explore the subject in more depth. In this discussion, we shall consider a system in which a single anode and cathode are present in a system known as the **basic wet corrosion cell**. Four essential components can be identified: anode, cathode, electrolyte and connections.

The part of a metal/electrolyte interface which behaves as an **anode** usually corrodes by loss of electrons from electrically neutral metal atoms in the solid state, forming discrete ions. These ions often enter the solution, but they may also react with other species at the interface to form insoluble solid corrosion products which usually accrue on the metal surface. This is a common anode reaction in neutral or alkaline environments and may block further metal dissolution, retarding the corrosion and resulting in passivation. As we have seen, the corrosion reaction of a metal M is usually expressed by the simplified equation:

$$M \rightarrow M^{z+} + ze^- \tag{2.7}$$

in which the number of electrons taken from each atom is governed by the valency of the metal. Commonly, $z = 1$, 2 or 3.

The cathode is an essential complement to the anode because it consumes the electrons generated by the anode. A cathode does not normally corrode, although it may suffer damage under certain conditions which will be discussed later. To determine the processes occurring at cathodes, we look for possible electron-consuming (reduction) reactions. Perhaps the most obvious is the reverse of eqn (2.7):

$$M^{z+} + ze^- \rightarrow M \tag{4.19}$$

commonly referred to as a **replating** reaction. Two other important reduction reactions may also occur at the cathode: a two-step process in which hydrogen gas is formed:

$$H^+ + e^- \rightarrow H \text{ (atom)} \tag{4.20}$$

$$2H \rightarrow H_2 \tag{4.21}$$

and a process that consumes dissolved oxygen and generates hydroxyl ions:

$$2H_2O + O_2 + 4e^- \rightarrow 4OH^- \tag{4.22}$$

As we saw in Section 2.3, hydrogen ions, H^+, are always present in water, to a greater or lesser extent, and therefore their reduction, represented by eqns (4.20) and (4.21), is always possible. The reaction varies with pH and is more likely at low pH, when $[H^+]$ is high.) The second reduction reaction, eqn (4.22), is dependent upon the level of dissolved oxygen in solution. Although in well-aerated solutions this is typically about 5–10 ppm and seems rather small, nevertheless, this amount is quite sufficient for this process to be important. When the level of dissolved oxygen varies locally in a corrosion cell, differential-aeration corrosion may result (see Chapter 7 for a full discussion). Other cathode reactions are possible; remember the only criteria are that the reaction must consume the electrons produced by the anode process and that the energy change must be favourable.

An **electrolyte** is an electrically conducting solution. Very pure water is not normally considered to be an electrolyte; the conductivity of typical commercial deionised water is about 1–10 mS m^{-1}. Under most practical conditions, however, an aqueous environment will have a sufficient conductivity to act as an electrolyte. 'Soft' tap water has conductivity typically about 10–20 mS m^{-1}, compared with a value for 3.5% sodium chloride solution of 5.3 S m^{-1}. At this stage, it is not intended to present an exact definition of an electrolyte; our current ideas will serve us well enough for the time being. In hot corrosion (Section 17.5) there need be no water present, for molten salts act as electrolytes.

The anode and cathode require an **electrical connection** for a current to flow in the corrosion cell. Obviously, a physical connection is not necessary when the anode and cathode are part of the same metal. In some case histories, it is the failure to identify the electrical connection that leads to unexpected corrosion, for example, Case 12.13.

All aqueous corrosion reactions can be thought of in terms of a simple wet corrosion cell, though they need not physically resemble it. Even when they are part of the same metal surface, anodes and cathodes are usually identifiable. Several of the experiments in Chapter 3 illustrate how this is possible. If we concede that, for corrosion to occur, the four components must be present, then

> **The removal of any one of the four components of the simple wet corrosion cell will stop the corrosion reaction.**

This is our own First Law of Corrosion Control. The control of corrosion is obviously a very important subject, to which much space is devoted in this book. Many of the measures adopted in the later chapters are just ways of applying this law.

Corrosion, then, arises when a chemical process is made possible by a net release of free energy across a metal/electrolyte interface. As expressed in eqn (4.13), this energy difference manifests as an electrical potential and a *potential* is precisely what it says, a *tendency* for corrosion.

We will now conduct a type of experiment known in physics as a thought experiment, an experiment we carry out in our minds only because we want to simplify a situation and there are practical difficulties which would distract from the process of explanation. Real systems will be introduced as we proceed.

Suppose we immerse an ideal metal, A, in pure water. We can imagine that, as in eqn (2.7), metal atoms may be converted into positively charged metal ions

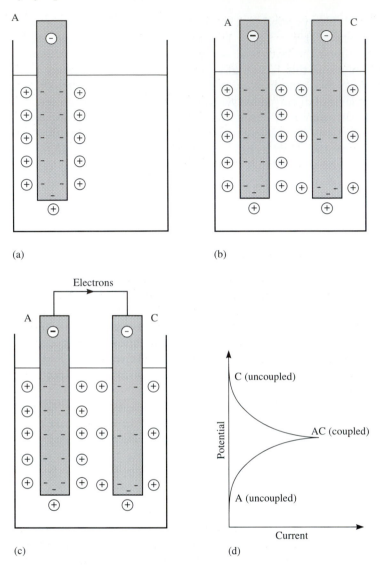

Fig. 4.2 (a) Atoms of metal A, immersed in a beaker of 'pure' water, corrode and form positive ions in the water, adjacent to the electrons they have left behind. (b) Atoms of metal C do the same as metal A, but to a lesser extent, leaving less negative charge in the metal. (c) A wire connection from metal A to metal C causes electron flow from the most negative to the least negative metal. (d) A graph of potential versus current for the situation in (c). As current flows, metal A becomes less negative and metal C becomes more negative.

which enter the electrolyte, leaving behind the electrons in the metal, as shown in Fig. 4.2(a). This is rather like the behaviour of electrical capacitors in that we have succeeded in bringing about a charge separation: the metal has become negatively charged overall and attracts the positive ions which are in solution.

As soon as there are metal ions in the electrolyte, there is a tendency for them to revert back to metal atoms in the solid phase once again. This tendency will increase as the concentration of metal ions increases.

It seems reasonable to assume that, with no other external influences, there may be some limit to these forward and reverse processes, and indeed there is. The system reaches what we call a **dynamic equilibrium** in which metal atoms and ions are continually entering and leaving the electrolyte. Overall, the observer sees no *net* change in the system, but the electrode has developed an **electrochemical potential**. We can investigate this with a real experiment. Students should realise that this is a simulation which really only tells part of the story at this stage; the full explanation will be given when we discuss mixed potentials.

Experiment 4.2(a)

To represent metal A, immerse a piece of zinc 100 mm × 50 mm × 4 mm in a 250 ml beaker containing 100 ml of 3.5% aqueous sodium chloride solution. Zinc is by no means an *ideal* metal, but it is fine for this experiment. The dimensions are not critical; use a piece of reasonable size to fit in the beaker if nothing else is available. Using a crocodile clip and a wire, connect the zinc to the black (negative) terminal of an electronic multimeter in volts mode. Immerse a saturated calomel reference electrode (SCE) in the beaker and connect it to the red (positive) terminal. At this stage you need only know that the SCE maintains a stable potential when immersed in a solution. Reference electrodes will be discussed in Section 4.4. Observe and note the potential reading for a few minutes.

Let us use the reference electrode (SCE) as an arbitrary datum of zero volts. Remember, when we refer to the electrochemical potential of an electrode, we do not mean just the excess electrons in the metal, but the potential of the complete assemblage of charge distributed over the solid/liquid interface. The arrangement and number of ions in the adjacent solution contribute as much to the potential as the excess electrons in the metal.

Now consider what would happen if we were to add a different metal, C, to the same solution, as shown in Fig. 4.2(b).

Experiment 4.2(b)

To represent metal C, substitute a piece of clean mild steel plate, approximately 100 mm × 50 mm × 4 mm for the zinc in Expt 4.2(a). Observe and note the potential reading for a few minutes. Next, remove the SCE and disconnect it from the multimeter. In its place, use the piece of zinc from Expt 4.2(a). Keep the meter set to volts and note the potential reading for a few minutes. Remember that multimeters used in volts mode have very high impedance (resistance) which stops current flowing between the metals.

We might expect exactly the same behaviour for electrode C as for electrode A, except that there is no reason to suppose that the amount of metal dissolution, and hence the amount of negative charge generated on C, will be the same as for A. Let us assume that fewer free electrons are created in electrode C. Thus our two different electrodes will have two different electrochemical potentials with electrode A being more negative than electrode C. Does this statement correlate with your observation in Expt 4.2(b)? You should have measured about –0.65 V for steel and about –1.05 V for zinc, i.e. about 0.4 V difference between the two metals.

Now, back in our thought experiment, in Fig. 4.2(b), let us make a wire connection to the two metals. What do we expect to happen?

Experiment 4.2c

Switch the multimeter to a current setting. This removes the very high impedance of the voltmeter and allows current to flow between the zinc and the iron. Observe and note the current. Which way is the current flowing?

Figure 4.2(c) shows the answer. The excess electrons in the two electrodes will try to minimise their potential energy by distributing themselves around the system. Since there are more free electrons in A than in C, electrons will flow from electrode A to electrode C. Electrode A loses some of its free electrons and becomes *less* negative overall. Electrode C gains free electrons and becomes *more* negative overall. The result of this is that the positive ions in the liquid phase of the interface of electrode A are now held more loosely. They may drift away into the bulk electrolyte, disturbing the equilibrium previously established and allowing further dissolution of metal A. Corrosion continues at A. On the other hand, the positive ions in the liquid phase of the interface at electrode C will be more strongly bound to the metal and may be converted back into metal atoms in the solid phase. Corrosion will be inhibited at electrode C.

This effect can be illustrated by means of an **Evans diagram** in which we plot electrical potential on the vertical axis, and electron current on the horizontal axis (Fig. 4.2(d)). In Fig. 4.2(b) we saw that, with metals A and C isolated, metal A is more negative than metal C. Also, because of the state of equilibrium existing on both metals, there is no measurable bulk electron current, so A and C are located accordingly on the vertical axis. We also said that when A and C are connected, an electron current flows *from* metal A *to* metal C. The electrochemical potentials of A and C are equal, as indicated by the potential coordinate of the point of intersection of the lines in Fig. 4.2(d). This system, in which two different metallic solids are in electrical contact and immersed in an electrolyte, is called a **bimetallic couple**. **Dissimilar metal corrosion** often results from bimetallic couples and is discussed in Chapter 5.

Compared with the situation *before* we connected them, metal A is now more positive and metal C more negative. The increase of (anodic) current along the horizontal axis for metal A, the anode, tells us that it is suffering enhanced corrosion. The increase of (cathodic) current for metal C, the cathode, tells us that it is protected. Notice that in this form of the Evans diagram, no sign is attributed to the current; the x-axis records the *magnitude* only.

In studying this system we have discovered an extremely important effect in corrosion which can be used to great effect in controlling the corrosion of large, expensive structures such as ships, buried pipelines or offshore oil rigs. When two metals are connected together, it is likely that the corrosion of one will be greater than if it were isolated. On the other hand, the other metal of the pair will be protected compared to its behaviour when isolated; this is the principle we adopt in the **sacrificial anode** method of **cathodic protection**. By using zinc to protect steel, we know that the zinc, which is more electronegative, will corrode sacrificially, protecting the steel. This important topic will be discussed fully in Chapter 16.

4.3   STANDARD ELECTRODE POTENTIALS

In the basic wet corrosion cell we saw that potential differences between anode and cathode can be measured by simply inserting a voltmeter into the circuit. While this is quite acceptable for most basic laboratory measurements, more accurate work requires the addition of a third electrode to the cell. This is necessary because of the need to define an absolute value for the electrode potential. The problem in the measurement of potentials is that only the potential *difference* can be measured. In Fig. 4.2(c), with an iron cathode and a zinc anode, the potential difference between the iron and the copper might be measured as a constant value, although the individual potentials of the iron and the zinc might be varying equally.

It would be very desirable to be able to predict the potential of a single metal electrode in an electrolyte, but it is impossible to measure potentials absolutely because all measurements are a comparison of one potential with another. This problem is overcome by defining a standard electrode against which all other measurements can be compared. Then, by defining its potential as zero volts, the measurement of an 'absolute' potential of a test electrode can be made.

In Expt 3.3, a simple galvanic series was built up by making a series of measurements using copper as a reference electrode, but copper was an arbitrary selection; in fact, hydrogen is chosen as a standard of reference for rigorous scientific work. By definition, it is given an electrode potential of exactly zero volts. The electrode potential of another element is then compared with the electrode potential for hydrogen and is called the **standard electrode potential** for that element. A table similar to the one obtained in Expt 3.3 is then compiled for different metals.

Question: *How can a gaseous element be used for a standard electrode?*

A cell is constructed according to Fig. 4.3. An inert metal electrode of platinum is used, over which is passed a steady stream of hydrogen gas at exactly 1 atmosphere pressure and 298 K. The electrode is immersed in a solution in which the concentration of hydrogen ions is exactly 1 M. This half of the cell is separated from the half-cell to be tested by a porous membrane designed to minimise the cross-contamination of the electrolytes which occurs as a result of the flow of ions. The test electrode is a combination of a metal and a solution of its ions, also of concentration 1 M. The whole cell must be at the standard temperature of 298 K.

The measurement of standard electrode potential can now be made. If, for example, we wish to measure the standard electrode potential of iron then the measured potential difference between the hydrogen electrode and the iron, under the exact conditions defined above, will be the required value. The half-cell in which the hydrogen reaction takes place is called the **standard hydrogen electrode (SHE)**.

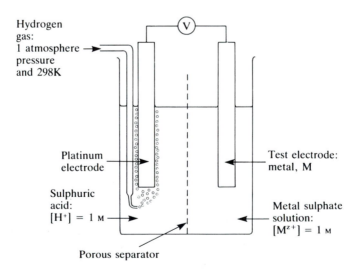

Fig. 4.3 Determination of standard electrode potentials.

Question: *How does the Nernst equation relate to this?*

Let us assume that we know nothing about the reactions for the moment, but since some of us might guess that iron dissolves in acid, the spontaneous reaction can be written:

Iron dissolves

$$Fe \rightarrow Fe^{2+} + 2e^- \tag{4.6}$$

Hydrogen gas is formed

$$H^+ + e^- \rightarrow H \text{ (atom)} \tag{4.20}$$

$$2H \rightarrow H_2 \tag{4.21}$$

overall reaction

$$Fe + 2H^+ \rightarrow Fe^{2+} + H_2(gas) \tag{4.23}$$

With reactants on the left of the equation and products on the right, substituting into the Nernst equation, eqn (4.17), we have

$$E = E^\circ - \frac{RT}{zF} \ln \frac{[Fe^{2+}][H_2]}{[Fe][H^+]^2} \tag{4.24}$$

Because of the experimental conditions, the terms $[H^+]$ and $[H_2]$ have been made equal to 1; $[Fe]$ has been approximated as unity. Equation (4.24) simplifies to

$$E = E^\circ - \frac{0.059}{2} \lg[Fe^{2+}] \tag{4.25}$$

Furthermore, in the correct setting-up of the cell, the concentration of iron ions is also made equal to 1 M. This reduces the lg term in eqn (4.25) to zero and leaves $E = E^\circ$. In other words, the *measured* potential difference *is* the electrode potential of the iron under standard conditions, and we say that, for the reaction, eqn (4.6), the standard electrode potential is given by E°. Notice that eqn (4.6) is an oxidation reaction. Note also that, because of Faraday's Law, eqn (4.13), if the measured value of potential is positive then $\Delta G < 0$, indicating a spontaneous reaction. This is, indeed, the case; E° is found to be $+0.44$ V, supporting the well-known property of iron that it dissolves spontaneously in acid. Notice also that the values substituted into eqn (4.17) need not be equilibrium values; they may take any reasonable values, here, usefully made unity.

In the same way as we have described the measurement of E° for iron, the standard electrode potential of other metals can be measured and a table of values compiled. The order in which the values are listed is known as the **electrochemical series**. It is conventional to list the entries as *reduction* reactions, and the table produced is then a table of **standard reduction potentials**. This means that the value measured for the *oxidation* of iron, $+0.44$ V, is listed as the *reduction* potential, -0.44 V; oxidation is the exact reverse of reduction. A

Table 4.1   Standard reduction potentials

Electrode reaction	$E°/V$
$Au^+ + e^- = Au$	+1.68
$Pt^{2+} + 2e^- = Pt$	+1.20
$Hg^{2+} + 2e^- = Hg$	+0.85
$Ag^+ + e^- = Ag$	+0.80
$Cu^{2+} + 2e^- = Cu$	+0.34
$2H^+ + 2e^- = H_2$	0.00
$Pb^{2+} + 2e^- = Pb$	−0.13
$Sn^{2+} + 2e^- = Sn$	−0.14
$Ni^{2+} + 2e^- = Ni$	−0.25
$Cd^{2+} + 2e^- = Cd$	−0.40
$Fe^{2+} + 2e^- = Fe$	−0.44
$Cr^{3+} + 3e^- = Cr$	−0.71
$Zn^{2+} + 2e^- = Zn$	−0.76
$Al^{3+} + 3e^- = Al$	−1.67
$Mg^{2+} + 2e^- = Mg$	−2.34
$Na^+ + e^- = Na$	−2.71
$Ca^{2+} + 2e^- = Ca$	−2.87
$K^+ + e^- = K$	−2.92

representative selection of standard reduction potentials is given in Table 4.1. Sometimes **standard oxidation potentials** are tabulated; these values are the *same in magnitude* as those in Table 4.1. but *of opposite sign.* Some texts show opposite signs for the values listed in Table 4.1. *This text uses the IUPAC convention, which ascribes the signs stated.* To avoid unnecessary confusion, the novice student is urged to avoid any texts using the opposite convention. The treatment we have developed has no need even to consider the possibility of using alternative signs.

Question: *How is the electrochemical series different from the galvanic series?*

The electrochemical series is determined under standard conditions. The values are absolute for each element and are independent of the electrolyte used. They may be substituted into the Nernst equation to obtain a prediction of the potential of a corrosion reaction under non-standard conditions.

The galvanic series, on the other hand, is true only for specified conditions of electrolyte, pressure or temperature. Unlike the electrochemical series, it can include the relative performance of alloys, a great advantage to engineers. It is therefore of considerable practical (field) significance, whereas the electrochemical series is more useful in laboratory or theoretical contexts.

Question: *The standard hydrogen electrode is a rather complicated arrangement to use in the laboratory. Is there not a simpler way of measuring potentials?*

Yes. You have already used a practical reference electrode in Expt 4.2, the SCE.

Scientists have measured the reduction potentials so accurately that there is now no practical need to use the hydrogen electrode. Robust and very stable, standard reference electrodes such as the one you used in Expt 4.2 can be used conveniently in most laboratory and field measurements. We shall now look at reference electrodes and ways of carrying out the measurements.

4.4 REFERENCE ELECTRODES

The most common reference electrode in the laboratory is the **standard calomel electrode**. The construction is designed to give a constant and well-defined potential against which other potential measurements can be made.

Calomel is an old name for mercury (I) chloride, Hg_2Cl_2. A combination of Hg_2Cl_2, mercury and a solution of chloride ions provides the very stable and reproducible potential. For convenience of use in laboratories, the chloride solution is a saturated solution of potassium chloride. In this case the standard calomel electrode is called the **saturated calomel electrode,** an example of which is shown in Fig. 4.4(a). Care should be taken not to confuse *standard* with *saturated*; the abbreviation SCE is often used to indicate the latter, though it could be taken to mean either. Since there are other standard electrodes which are not saturated and which have different potentials (see below) it is safest to be quite clear which electrode is being used.

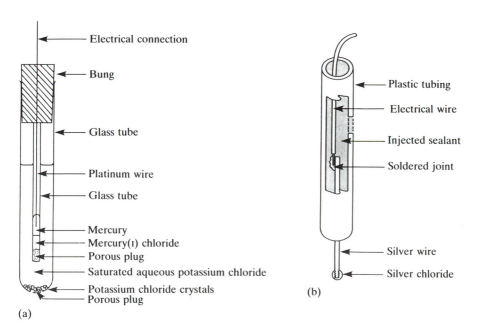

(a)

(b)

Fig. 4.4 Two designs of laboratory reference electrode: (a) a saturated calomel reference electrode and (b) a silver/silver chloride reference electrode.

The saturated calomel electrode consists of a platinum wire in contact with a small sample of mercury and mercury (I) chloride. This is held within a thin glass tube by a porous plug; the glass tube is contained inside a larger glass tube, again with a porous plug. The larger tube contains the saturated potassium chloride solution. The porous plugs allow the passage of ions (and hence current) without causing significant cross-contamination of the potassium chloride and the electrolyte of the test cell. If this were to happen, the potential of the calomel cell would not be constant and the potential of the test electrode might also change. A simple electrical connection to the platinum wire allows the electrode to be introduced into a corrosion cell such as will be considered in Section 4.11.

Variations of the calomel electrode exist in which different concentrations of potassium chloride are used. This is because the solubility of the salt is not constant over a range of temperature, and a solution which is saturated at one temperature is not saturated at a higher temperature. However, since most experimental work is performed at 298 K, there is little error involved with the use of the saturated solution and the electrode described here is probably the most widely used. It has the advantage that, in use, it is a simple matter to inspect the electrolyte visually. If crystals of potassium chloride are present in the tube, the electrode is probably functioning correctly. If crystals are absent, the electrolyte is possibly diluted and the electrode may not be providing the required reference potential. The porous plugs are not infallible and regular checks should be made to ensure they are still allowing free passage of ions.

A second type of reference electrode is gaining favour for laboratory use because it can be made quite small, smaller than is convenient for the calomel electrode. It is made with a silver wire, the end of which carries a coating of silver chloride (Fig. 4.4(b)). This robust coating can be easily applied by dipping the wire into a melt of the silver salt. The wire is then enclosed in a suitable glass tube or capillary and an electrical connection is made.

Silver/silver chloride electrodes are particularly suitable for use in sea-water environments, and are often used as reference electrodes in cathodic protection systems (see Chapter 16). An electrode suitable for laboratory use has a length of 0.25 mm silver wire wound about 8–10 times around a narrow cylindrical former such as a wooden cocktail stick, making a helical basket. The silver is cleaned thoroughly by dipping into nitric acid and washing in deionised water. A clean crucible is prepared containing a small quantity of molten silver chloride. Next the silver basket is dipped several times in the molten salt, coating the whole of the silver basket in silver chloride. When the salt is dry, the basket is soldered to a length of insulated electrical wire and the soldered joint is completely sealed with a proprietary epoxy resin sealant. This is left to dry overnight. Next the assembly is threaded into a suitable length of glass tube, say 5–8 mm diameter, so that the basket is close to the end of the tube and only silver or coated silver is protruding. This working end of the glass tube is also filled with sealant, or sometimes the whole tube is filled. The electrode is finally activated by connecting it to a piece of aluminium and immersing both in a strong electrolyte for several minutes. A high proportion of the electrodes made

Table 4.2 Standard reference electrode potentials

Electrode	Electrolyte	Potential/V
Calomel (SCE)	Saturated KCl	+0.2420
Calomel (NCE)	1.0 M KCl	+0.2810
Calomel	0.1 M KCl	+0.3335
Silver/silver chloride (SSC)	1.0 M KCl	+0.2224
SSC	Sea-water	+0.25 approx
Copper/copper sulphate (CSE)	Sea-water	+0.30 approx
Platinised platinum	0.1 M NaCl	−0.12 approx
Gold	0.1 M NaCl	−0.25 approx
Zinc	Sea-water	−0.79 approx

in this way should be within 5 mV of each other and are stable for periods of 6–12 months, depending upon the environment in which they work.

Many modern applications of corrosion science involve the use of micro-electrodes for investigating localised forms of corrosion such as pitting. A reference electrode made from platinised platinum is one such micro-electrode which can be used for investigating tiny local potential variations. Gold, as well as platinum, can be used; the advantage of both metals is that they are available in wire of very fine diameter. More on these techniques is given in Chapter 7, and details of the probe manufacture can be found elsewhere. [1]

The reference potentials of some common standard electrodes are given in Table 4.2.

4.5 CELL POTENTIAL

Modern batteries are merely well-designed corrosion cells in which the electrical current produced by the corrosion reaction is used to drive an external device. The **Daniell cell** was a very early form of battery which consisted of copper and zinc immersed in solutions of their salts (Fig. 4.5). The external circuit shown in the figure is not part of the cell, but will be used in an experiment below. The use of the Daniell cell as an example to aid development of our theory is convenient because it is simple in construction and common in teaching laboratories. Although its practical uses as a battery are now few because of the availability of more modern designs, it can be thought of as a wet corrosion cell, described in Section 4.2.

Let us begin by adopting a standard form of symbols by which the cell can be represented, rather than in a diagram:

$$Zn \mid Zn^{2+} \parallel Cu^{2+} \mid Cu$$

The two electrodes are written at the extreme left and the extreme right; between them are the ions needed for the redox processes. The double vertical line

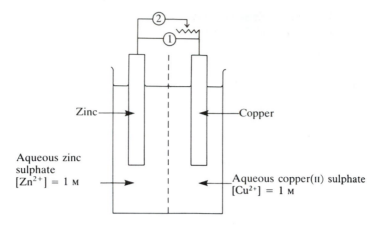

Fig. 4.5 The Daniell cell and associated apparatus for Expt 4.2: (1) and (2) represent digital multimeters.

represents some form of separator for the two different ionic species. Notice that we have written it in the same sense as the figure, i.e. with the zinc on the left.

Our analysis begins with no preconceptions about what actually happens. Let us take a guess that zinc is the anode. If you have done the experiments in Chapter 3 you will know this guess is correct. The zinc anode (oxidation) reaction is

$$Zn \rightarrow Zn^{2+} + 2e^- \tag{4.26}$$

and the copper cathode (reduction) reaction is

$$Cu^{2+} + 2e^- \rightarrow Cu \tag{4.7}$$

so the overall reaction is

$$Zn + Cu^{2+} \rightarrow Zn^{2+} + Cu \tag{4.27}$$

We can use these reactions to find the theoretical potential obtainable from the cell. Using the Nernst equation, eqn (4.18), for each of the cell half-reactions we can write:

$$E_{(Zn/Zn^{2+})} = E^{\circ}_{(Zn/Zn^{2+})} - \frac{0.059}{2} \lg [Zn^{2+}] \tag{4.28}$$

with the reactants on the left and products on the right of the equation. Note that $E_{Zn/Zn^{2+}}$ and $E^{\circ}_{Zn/Zn^{2+}}$ are *oxidation* potentials because we have written an oxidation reaction eqn (4.26). Similarly,

$$E_{(Cu^{2+}/Cu)} = E^{\circ}_{(Cu^{2+}/Cu)} - \frac{0.059}{2} \lg \frac{1}{[Cu^{2+}]} \tag{4.29}$$

where, this time, $E_{Cu^{2+}/Cu}$ and $E^{\circ}_{Cu^{2+}/Cu}$ are reduction potentials, in accordance with eqn (4.7). Logically, the cell potential is the sum of the oxidation potential of the anode and the reduction potential of the cathode:

$$E_{(cell)} = E_{(Zn/Zn^{2+})} + E_{(Cu^{2+}/Cu)} \tag{4.30}$$

and, similarly for standard conditions:

$$E^{\circ}_{(cell)} = E^{\circ}_{(Zn/Zn^{2+})} + E^{\circ}_{(Cu^{2+}/Cu)} \tag{4.31}$$

Then by adding eqns (4.28) and (4.29) we get

$$E_{(cell)} = E^{\circ}_{(cell)} - \frac{0.059}{2} \lg \frac{[Zn^{2+}]}{[Cu^{2+}]} \tag{4.32}$$

If we simplify matters by using ion concentrations of 1 M, the lg term vanishes, leaving $E_{(cell)} = E^{\circ}(cell)$. Remember, by using this form of the Nernst equation, eqn (4.18), we have decided to consider only standard conditions. Thus, for 1 M electrolyte,

$$E^{\circ}_{(cell)} = E^{\circ}_{(Zn/oxidation)} + E^{\circ}_{(Cu/reduction)} \tag{4.33}$$

From Table 4.1, $E^{\circ}_{(Zn\ oxidation)} = -(-0.76)$ V and $E^{\circ}_{(Cu\ reduction)} = +0.34$ V, therefore

$$E^{\circ}_{(cell)} = (+0.76\ V) + (+0.34\ V)$$
$$= +1.10\ V$$

In many texts, a convention has been adopted which should be explained. Whenever we write down the symbols for a cell, or sketch one in a diagram, there is always one electrode on the left and one on the right. It has been agreed to subtract the potential of the electrode on the left from the potential of the electrode on the right. In Figure 4.5 we drew the zinc on the left and the copper on the right. So we can write

$$E^{\circ}_{(cell)} = E^{\circ}_{(copper)} - E^{\circ}_{(zinc)} \tag{4.34}$$

This convention requires that we substitute *reduction* potentials into eqn (4.34). Writing the two half-reactions as reduction reactions:

$$Zn^{2+} + 2e^- \rightarrow Zn;\ E^{\circ} = -0.76V \tag{4.35}$$

$$Cu^{2+} + 2e^- \rightarrow Cu;\ E^{\circ} = +0.34V \tag{4.7}$$

whereupon, according to the *convention*,

$$E^{\circ}_{(cell)} = (+0.34) - (-0.76)$$
$$= +1.10\ V$$

The equivalence of the two approaches has been demonstrated, although the second is less intuitive. The cell reaction is found by eqn (4.7) – eqn (4.34). This gives

$$Cu^{2+} - Zn^{2+} \rightarrow Cu - Zn \qquad (4.36)$$

which, on rearrangement, gives eqn (4.27).

You will notice that our method gives exactly the same results as those obtained using the convention. Students using our method simply express the equations for the reactions and substitute standard potentials (oxidation or reduction) according to whether an oxidation or reduction reaction has been used. An important check is now made to see if the cell reaction is possible in the direction in which we have written it. By simply substituting the value +1.10 V into Faraday's Law, eqn (4.13), we see that a negative value of $\Delta G°$ is obtained and the way that we have described the cell leads to the prediction of a spontaneous reaction. If in our method we had selected copper instead of zinc for the anode, or if by convention we had written the cell with the copper on the left, we would have obtained a negative cell potential,, leading us to conclude that the cell reaction was not spontaneous. Thus we always know, by inspection, which is the anode and which is the cathode.

Question: *What if I did not use ion concentrations which were exactly 1 M?*

The potential of the cell will be modified by the Nernst equations and the lg terms will no longer be zero. Note that the Nernst equation, as we have used it, becomes inaccurate at high concentrations of electrolyte. This is because of our use of concentrations rather than activities. A further small problem is the presence of a so-called **junction potential**, a potential difference across the interface of the two different electrolytes. At this stage it is sufficient to be aware of it; we shall not discuss junction potentials any further.

Experiment 4.3

Set up a Daniell cell and circuit as shown in Fig. 4.5. As in Expt 4.2, it is useful to use two modern digital meters (labelled 1 and 2 in Fig. 4.5) because they can be used to measure both potential and current at the flick of a switch. First, switch meter 1 to the volts mode with meter 2 switched off. No current will flow through meter 2, so meter 1 will measure a potential quite close to +1.1 V, confirming our calculations. Next, switch meter 2 to the current mode. Instantly, you will see that the potential measured by meter 1 has dropped as current flows in the external circuit. You can use the rheostat to alter the resistance of the circuit and show the greater the current flow, the greater the drop in potential.

Question: *Why does current flow reduce the potential?*

The potentials used in eqn (4.33) are for equilibrium (no net current) conditions and are calculated using the Nernst equation. When a net current flows (see Fig. 4.2(d)), the potential of the anode rises; it becomes less negative. Similarly, the potential of the cathode falls as it becomes less positive. The cell potential (the magnitude of the difference between anode and cathode) is thus reduced. This is another consequence of thermodynamics, known in chemistry as **Le Chatelier's Principle:**

> **A system will always react to oppose a change imposed upon it.**

The cell potential can be thought of as an ability to supply current. As soon as current is drawn, the ability is reduced. If this were not so then we could go on drawing as much current as we liked from a battery without penalty, effectively creating a limitless supply of free energy.

The answer to the question is more fully explained in later sections, in particular, Section 4.9.

4.6   THE KINETICS OF CORROSION REACTIONS

We have until now been considering the thermodynamic implications of corrosion reactions, as evidenced by electrochemical potentials, but we have discovered that when a current flows the potentials change. This brings us back to the point that thermodynamics tells us only about the *tendency* of a system to corrode. Corrosion reactions not in equilibrium cause current to flow and we must fully investigate the relationships between potential and current to appreciate corrosion kinetics.

Consider two pieces of metal, areas 10 mm^2 and 1 mm^2 such that they both corrode in separate cells and each produces a current of 10 electrons per second. It is easy to see that the smaller piece will suffer corrosion damage 10 times worse (if $z = 1$) than the larger piece because the surface mass affected by corrosion is directly proportional to the rate of generation of electrons. When measuring corrosion currents we eliminate the effects of area by considering **current density**. Throughout the discussions which follow we shall use I to represent an absolute current (A) and i to represent current density (A m^{-2}). We shall always imply a flow of electrons as being the current; conventional currents will not be used. We shall also use the symbols i_a and i_c to represent the magnitudes of anodic and cathodic current densities. When they need to be added, it is necessary to treat them as having opposite signs because the currents flow in opposite directions.

Let us return to the thought experiment of Section 4.2. This time we place a piece of pure copper of unit area in a beaker of pure water. Immediately there applies a situation parallel to the energy profile of Fig. 4.1 and redrawn in Fig. 4.6(a). Note that both ΔG and $\Delta G^{\ddagger}$ are treated as variable in this discussion

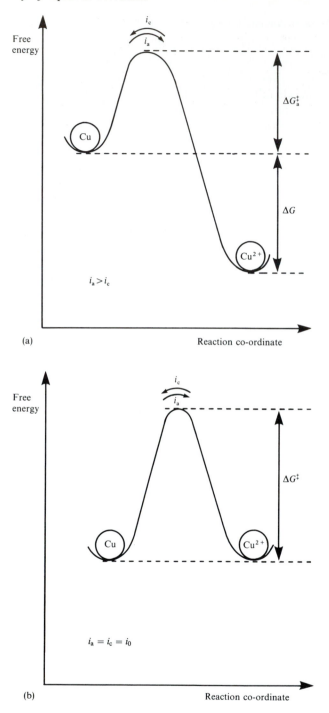

Fig. 4.6 (a) An energy profile for copper in pure water $i_a > i_c$. (b) An energy profile for copper in equilibrium with a solution of its divalent ions; $i_a = i_c = i_0$.

because they are both dependent upon the nature of the materials on each side of the metal/electrolyte interface and these materials change.

There is sufficient available energy in the environment for a steady flow of copper atoms to 'pass over the energy barrier', $\Delta G_a^{\ddagger}$, and proceed to the Cu^{2+} ionic form. The copper begins to dissolve (corrode) and the concentration of copper ions in the water, initially zero, will slowly increase.

$$Cu \rightarrow Cu^{2+} + 2e^- \tag{4.37}$$

Remember that a single piece of metal placed in an electrolyte can act as its own anode, cathode and electrical connection. Individual areas of the metal can be anodic to others because of microvariations in the solid structure of the metal, or environmental differences over the surface as a whole (see Chapter 6).

The tendency of the copper to corrode decreases as the current increases from zero, and the value of ΔG diminishes, together with the potential, in accordance with Faraday's Law. The thermodynamic energies of metal atoms and the assembly of adjacent ions tend to approach each other.

We have already discussed in the early sections of this chapter how the rate of reaction diminishes as the activation free energy barrier, $\Delta G^{\ddagger}$, increases. As soon as copper ions are present in solution, there is a possibility for them to 'pass back over the energy barrier' and replate onto the metal. The rate of this process is governed by the *activation free energy in the reverse direction*, $(\Delta G + \Delta G_a^{\ddagger})$, a quantity initially greater than that for the forward reaction, $\Delta G_a^{\ddagger}$. However, this free energy barrier is reduced in magnitude as the energies of the two species approach each other, increasing the extent of the backward reaction of copper ions plating out. On the other hand, the rate of the forward reaction decreases because its activation free energy increases. The situation is thus obtained that the rate of the decreasing forward reaction becomes equal to the rate of the increasing backward reaction and equilibrium is established (Fig. 4.6(b)) at an equilibrium value of free energy of activation $\Delta G^{\ddagger}$, and with $\Delta G = 0$. For a divalent metal, M, we can rewrite eqn (4.37) as

$$M \overset{i_a}{\underset{i_c}{\rightleftharpoons}} M^{2+} + 2e^- \tag{4.38}$$

When the state of equilibrium is reached, $i_a = i_c$; the measured current density, $i_{meas} = (i_a - i_c)$ and no *net* current flows. There *is* current flowing, but it is *equal and opposite and cannot be measured*. It is called the **exchange current** and is denoted by I_0, or i_0 when divided by the area.

The non-homogeneous distribution of ions which has resulted from the immersion of a metal in an aqueous electrolyte is commonly referred to as the **double layer**. It is illustrated schematically in Fig. 4.7. The double layer consists of two parts: a compact layer and a diffuse layer.

The compact layer, or **Helmholtz layer**, is closest to the surface in which the distribution of charge, and hence potential, changes linearly with the distance from the electrode surface. The more diffuse outer layer, the **Gouy–Chapman layer**, occurs where the potential changes exponentially.

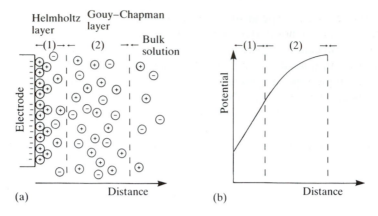

Fig. 4.7 The double layer: (a) distribution of ions as a function of distance from an electrode behaving as an anode; and (b) variation of potential with distance for the model in (a).

The constitution of the double layer will parallel the changes of potential occurring at the electrode. It may also reach an equilibrium condition corresponding to the energy profile of Fig. 4.6(b), or when the equilibrium is destroyed by an increase in either the forward or backward reactions, a new condition may be established in which a continuous flow of anions and cations in the electrolyte adjacent to the electrodes performs the bulk current-carrying requirement of the electrode reactions.

Faraday's Law of Electrolysis states that

$$Q = zFM \tag{4.39}$$

where Q is the charge created by the ionisation of M mol of material. Differentiating with respect to time we get

$$\frac{dQ}{dt} = zF\frac{dM}{dt} \tag{4.40}$$

Now the rate of flow of charge is current, I, and if we consider the passage of charge across unit area of cross-section then we can use current density, i. Then, dM/dt becomes J, the flux of substance, and eqn (4.40) becomes

$$i = zFJ \tag{4.41}$$

The flux of substance is another name for corrosion rate per unit area. *Hence we confirm the important concept that current density and corrosion rate can be equated.*

The ability to determine a corrosion rate by measurement of current density is a most important finding. However, in practical terms, to say that a metal is corroding at a rate of 0.003 $A\,m^{-2}$ is less helpful than an average rate of deterioration per unit area expressed as the average depth of corrosion over a given area in a given time. Thus, engineers prefer, for example, a corrosion rate of 2.5 mm per year, meaning that in one year the metal will have corroded on average across the whole of its exposed area to a depth of 2.5 mm. Millimetres per year is sometimes written mmpy. In the United States a corrosion rate expressed in mpy means milli-inches per year, sometimes referred to as mils. Worked example E18.4 shows how the conversion from current density to mm per year is carried out.

A variety of other practical units are used to state corrosion rates. [2] One commonly used unit is milligrams weight lost per square decimetre per day (mdd). In certain forms of corrosion, such as crevice or pitting corrosion, these methods of considering corrosion rate are dangerous because an average corrosion rate is meaningless; corrosion can be very rapid and penetrating over very small areas of a large exposed surface; it can also vary considerably over long periods such as a year.

4.7 POLARISATION

When a metal is not in equilibrium with a solution of its ions, the electrode potential differs from the equilibrium potential by an amount known as the **polarisation**. Other terms having equivalent meaning are **overvoltage** and **overpotential**. The symbol commonly used for polarisation is η. Polarisation is an extremely important corrosion parameter because it allows useful statements to be made about the *rates* of corrosion processes. In practical situations, polarisation is sometimes defined as the potential change away from some other arbitrary potential, and in mixed potential experiments, this is the free corrosion potential (see Section 4.9).

Consider the process described in eqn (4.38). In the last section we saw that corrosion rate and current density are directly related. At the beginning of this chapter we said that corrosion rate, v, could be expressed as

$$v = k_{(corr)}[\text{reactants}] \tag{4.1}$$

where

$$k_{(corr)} = A \exp(-\Delta G^{\ddagger}/RT) \tag{4.2}$$

A is constant. From these two equations we see that

$$v = A \exp(-\Delta G^{\ddagger}/RT) \cdot [\text{reactants}] \tag{4.42}$$

At equilibrium, the rate of the forward (anodic) reaction is i_a, and equals the rate of the reverse (cathodic) reaction, i_c. (Remember that $i_0 = i_a = i_c$ at equilibrium.) It is usually possible to treat the concentration of reactants (e.g. the solid metal, for the forward reaction) as constant, and we shall incorporate the term into a new constant, A_0. Thus, if we consider the rate of the forward

95

reaction for which the activation free energy is $\Delta G^{\ddagger}$, we can write eqn (4.42) as

$$i_a(\text{at equilibrium}) = i_0 = A_0 \exp\left(\frac{-\Delta G}{RT}\right) \tag{4.43}$$

When the forward reaction is faster than the reverse reaction ($i_a > i_c$) and an overall corrosion process occurs, equilibrium is destroyed and the free energies of the metal and its ions are at different levels (Fig. 4.6(a)).

The deviation from the equilibrium potential, the polarisation, is the combination of an anodic polarisation on the metal and a cathodic polarisation of the environment. Compare parts (a) and (b) of Fig. 4.6; the energy of the metal has increased and the energy of the environment has decreased. These potential deviations away from the equilibrium value may or may not be equal, and we shall assume for the moment that they are not. The changes are redrawn in Fig. 4.8.

If we call the total polarisation η, then we can define the anodic polarisation as $\alpha\eta$ and the cathodic polarisation as $(1 - \alpha\eta)$. Note that in Fig. 4.8, the polarisations have been converted into free energies by multiplication by the factor, zF, as in eqn (4.13). This enables us to determine the new activation energy for the anodic reaction, which can be seen to be $(\Delta G^{\ddagger} - \alpha\eta zF)$ because the energy state of the metal has increased and the activation energy reduced. Thus we can write

$$i_a = A_0 \exp\left(\frac{-\Delta G^{\ddagger} + \alpha\eta zF}{RT}\right) \tag{4.44}$$

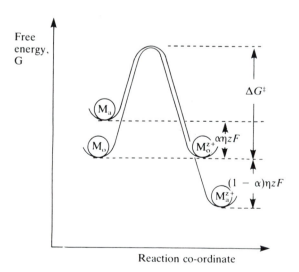

Fig. 4.8 An energy profile for an anode at equilibrium, represented by the curve M_0/M_0^{z+} and a similar profile for an anodic activation polarisation of η represented by the curve M_a/M_a^{z+}.

$$i_a = A_0 \exp\left(\frac{-\Delta G^{\ddagger}}{RT}\right) \exp\left(\frac{\alpha\eta zF}{RT}\right) \tag{4.45}$$

Note that, similarly, the activation energy of cations, M^{z+}, being converted into M, has increased by an amount $(1 - \alpha)\eta zF$. Substituting eqn (4.43) into eqn (4.45) gives

$$i_a = i_0 \exp\left(\frac{\alpha\eta zF}{RT}\right) \tag{4.46}$$

and a comparable expression for the cathodic reaction is

$$i_c = i_0 \exp\left(\frac{(1 - \alpha)\eta zF}{RT}\right) \tag{4.47}$$

from which the bulk current flow, $i_{meas} = (i_a - i_c)$, i.e.

$$i_{meas} = i_0\left[\exp\left(\frac{\alpha\eta zF}{RT}\right) - \exp\left(\frac{(1 - \alpha)\eta zF}{RT}\right)\right] \tag{4.48}$$

This equation is known as the **Butler–Volmer equation** and is much used in this subject.

Now let $A' = \alpha zF/RT$; then

$$i_a = i_0 \exp(A'\eta) \tag{4.49}$$

Taking logarithms we get

$$\ln i_a = \ln i_0 + A'\eta \tag{4.50}$$

Rearranging eqn (4.50) gives

$$\ln\left(\frac{i_a}{i_0}\right) = A'\eta \tag{4.51}$$

Converting to base 10 logarithms and rearranging eqn (4.51)

$$\eta = \frac{2.303}{A'} \lg\left(\frac{i_a}{i_0}\right) \tag{4.52}$$

Letting $\beta = 2.303/A'$, we now have the important result that for the anodic process

$$\eta_a = \beta_a \lg\left(\frac{i_a}{i_0}\right) \tag{4.53}$$

where

$$\beta_a = \frac{2.303\,RT}{\alpha zF} \tag{4.54}$$

Written in the general form:

$$\eta = C \lg i + D \tag{4.55}$$

the equation is known as the **Tafel equation**, while more specifically for the anode process:

$$\eta_a = \beta_a \lg i_a - \beta_a \lg i_0 \tag{4.56}$$

and for the cathode process:

$$\eta_c = \beta_c \lg i_c - \beta_c \lg i_0 \tag{4.57}$$

where, similarly,

$$\beta_c = \frac{2.303 RT}{(1 - \alpha)zF} \tag{4.58}$$

Constants β_a and β_c are called the anodic and cathodic **beta** or **Tafel constants**.

Examination of the Tafel equation in the form of eqn (4.55) tells us immediately that a graph of η against $\lg i$ for either of the two processes gives a straight line with a slope equal to the respective β constant. The intercept, D, is given by $-\beta \lg i_0$. In the context of the Nernst equation, eqn (4.17), we found that $2.303 RT/F$ has the value 0.059 and, modified by αz, sensible values for Tafel constants can readily be evaluated. As we have seen, z takes values 1, 2 or 3 and α is usually about 0.5. In practice, a value of ± 0.03 to ± 0.1 V (or ± 30 to ± 100 mV) per decade of current density is common.

Question: *Where is all this leading?*

We are working towards the goal of a practical electrochemical technique for measuring corrosion. We next need to plot the variation of polarisation with $\lg i$ for both anodic and cathodic reactions. We need to decide upon representative values for the Tafel constants and the exchange current density. If we choose as representative values $\beta_a = +0.1$ V per decade, $\beta_c = -0.1$ V per decade, and $i_0 = 0.01$ A m^{-2}, substitution into eqn (4.56) leads to data which, when plotted, take the form of Fig. 4.9(a). For example, when $i_a = 10^{-2}$ then $\eta = 0$. This is true also for $i_c = 10^{-2}$. Remember that we are plotting η versus $\lg i$, not $\lg i/i_0$. The anodic polarisation varies as line (i_a) in Fig. 4.9(a) and the cathodic polarisation as line (i_c).

Examination of Fig. 4.9(a) shows that when the electrode is anodically polarised to $+0.1$ V, the magnitude of the anodic current density is 10^{-1} A m^{-2} whereas the magnitude of the cathodic current has fallen to 10^{-3} A m^{-2}. Since we can only measure the difference between the anodic and cathodic currents, we obtain

$$\begin{aligned} i_{meas} &= i_a - i_c \\ &= 10^{-1} - 10^{-3} \\ &= 0.099 \,\text{Am}^{-2} \end{aligned} \tag{4.59}$$

As the polarisation is increased, so i_a increases, i_c decreases and $i_{meas} \rightarrow i_a$.

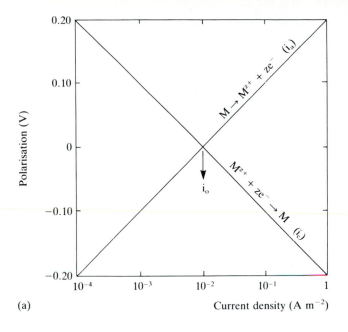

(a)

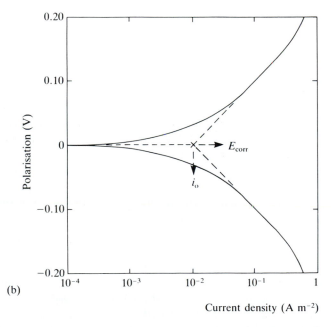

(b)

Fig. 4.9 Two Tafel plots: (a) theoretical Tafel plot; and (b) an idealised practical Tafel plot.

Using eqn (4.59) and substituting for i_a in eqn (4.53), we obtain

$$\eta_a = \beta_a \lg \left(\frac{i_{meas} + i_c}{i_0} \right) \tag{4.60}$$

Obviously when $i_{meas} \gg i_c$ then linear Tafel behaviour will be observed experimentally. However, at polarisations close to equilibrium, where i_a is comparable with i_0, the measured value of the current density will be far removed from the true value of i_a and substantial deviations from linearity will be obtained. The same arguments apply whether anodic or cathodic polarisations are being used. Thus if we try to obtain the data experimentally, the graph of Fig. 4.9(a) will become Fig. 4.9(b). Extrapolation of the linear portions of the polarisation plots allows a determination of i_0.

4.8   DIFFUSION PROCESSES AND THE DOUBLE LAYER

When the rate of a corrosion process is examined, several different stages must be analysed. So far we have discussed the kinetics of the reactions occurring at the solid/liquid interface. The transport of the species through the solution is also important because of the time it takes. Unlike the passage of electrons as current, which is deemed to occur in a negligible time compared with other processes, the current through the solution is carried by species much more massive than electrons.

In the analysis of rates of reaction there is an important principle:

The rate of a reaction is determined by the slowest step.

A simple analogy is to consider what happens to the traffic on a highway at the scene of roadworks. When the traffic is light, the roadworks pose no obstruction and the rate of passage of vehicles along the road is the same as the rate at which they join it. However, when the traffic is heavy, the bottleneck caused by the restriction creates a long tailback, while the traffic which has passed by flows freely. The rate of flow of cars along the highway as a whole is the rate at which they pass the obstruction; this step is called the **rate-determining step.**

In a corrosion cell, the step which is the slowest can vary at any given time, just as on our highway. When small currents are involved the transport of cathode reactant, e.g. dissolved oxygen, through the solution is relatively easy and the activation process is the rate-determining step. However, when large currents flow, the cell demands a greater charge transfer than can be accommodated by the electrolyte. The speed of passage of the dissolved oxygen species becomes the slowest step and is thus rate-determining. Under these conditions we refer to the process as **diffusion-controlled**.

Figure 4.10 represents the variation of cathode reactant concentration with distance from the cathode. Under zero-current conditions, labelled 1 in the figure, the concentration of species, c_0, will be uniform throughout the

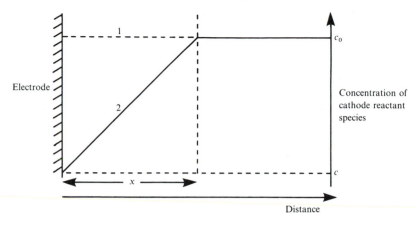

Fig. 4.10 Variation of anion concentration with distance from a cathode.

electrolyte. When the cell is connected and current flows, labelled 2 in the figure, a drop in concentration at the cathode results. A concentration gradient is established such that for concentrations, c, and distances, x, the gradient is dc/dx. Fick's First Law of Diffusion states that

$$J = -D\frac{dc}{dx} \tag{4.61}$$

where J is the flux of substance and D is a diffusion coefficient. We have already seen from Faraday's Law of Electrolysis that

$$i = zFJ \tag{4.41}$$

thus, substituting eqn (4.61) into eqn (4.41) gives

$$i = -zFD\frac{dc}{dx} \tag{4.62}$$

To simplify the situation, assume that the concentration gradient is linear and can be replaced by $(c_0 - c)/x$. Equation (4.62) becomes

$$i = -zFD\frac{(c_0 - c)}{x} \tag{4.63}$$

The negative sign is conventional; it tells us the current is being carried away from the cathode (as hydroxyl ions). We are interested only in the magnitude of the current, and since c can never be negative, the magnitude of the current is greatest when $c = 0$. The maximum or **limiting current**, i_L, is given by

$$i_L = -zFD\frac{c_0}{x} \tag{4.64}$$

Notice that when $c > c_0$ the implication is that the concentration of species in the region of the electrode is increasing. The resulting sign change of i from negative to positive tells us the current is reversed.

Using the Nernst equation for condition 1 (no current):

$$E_1 = E° + \frac{0.059}{z} \lg c_0 \tag{4.65}$$

and for condition 2:

$$E_2 = E° + \frac{0.059}{z} \lg c \tag{4.66}$$

We have previously defined the polarisation as the change of potential away from the equilibrium (no net current) condition. Thus $\eta = E_2 - E_1$ and subtracting eqn (4.65) from eqn (4.66) we get

$$\eta = \frac{0.059}{z} \lg\left(\frac{c}{c_0}\right) \tag{4.67}$$

Using eqn (4.63) and eqn (4.64) it can be shown that

$$\frac{c}{c_0} = \left(1 - \frac{i}{i_L}\right) \tag{4.68}$$

Substituting eqn (4.68) into eqn (4.67) we obtain

$$\eta = \frac{0.059}{z} \lg\left(1 - \frac{i}{i_L}\right) \tag{4.69}$$

From this equation it can be seen that as $i \to i_L$ then $\eta \to -\infty$. The overall effect of diffusion polarisation on the cathodic part of the $E/\lg i$ plot is easy to define and should be illustrated in the results you obtained in the experiment. For small currents, $i_{meas} \to 0$ because $i_c \to i_a$; non-linearity of the $E/\lg i$ plot is obtained. For intermediate currents, $i_{meas} \to i_c$, the Tafel equation holds and linearity is observed. As the current increases still further the plot begins once more to deviate from linearity towards more negative values, approaching the limiting current density asymptotically.

4.9   MIXED POTENTIAL THEORY

In Section 4.5 the Daniell cell was considered as a corroding system in which, as current is drawn, the cell potential falls from its maximum theoretical value. We can use a graph of potential versus current density to illustrate diagrammatically the typical polarisation of both electrodes in a corrosion cell. By plotting current density on a logarithmic scale, the polarisation lines will be linear, in accordance with the Tafel equation. The diagrams drawn in the way described below are commonly called **Evans diagrams**, after one of the founders of corrosion science, Ulick Evans.

Figure 4.11(a) shows such a diagram using a zinc anode (A) and a copper cathode (C), as we were using in the Daniell cell. First, the equilibrium

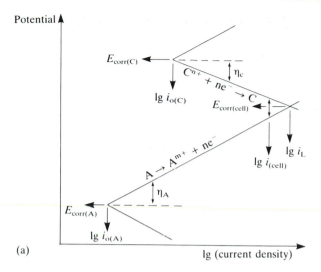

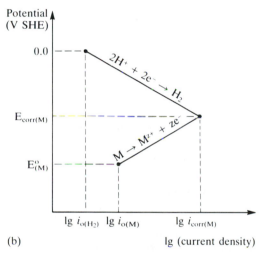

Fig. 4.11 Two mixed potential plots (Evans diagrams). (a) The Daniell cell: the individual free corrosion potentials of the copper (C) and zinc (A) are changed to a cell potential when the metals are short-circuited. The limiting current density, i_L, is never achieved because of the finite internal resistance of the cell. (b) A metal, M, corroding in an acid solution with the evolution of hydrogen. The microstructure of the metal causes it to act as its own anode and cathode; the cathode reaction is the reduction of hydrogen ions.

potentials of the individual metals in their solutions, $E_{(A)}$ and $E_{(C)}$, are recorded at their respective exchange current densities; the cathode has the more positive value. Note how this diagram is more sophisticated than the simple one in Fig. 4.2(d) because it now includes exchange current densities. As the zinc corrodes, according to

$$Zn \rightarrow Zn^{2+} + 2e^-$$ (4.26)

it is polarised upwards to more positive values by an amount η_a, the anodic polarisation. At the copper electrode,

$$Cu^{2+} + 2e^- \rightarrow Cu \tag{4.7}$$

causes a cathodic polarisation, η_c, downwards to more negative values. For completeness it may be desirable to draw both the anodic and cathodic polarisations for each electrode. The slopes of these lines, the anodic and cathodic beta (Tafel) constants, are not necessarily the same. The steady-state current, i_{corr}, will be obtained at the potential of intersection, E_{corr}, of the anodic polarisation line of the anode and the cathodic polarisation line of the cathode. These parameters, commonly called the **free corrosion current density** and the **free corrosion potential**, are of considerable practical importance because they are most commonly determined in corrosion cell measurements.

Question: *If the potential of the cathode is more positive than for the anode, why do the anode reactions have positive slopes in the diagram? Surely, as the anode corrodes it will become more anodic and the line will have a negative slope.*

Remember the simple explanation given towards the end of Section 4.6. When a piece of metal is placed in an electrolyte, metal ions enter the solution and the metal is left with an excess of negative electrons. This is the source of the electrochemical potential. If the metal is now connected to a more cathodic metal (a more positive metal), the electrons will flow away from the anode and towards the cathode, so the anode will become less negative and more positive. Le Chatelier's Principle says the system reacts to oppose any change we try to impose upon it. When a metal corrodes it loses some of its thermodynamic desire for corrosion.

The use of Evans diagrams such as that shown in Fig. 4.11(a) is not restricted to simple cases in which the cathode reaction is one of replating metal ions. Section 4.2 described two other common cathode reactions which occur when a metal corrodes in aqueous solution: hydrogen evolution and reduction of dissolved oxygen. The well-known dissolution of metals in acids, accompanied by the evolution of hydrogen, is itself a corrosion reaction and can be described by an Evans diagram such as Fig. 4.11(b). The cathode reaction is now represented by eqns (4.20) and (4.21), and at the point where the polarisation line intersects the metal dissolution line, the corrosion rate is $i_{corr(M)}$. The value of $i_{0(H_2)}$ occurs at 0.0 V SHE, but at different current densities according to the metal. Obviously, the rate of corrosion of a metal in acid is governed by such factors as

The anodic polarisation line of the metal

The exchange current density of hydrogen evolution on the metal

If $i_{0(M)}$ or $i_{0(H_2)}$ occurs at higher current densities the metal will corrode much faster; a good example is iron and zinc. In hydrochloric acid under identical conditions iron corrodes much faster than zinc, even though zinc is much more active according to its position in the galvanic series (Section 5.1). This is because $i_{0(H_2)}$ is much greater on iron than on zinc. Explanation by Evans diagram is shown in Fig. 4.12. Noble metals such as platinum and palladium have very high values of $i_{0(H_2)}$ but their greatly positive electrode reduction potentials are responsible for their lack of corrosion.

The hydrogen evolution line for zinc is shown as line a, while the anodic dissolution of zinc is line b. The free corrosion potential and corrosion current density of zinc are shown at the intersection of these two lines. The equivalent lines for iron (corroding independently of the zinc) are shown as a′ and b′ respectively. The corrosion potential and corrosion current density for the zinc/iron couple occur at the intersection of the cathodic and anodic lines for the couple. To obtain them, it is necessary to sum the two cathodic processes a + a′ and the two anodic processes b + b′. These are shown as the dashed diagonal lines in Fig. 4.12.

This type of diagram is extremely useful in both explaining and predicting corrosion rates in different environments. This example (in acid solution) was chosen so that the cathodic lines were representative of hydrogen evolution. In neutral or alkaline solutions, the cathode reaction is represented by eqn (4.22), but is strongly affected by the rate of oxygen diffusion to the metal surface and by stirring. As we shall see in Chapter 5, a different environment can

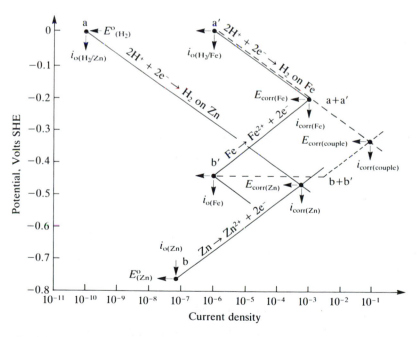

Fig. 4.12 A mixed potential plot for the bimetallic couple of iron and zinc. The diagram also explains the higher corrosion rate of iron than zinc in hydrochloride acid solution. Despite the more positive reduction potential of iron, the evolution of hydrogen on iron has a high exchange current density.

considerably alter the plot. In particular, experimental effects can sometimes cause complications (Section 5.2). However, this should not detract from the usefulness of Evans diagrams, especially in situations where two or more metals may be coupled together. The success of Evans diagrams is that they combine thermodynamics and kinetics to form a whole picture; the potential axis is the thermodynamic factor and the current axis is the kinetic factor. Modelling of experimentally derived information is now likely to be done with micro-computers instead of Evans plots because microcomputers give rapid access to the component parts of the curves. This technique will be discussed in detail in Section 4.12.

4.10 POURBAIX DIAGRAMS

Marcel Pourbaix added a new dimension to electrochemical studies by specifying a most valuable diagram designed to illustrate how the stability of different species were affected by potential and pH. He arbitrarily distinguished a corroding condition from a non-corroding condition by means of a threshold potential.

A metal is deemed to be in a **corroding** condition when the concentration of its ions in solution $\geq 10^{-6}$ M.

If the concentration of ions is less than this value then the metal is deemed to be in a condition of **immunity**.

Until now we have considered all corrosion reactions to be represented by

$$M \rightarrow M^{z+} + ze^- \qquad (2.7)$$

where the ionic species has been assumed to enter the solution as a soluble species. This may not be the case; many corrosion products are insoluble in the electrolyte, e.g. oxides. Often, the insoluble product forms a film on the corroding surface which prevents the electrolyte from coming into contact with the metal and greatly reduces the corrosion rate. This is the third electrode condition and is called **passivity.** Experiments 3.12, 3.13 and 3.14 illustrate this property. The corrosion of many metals, even in simple electrolytes, may consist of more than one reaction. These reactions are often dependent on the pH of the electrolyte.

Pourbaix succeeded in correlating the dependence of pH and the potential of the electrode with the condition of the electrode. The result of his work is a chart for each metal which shows the conditions under which a metal will be corroding, immune or passivated. The charts are called *E*/**pH** or **Pourbaix diagrams**.

The construction of an *E*/pH diagram, rather like a table of logarithms, is based upon quite simple principles but requires many calculations. With a microcomputer *E*/pH diagrams can now be calculated for almost any given set of conditions in a matter of moments. We shall consider the construction of the E/pH diagram for zinc in water, used in the next section to create a polarisation curve model. The description has been much simplified to assist the explanation, but the principles involved in the construction of the diagram are the same for all metals.

When zinc corrodes in pure water, up to four species can be present over the complete range of potential and pH: Zn, Zn^{2+}, $Zn(OH)_2$ and ZnO_2^{2-}. Therefore five reactions must be written to describe the interconversion of each of the species.

The usual anode reaction

$$Zn = Zn^{2+} + 2e^- \tag{4.26}$$

The formation of insoluble zinc hydroxide, $Zn(OH)_2$

$$Zn + 2H_2O = Zn(OH)_2 + 2H^+ + 2e^- \tag{4.70}$$

The formation of a soluble zincate ion, ZnO_2^{2-}

$$Zn + 2H_2O = ZnO_2^{2-} + 4H^+ + 2e^- \tag{4.71}$$

The dissolution of zinc hydroxide by acid

$$Zn(OH)_2 + 2H^+ = Zn^{2+} + 2H_2O \tag{4.72}$$

The formation of zincate from zinc hydroxide

$$Zn(OH)_2 = ZnO_2^{2-} + 2H^+ \tag{4.73}$$

Those reactions which involve the generation of electrons—(4.26), (4.70) and (4.71)—must be influenced by variations of electrode potential, whereas those in which hydrogen ions are formed—(4.70), (4.71), (4.72) and (4.73)—will be controlled by pH. Reactions (4.70) and (4.71) are controlled by both potential and pH.

Reaction (4.26) has already been studied in detail. It is independent of pH variation and influenced only by variation of E in accordance with the Nernst equation. Substituting the value of $[M^{z+}] = 10^{-6}$ M, we can calculate the threshold corrosion potential as

$$E = -0.76 + \frac{0.059}{2}\lg(10^{-6})$$

i.e. $E = -0.76 - 0.177$
$$E = -0.937 \text{ V}$$

Remember that all potentials calculated from the Nernst equation are relative to the standard hydrogen electrode and are more accurately quoted as -0.937 V SHE.

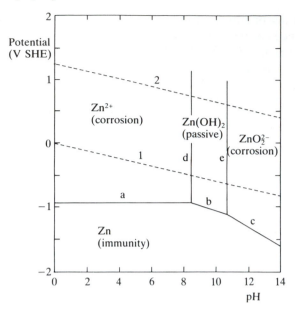

Fig. 4.13 E/pH diagram for zinc in water.

We have thus established that the threshold corrosion potential occurs at -0.937 V when $[Zn^{2+}] = 10^{-6}$ M. On a graph of E as ordinate and pH as abscissa we can draw a horizontal line at -0.937 V.

Figure 4.13 shows an E/pH diagram for the system, and the line at -0.937 V is labelled line a.

Question: *How are the other lines obtained?*

For small values of pH, (4.26) is the only significant reaction, but as the electrolyte becomes more alkaline, the concentration of hydroxide becomes large enough to cause reaction (4.70) to predominate. If we take the equilibrium condition and substitute into the Nernst equation we can write

$$E = E^\circ + 0.0295 \lg \frac{[Zn(OH)_2][H^+]^2}{[H_2O]^2[Zn]} \tag{4.74}$$

As has been explained earlier, our simplified treatment using concentrations rather than activities requires that we substitute unity for terms other than ions in solution. This simplifies eqn (4.74) to

$$E = E^\circ + 0.0295 \lg [H^+]^2 \tag{4.75}$$

E° is the standard electrode potential for the $Zn/Zn(OH)_2$ combination. This value has been measured as -0.439 V, and as we saw in Section 2.3, $-\lg [H^+] =$ pH. Equation (4.75) therefore becomes

$$E = -0.439 - 2 \times (0.0295) \times \text{pH} \tag{4.76}$$

Equation (4.76) is the equation of the line representing reaction (4.70) and labelled as such in Fig. 4.13. We can confirm its point of intersection with line a by substituting the value of -0.937 V for E in eqn (4.76):

$$\text{pH} = \frac{-0.937 + 0.439}{-0.059} = 8.44 \tag{4.77}$$

This tells us the equation of line d, for at this value of pH, $Zn(OH)_2$ becomes the most stable species instead of Zn^{2+}. A vertical line is thus drawn at pH 8.44 (shown as line d in Fig. 4.13) to represent reaction (4.72), which is independent of potential.

Line d can be found by another method, for if the equilibrium constant for reaction (4.72) is known, we can write

$$K = \frac{[Zn^{2+}][H_2O]^2}{[Zn(OH)_2][H^+]^2} \tag{4.78}$$

$$\lg K = \lg [Zn^{2+}] + 2\text{pH} \tag{4.79}$$

K has been measured by chemical means and found to have the value 7.58×10^{10}, and we have defined $[Zn^{2+}] = 10^{-6}$, therefore substituting into eqn (4.79):

$$10.88 = -6 + 2\text{pH}$$
$$16.88 = 2\text{pH}$$
$$8.44 = \text{pH}$$

As pH is increased still further it is found that the insoluble $Zn(OH)_2$ dissolves to form the so-called zincate ion. This occurs at pH = 10.68 and is vertical line e in Fig. 4.13. Line c can be determined in a manner identical to that described for line b.

The four regions of the diagram can be considered as domains of immunity, corrosion or passivation. Each domain indicates a region in which one species is the most thermodynamically stable. If the metal is the most stable species then it is considered immune to corrosion, but if a soluble ion is most stable then the metal should corrode. A region in which an insoluble corrosion product is the most stable species is considered passive. Here it is not so easy to predict whether the metal will corrode or not. There are many factors which determine whether a solid corrosion product will form a sufficiently protective film on the surface of the metal to prevent corrosion from taking place. These factors will become more apparent as the corrosion properties of different materials are discussed in other chapters.

Question: *What do the dashed lines in Fig. 4.13 mean?*

The dashed lines represent two other reactions which are possible in aqueous solutions:

1. The reduction of hydrogen ions to liberate hydrogen gas

$$2H^+ + 2e^- \rightarrow H_2 \tag{4.20/21}$$

2. The oxidation of water to liberate oxygen gas

$$2H_2O \rightarrow O_2 + 4H^+ + 4e^-$$ (4.80)

Reaction 1 gives a Nernst equation:

$$E = E° - 0.059pH$$ (4.81)

assuming that the pressure of hydrogen is 1 atmosphere. $E°$ for hydrogen is 0.00 V, so the line intersects the y-axis at $E = 0$ for pH = 0 and at pH = 10 has a value of $E = -0.59$ V. Below line 1, hydrogen gas is the most stable species, while above it, the hydrogen ion is stable. Hydrogen gas is always liberated at a cathode and this will occur when the cathode is in the domain of potential and pH below line 1.

Reaction 2 gives the same Nernst equation (4.81) where $E°$ has been measured as +1.228 V. Line 2 therefore intersects the potential axis at this value and at pH = 10 has the value $1.228 - 0.59 = 0.638$ V. Above this line, oxygen gas is the most stable species and will be liberated on an anode which lies in this region of potential and pH.

Question: *How can I make water either strongly acid or strongly alkaline without it, in effect, becoming another electrolyte?*

The question has highlighted one weakness of this theoretical approach to corrosion. In short, in order to cause an imbalance in either the hydrogen or hydroxide ions, it is necessary for another counter-ion to be present and the properties of water may be modified as a result. This may invalidate some of the calculations and their predictions in this simple form of the diagram. Often, it is possible to have a counter-ion present which does not significantly interfere, but this is not always so. In the case of copper and water, for example, the presence of chloride ions in the electrolyte does have a significant bearing on the electrode processes. However, it is still possible to construct an E/pH diagram and modern software is readily capable of this, provided all the relevant thermodynamic data are available.

A further problem is that the domains have been calculated using thermodynamic data and, as we have seen, the actual reactions are kinetically as well as thermodynamically controlled. Thus, unlike Evans diagrams, E/pH diagrams give no information about corrosion rates. There are many environmental factors, too, which cannot easily be encompassed by the E/pH diagram. Changes of flow rates, oxygen concentration, temperature and pressure are just some of the ways in which the environment continually influences the course of corrosion reactions, and what may be a simple system on paper, or even in the laboratory, becomes highly unpredictable in nature. Although the E/pH diagram would seem, at first sight, to be exceedingly useful to predict the course of corrosion reactions, there are some limitations which must be borne in mind before it can be used to effect.

4.11 THREE-ELECTRODE CELLS AND *E*/lg *i* PLOTS

The three-electrode cell is the standard laboratory apparatus for the quantitative investigation of the corrosion properties of materials. It is a refined version of the basic wet corrosion cell and a typical example is illustrated in Fig. 4.14. It can be used in many different types of corrosion experiments. First we shall examine the components in more detail.

The **working electrode** is the name given to the electrode being investigated. It is useful, though not essential, if the electrode is designed to have a surface area of at least 100 mm^2 (1 cm^2); current measurements can then be more readily converted into current densities, which should be used in calculations. Laboratory and field experiments are, however, often better performed using larger electrodes, if convenient. We use the term 'working electrode' rather than 'anode' because we are not limited to investigations of anodic behaviour alone; cathodic behaviour can also be examined.

Practical working electrodes can be constructed in a variety of ways. A simple method is to mount a small specimen in cold-setting resin (Fig. 4.15(a)). Electrical connection must be made to the specimen, and this can be done with solder or spot weld on the reverse side before mounting. After mounting, specimens are often ground and polished, as for metallographic examination. If this technique is used, the surface will be activated, in other words, passive films may have been either removed entirely or just changed from the as-received condition. This should always be borne in mind. Obviously, if the original passive film is part of the corrosion investigation, no pretreatment should be used. In fact, this is the biggest single reason for discrepancies between lab and field data. Surfaces in real engineering systems are most often as received from manufacture and are not similar to specimens prepared for metallography.

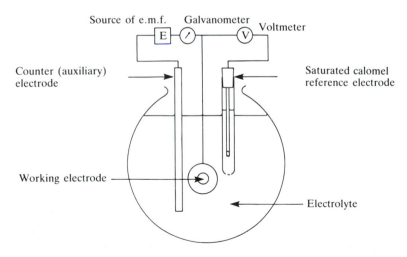

Fig. 4.14 The three-electrode cell.

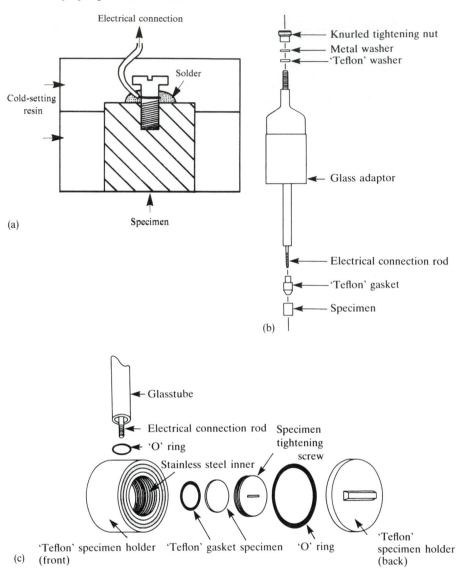

Fig. 4.15 Designs for working electrodes: (a) simple laboratory mount in cold-setting resin; (b) cylindrical specimens; and (c) disc specimens. Designs (b) and (c) ©1984 EG&G Princeton Applied Research Inc.

Although these specimen configurations are adequate for most purposes, accurate work may require a more carefully designed method of mounting. Two designs which are commercially available are illustrated in Fig. 4.15(b) and (c). Specimens in the form of thin discs of approximately 15 mm diameter, or cylinders of about 10 mm diameter × 10 mm length, can be quickly assembled into working electrodes with well-defined characteristics.

The **counter (auxiliary) electrode** is the name given to the second electrode. The counter electrode is present specifically to carry the current created in the circuit by the investigation; it is not required for measurements of potential. Usually, a carbon rod is used, but it can be any material that will not introduce contaminating ions into the electrolyte. Platinum or gold can also be used with success, especially if space is at a premium, when smaller electrodes can be used; titanium is also suitable.

The **reference electrode** is present to provide a very stable datum against which the potential of the working electrode can be measured. It cannot itself carry any more than the most negligible current. If it did, it would participate in the cell reactions and its potential would no longer be constant, hence the requirement for the counter electrode. By far the most convenient reference electrode to use in such an experiment is a saturated calomel electrode (SCE).

The external circuit can be varied considerably. The essential components are a current-measuring device, a potential-measuring device and a source of potential. The **current-measuring device** should be capable of reading microamps at least. The **potential-measuring device** should draw no current during the act of measurement; traditionally potentiometers have been used for this purpose. The modern digital meter, however, can have an impedance of the order of gigohms, and may be used with as good an accuracy as a potentiometer.

The **source of potential** must 'drive' the working electrode to produce the desired cell reactions. Typical potentiostats are readily available commercially and have been used extensively by corrosion scientists. Potentiostats apply predetermined potentials to the working electrode so that measurement of the cell current can be made. This is done by altering the current at the counter electrode to maintain the set value of working-to-reference potential. A simple constant voltage source is not suitable.

Many computer-driven instruments are now available offering a wide range of capabilities, but a very simple and inexpensive rig is quite capable of accurate measurements.

The three electrodes are usually placed in a suitable glass vessel of capacity about 0.5 to 2 litres containing the chosen electrolyte.

Question: *Does the electrolyte I choose affect any of the measurements?*

Yes, considerably, because environment is a fundamental part of the whole corrosion process. It is extremely important to consider the conductivity of the electrolyte, since by carrying the ionic current, it plays such an important role in corrosion reactions. The use of a reference electrode is to enable the potential of a working electrode to be measured and it should be placed as close to the electrode surface as possible. This is because the measured potential will always include the potential difference across the electrolyte occupying the space between the working electrode surface and the reference electrode. Most corrosion measurements involve the use of direct currents, so Ohm's Law applies and the potential difference across the electrolyte can be estimated by means of eqn (2.19), i.e. $V = IR$. Not surprisingly, this potential is often referred to as the **Ohmic** or *IR* **drop**, and may be large if either the current or the resistance of the electrolyte is large. It is usually preferable to make the *IR* drop as small as possible, otherwise its contribution to the overall cell potential may be difficult to quantify. When using a high conductivity electrolyte, such as 3.5% sodium chloride solution or sea-water, the effect will be small, so the experimental apparatus described in Fig. 4.14 is adequate for most investigations. If more dilute solutions are necessary for the experiments then it is essential to use a more sophisticated reference electrode measurement. This is achieved by means of a device called a **Luggin capillary** (Fig. 4.16(a)).

The traditional Luggin design utilises a glass capillary with a very fine tip placed as close to the metal surface as is practical. The capillary leads away from the reaction vessel to a separate small receptacle containing an SCE and filled with saturated potassium chloride solution. The tube which carries the electrolyte between the reference and test electrodes is known as the **salt bridge**.

The biggest problem with such an arrangement is from contamination of the low conductivity electrolyte by diffusion of the saturated potassium chloride. A better method is to use the equipment of Fig. 4.16(b) in which a special Vycor tip considerably reduces the leakage rates and the *IR* drop through the tip.

Experiment 4.4

Set up the equipment described in Fig. 4.14 using pure copper of measured exposed surface area as a working electrode, and a solution of 3.5% sodium chloride as an electrolyte. Adjust the potentiostat to read -0.400 V. Scan through the potentials to $+0.400$ V and take measurements of current at 0.010 V (10 mV) intervals. Allow the cell to settle at each value of potential for one minute before moving on to the next. Observe the surface of the specimen and note any changes that occur.

Question: *What shall I do with the data I get from this experiment and what does it mean?*

Compile a table of applied potential (V) and current density $(A\,m^{-2})$, treating all current values as positive. Plot a graph of E as ordinate and $\lg i$ as abscissa. You should obtain a graph similar to Fig. 4.17.

The *E*/$\lg i$ **plot**, or **potentiodynamic polarisation curve**, is one of the most common methods of examining the corrosion behaviour of materials. It has become common practice in *E*/$\lg i$ plots for all current densities to be treated as positive. This is really just a convenience for it reduces the size of the graphs and gives a much clearer indication of the value of potential when the current density changes from negative to positive. The portion of the graph for which you measured negative currents, that is from -0.400 V to about -0.240 V, represents the copper behaving as a cathode. From -0.240 V to more positive

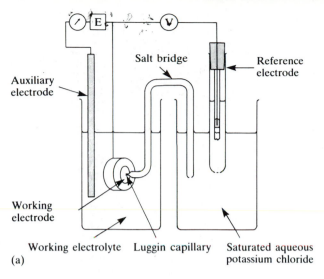

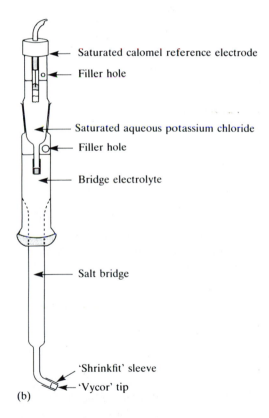

Fig. 4.16 Designs of Luggin capillary for minimising the IR drop in a corrosion cell: (a) simple capillary made from glass tube salt bridge; and (b) specialist design incorporating ball-and-socket ground glass joint. ©1984 EG&G Princeton Applied Research Inc.

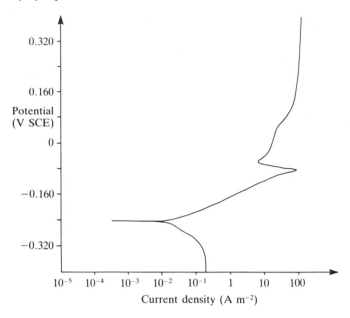

Fig. 4.17 Potentiodynamic scan for copper in 3.5% sodium chloride solution.

values of potential, the copper is behaving as an anode, and it is during this part of the scan that you will have observed visible changes to the specimen because of the numerous reactions which occurred as the copper corroded.

Question: *But so far you have described only one copper corrosion reaction, eqn (4.37). Are there more?*

Yes. The reaction we have just been considering, in which copper corrodes to its divalent ion, Cu^{2+}, has been much simplified. The corrosion reactions of copper, especially in a solution containing chloride ions, are quite complicated, but the reactions may briefly be described as the formation of copper (I) (cuprous) and copper (II) (cupric) oxides, together with insoluble hydrated chlorides.

You should have observed the formation of surface corrosion products while the specimen was in the anodic regime, i.e. −0.240 V and upwards. At potentials in the region of −0.050 to −0.080 V, a marked reduction in the corrosion current is observed. This represents partial passivation of the metal because of the presence of the corrosion products, in this case quite small, as evidenced by the relatively small reduction in current. Much more will be said about passivation in Sections 7.3 and 16.6 where examples of strong passivation will be given.

Question: *What is the significance of the changeover from negative to positive currents?*

Considerable. In practical terms it represents the state the metal assumes under freely corroding conditions. The value of potential is commonly called the **free**

corrosion potential and the symbol used is E_{corr}. In your experiment you should have measured E_{corr} at about -0.24 V SCE. At this potential the specimen could be described as being in a steady-state condition, equivalent to the condition it would have achieved with the potentiostat switched off. Remember that the purpose of the instrument is to perturb the potential of the specimen away from its rest potential. In practice, equilibrium conditions are impossible to achieve; the metal surface acts as an assembly of many tiny anodes and cathodes, and corrosion occurs at a rate given by the theoretical anode current density, i_a. At the free corrosion potential i_a is replaced by i_{corr}, **the corrosion current density**. This is the simplest way of quantifying the actual corrosion rate under freely corroding conditions, and it emphasises the importance of carrying out polarisation scans and Tafel plots of the data obtained.

Question: *So what actually is the corrosion rate?*

Corrosion rate is always equivalent to i_a at the prevailing potential for a metal dissolution reaction such as eqn (4.36). When only a single reversible metal redox reaction is being considered and is at equilibrium, $i_0 = i_a = i_c$. When more than one redox reaction occurs, as in this case, there is a separate exchange current density for each redox process, and the corrosion rate is now symbolised by i_{corr} (still equal to i_a for the metal dissolution).

Question: *Are there any other ways of using this theory to find corrosion rates?*

Yes. Suppose we take the data of Fig. 4.9(b). Instead of plotting a logarithmic graph, we plot a linear graph of E versus current density, and we plot net cathodic current density as negative and net anodic current density as positive. The resulting graph is shown in Fig. 4.18(a). It can be seen that in a narrow range of polarisation of about 30 mV either side of E_{corr}, the graph is approximately linear. The same data is redrawn over a ± 50 mV range in Fig. 4.18(b).

It can be shown that, for small values of polarisation,

$$\frac{\Delta\eta}{\Delta i} = \frac{\beta_a\beta_c}{2.3 i_{corr}(\beta_a + \beta_c)} \tag{4.82}$$

where $(\Delta\eta/\Delta i)$ is the slope of the graph. It is called the **polarisation resistance**, since it has units of ohms, and is often denoted by the symbol R_p. Thus,

$$i_{corr} = \frac{\beta_a\beta_c}{2.3 R_p(\beta_a + \beta_c)} \tag{4.83}$$

Using the values arbitrarily chosen above for both β_a and β_c as 0.1, and taking R_p from the curve (Fig. 4.18(b)) as $0.1/0.04 = 2.5$, we calculate i_{corr} to be 0.0087 A m^{-2}. Compare this to i_{corr} ($= i_0$ in this special case) $= 0.01$ A m^{-2} as we have drawn the graph in Fig. 4.9(b).

This method is called the **linear polarisation technique**. [3, 4] As we have shown in our analysis, it is approximate and the reliability of the value obtained

117

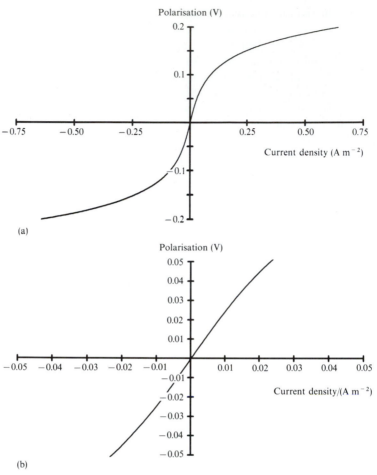

Fig. 4.18 (a) Data from Fig. 4.9(b) redrawn on a linear scale of current density and with negative cathodic current density and positive anodic current density. (b) Expanded graph of (a) showing the linear portion of the graph for polarisations of ± 30 mV. The slope of the graph in this region is called the polarisation resistance, R_p.

for the corrosion rate depends very much on the system. Many experimental determinations do not yield a linear relationship, making it difficult to determine R_p. There are many examples in which the polarisation kinetics around E_{corr} are not closely determined by the Tafel equation (see Section 4.12). In such cases, this analysis does not apply because it is not possible to determine values of the Tafel constant. A full analysis of the polarisation curve is necessary to determine the kinetic parameters. The technique is often used throughout industry as a quick method of measuring corrosion rate, but in our opinion there is much scope for abuse, and wildly inaccurate or unreliable values of a very important corrosion parameter can easily be obtained by untrained people when they fail to appreciate that the Tafel relationship does not apply for the situation being measured.

Question: *How do I calculate values for* i_0?

There is no theoretical method for calculating i_0. One way is to perform E/lg i measurements very carefully for each metal, both anodically and cathodically polarised under conditions in which the free corrosion potential is a result of only the metal redox pair of reactions. If good linearity of the resulting Tafel plot is obtained, then the extrapolated linear portions of the anodic and cathodic polarisations should intersect at the value corresponding to lg i_0 and E_{corr} (refer to Fig. 4.9(b)). However, not all such experiments result in pure Tafel behaviour; see the discussion on i_0 in Section 4.12.

Question: *If I carry out the scan in Expt 4.4 in the reverse direction, from +0.40 V to −0.40 V, will I achieve the same result?*

No, not in this case, because the formation of surface corrosion products affects the currents observed. You should not expect to observe the same behaviour by starting with a surface that is already heavily corroded (as it is at +0.40 V) and scanning to more negative potentials. It is frequently best to begin all such scans at E_{corr} (when the surface will be as close as possible to naturally occurring conditions) and then to consider deviations from this value, whether anodic polarisations (more positive) or cathodic polarisations (more negative).

Question: *When the copper is in the region of potential from −0.24 to −0.40 V, it is acting as a cathode in the aqueous corrosion cell. What is the anode and what corrodes in this cell at these potentials?*

The auxiliary electrode, in this case the carbon rod, becomes the anode. There is no corrosion in the manner of eqn (2.7), because the auxiliary electrode is chosen so that it will not contaminate the solution with ions which are not part of the investigation. Other electron-producing reactions must take place at the electrode surface if the electrons required by the copper cathode are to be produced. A common reaction in the electrolyte is the generation of oxygen gas by the oxidation of water:

$$2H_2O \rightarrow O_2 + 4H^+ + 4e^- \tag{4.80}$$

Another possibility, again in the chloride-containing electrolyte, is the generation of chlorine gas:

$$2Cl^- \rightarrow Cl_2 + 2e^- \tag{4.84}$$

Question: *Is there a simpler way of measuring free corrosion potentials?*

Yes. When you measured a simple galvanic series in Expt 3.3 you were actually comparing the free corrosion potentials of the specimens. You used a copper electrode as a crude reference electrode and assumed that its potential would not change significantly. A better idea would be to carry out the same experiment using an SCE, as was done in Expts 4.2(a) to 4.2(c). Here, the measured potentials are not equilibrium potentials, as we first suggested, in fact they are mixed potentials because the cathode process is the consumption of oxygen in the solution.

Question: *You explained in detail that the Tafel equation predicts a linear relationship between E and lg i. My E/lg i plot is not at all linear.*

There are several reasons why an E/lg i plot may be non-linear. We have already discussed the deviations from linearity which occur in the vicinity of E_{corr} when i_{meas} is significantly different from i_a or i_c. Four other reasons can be offered.

If there is more than one electrode reaction
The corrosion products which form on the surface of the specimen during your experiment serve to reduce the conductivity of the anode. This results in lower current densities than would be predicted by the Tafel equation and a deviation upwards on the E/lg i plot. This is a good reason why the determination of cathodic polarisation data is made by scanning either from the free corrosion potential to more negative potentials, or from the cathodic to the anodic regime. There is little point in obtaining cathodic data on a specimen which has already been badly corroded, unless there is a special reason for doing so.

If the scan rate is too fast
If the specimen is not allowed time to react to each potential step before the next one is imposed then linearity cannot be expected.

Because of diffusion polarisation
The theory which we have examined so far has considered only the kinetics of the electrode processes. Another major factor in the kinetics of corrosion is the transport of charge (diffusion) through the electrolyte and this will be discussed next.

If the Tafel relationship is not obeyed
The free corrosion potential occurs at a point of mixed potential and either the anodic or cathodic lines at this point of intersection are not under activation control, i.e. they are not in the potential domain in which the Tafel relationship is valid. This is usually because the cathodic polarisation line is determined by the diffusion of oxygen.

4.12 POLARISATION CURVES

In Section 4.7, we developed the theory of the E/lg i plot in which we saw how a pair of reactions (Fig. 4.9(a)), one for an anodic process (i_a) and the other for a cathodic process (i_c) were combined to create an idealised experimental curve (Fig. 4.9(b)). In Section 4.8 we discussed the polarisation curve for copper in sodium chloride and pointed out that the experimentally measured polarisation curve (Fig. 4.17) was not at all like a Tafel plot of the kind shown in Fig. 4.9(b), because the experimental curve contained information for other reactions which had not been considered. In the understanding of corrosion processes, and in the actual determination of corrosion rates for industrial processes, it is important to explain the shape of polarisation curves and to be able to extract the relevant information from them. It is the case, unfortunately, that polarisation curves are frequently misunderstood, measured incorrectly and poorly interpreted so we will make some effort to explain how they arise and what we can get from them.

In the past, it was difficult to disaggregate polarisation curves into their component parts because there were too many variables and the work required was too laborious. As a result, great emphasis was placed upon schematic diagrams of the kind discussed in Section 4.9. Figures 4.11 and 4.12 are diagrams only to *represent* the reactions which the corrosion scientist *thinks* are the ones to consider. Rarely do the diagrams have any real basis in experiment because many of the parameter values are unknown, parameter values needed for their accurate construction.

Microcomputers have enabled corrosion scientists to analyse behaviour measured in the laboratory, and today software exists for synthesis of experimental curves and for estimation of electrochemical parameters not previously available, except by very much more painstaking electrochemical measurements in precisely controlled conditions. In this way, the microcomputer has allowed us to substitute schematic Evans diagrams with modelled curves. With the aid of E/lg i plots produced by the Simpler software package, let us examine some techniques.

In the last section, we constructed an E/pH diagram for zinc in aqueous solution; let us now construct a real polarisation curve for zinc in aqueous sodium chloride at pH 3. Turn back to Fig. 4.13. A polarisation curve with potential as the y-axis occurs as a slice through the E/pH curve in the third dimension at the pH of the electrolyte, 3.0 in this case. Thus, the lg i axis would be a z-axis perpendicular to the plane of the page with its origin cutting the pH axis at 3.0. If we were to investigate the polarisation curve at pH 10, the curve would be quite different because, at potentials above -1.1 V SHE, the metal would be passive rather than corroding and coated in an insoluble layer of zinc hydroxide. In order to keep this discussion simple, we selected pH 3 so the film is not present.

In the laboratory, as we saw in Expt 4.4, the potentiostat applies potentials, as a result of which we measure currents and then, knowing the area of the specimen, convert to lg current densities. At pH 3.0, Fig. 4.13 shows that the redox process for the metal is given by:

$$Zn \rightleftharpoons Zn^{2+} + 2e^-$$
(4.26)

In a real experimental situation, we must also consider the oxidation and reduction of water, i.e. hydrogen ions (eqns (4.20 and 4.21)), line 1 in Fig. 4.13, and hydroxyl ions (eqn. (4.22)), line 2. Again, note that the equilibrium potential of both of these processes varies with pH. Reasonably, in this case, we will assume that the sodium and chloride ions have no influence, although this is not always true (copper, for example). Using the Nernst equation, eqn (4.17), we must compute the equilibrium potential for each of the three redox processes selected. This was done for zinc in the previous section and was found to be -0.937 V SHE, i.e. -1.179 V SCE. For a calculation involving copper, see worked example E18.3. This calculated value of potential is then used to calculate polarisation.

Next we need to calculate the polarisation/current density relationship. We write an expression for the total polarisation of an electrode:

$$\eta_{\text{total}} = \eta_{\text{activation}} + \eta_{\text{diffusion}} + \eta_{\text{resistance}} \tag{4.85}$$

where the polarisations due to activation, diffusion and ohmic resistance across the electrolyte are summed to give the total polarisation. Using eqns (4.51) and (4.52), together with eqn (4.69) and eqn (2.19) we get

$$\eta_{\text{total}} = \frac{2.303RT}{\alpha zF} \ln\left(\frac{i}{i_0}\right) - \frac{2.303RT}{zF} \ln\left(1 - \frac{i}{i_L}\right) + i(R_s + R_f) \tag{4.86}$$

in which R_s is the resistance of the solution and R_f is the resistance of any film present on the working electrode surface. For the anode reaction we use i_a and for the cathode reaction we use i_c. R and F are constants; R_s, R_f and T are system constants; and i_L, i_0 and z are reaction constants. Of these, i_L, i_0, R_s and R_f are not necessarily known, but sensible values can be substituted from examination of a laboratory-determined curve or from other sources. For example, with knowledge of solution conductivity and cell geometry it is easy to make a guestimate for a cell such as the one in Fig. 4.14 when the metal surface is considered to be active. According to our guestimate, R_s is 10–30 Ω and R_f can be taken as zero. If the surface is in some way passivated, we may need to choose R_f as 100 Ω or perhaps 1,000 Ω.

Taking each reaction separately, the computer calculates the i_a and i_c values for each applied potential over a given range. The mathematics is not difficult, but because eqn (4.85) cannot be rearranged for i in terms of η, an interpolation method must be used.

To create a polarisation curve, it is then necessary to sum, at each potential, all anodic and cathodic current densities for the N selected processes:

$$i_{\text{meas}} = \sum_1^N i_a - \sum_1^N i_c \tag{4.87}$$

$$i_{\text{meas}} = i_{a(H_2 \to H^+)} + i_{a(OH^- \to O_2)} + i_{a(Zn \to Zn^{2+})} \\ - i_{c(H^+ \to H_2)} - i_{c(O_2 \to OH^-)} - i_{c(Zn^{2+} \to Zn)} \tag{4.88}$$

Figure 4.19 shows the results of such a calculation, together with an experimental curve for comparison. Table 4.3 lists the parameter values used in the model. For this model we include also $T = 20°C$ and $(R_s + R_f) = 300$ Ω. The experiment used an air-saturated 3.5% aqueous sodium chloride acidified to pH 3 with HCl, together with a scan rate of 0.1 mV s^{-1}. It can be seen that, for an applied potential range of 2000 mV, the fit of experimental and model curves is excellent.

Table 4.3 Parameter values used to calculate the model polarisation curve in Fig. 4.18

Reaction	Parameter				
	$E°/V$ SHE	$i_0/nA\ cm^{-2}$	$i_{La}/nA\ cm^{-2}$	$i_{Lc}/nA\ cm^{-2}$	α
$2H^+ + 2e^- \rightleftharpoons H_2$	0.00	1.0	1.0×10^2	2.0×10^6	0.6
$O_2 + 2H_2O + 4e^- \rightleftharpoons 4OH^-$	1.23	10	1.0×10^3	6.0×10^4	0.5
$Zn \rightleftharpoons Zn^{2+} + 2e^-$	-0.76	1.0×10^3	5.0×10^6	2.0×10^3	0.5

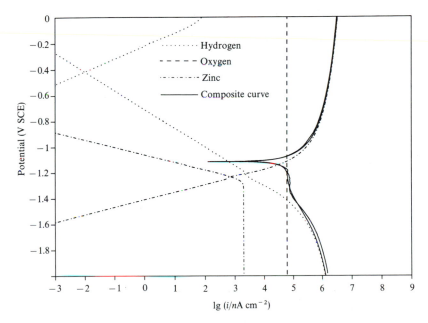

Fig. 4.19 E/lg i plot (polarisation curve) for zinc in 3% sodium chloride solution acidified to pH 3. Experimental and model curves are shown (solid lines, almost superimposed), together with the components of the model; zinc dissolution, hydrogen evolution and oxygen consumption.

A number of valuable points are clear in this model.

1. The free corrosion potential, E_{corr}, occurs at the mixed potential point of intersection of the metal anodic dissolution line and the cathodic oxygen consumption line.

2. This potential regime is *not* governed by a Tafel relationship, because the oxygen process is under diffusion control; the current density is determined by the rate at which oxygen can diffuse to the surface through the electrolyte (see Section 4.8.) It would therefore be a mistake to try to invoke a Tafel slope along the cathodic part of the experimental curve in the region of E_{corr} in order to find

i_{corr}. In this case, the corrosion rate is the value of i_a at the point where the (vertical) oxygen consumption line intersects, i.e. 6.0×10^4 nA cm^{-2}.

3. Although the hydrogen evolution line is evident at lower potentials it does not influence the free corrosion potential.

4. At potentials above E_{corr}, there is evidence of a limiting current density, just as for the cathodic lines.

5. Also affecting the curves at higher current densities is the quite high value of the ohmic drop term, 300 Ω in this case. Study the last term in eqn (4.85). You will see that when i and $(R_s + R_f)$ are both large, the polarisation is greater. Using very high conductivity solutions and highly active surfaces will reduce the effects of this term.

6. When you know which reactions are important—zinc oxidation, hydrogen ion reduction, and oxygen reduction—you can ignore those which are not important; zinc ion reduction, hydrogen oxidation and hydroxyl ion oxidation; some would say these need not have been calculated in the first place. Our advice is that until you know which lines to ignore, it is better to include both members of the chosen redox pairs.

In Chapter 7 we shall use the modelling techniques again to examine the polarisation behaviour of passivating materials such as stainless steels.

4.13   ELECTROCHEMICAL IMPEDANCE

Throughout this chapter, we have dealt exclusively with direct current. Only in recent years have ac impedance techniques been used in corrosion science, restricted in the past by the lack of the necessary sophisticated electronics. Today, these limitations have been overcome and modern techniques employ the latest digital electronics and computer control.

In the last section, the potential/current variation was expressed by eqn (4.86), based upon three components of polarisation for activation (charge transfer), diffusion and resistance, eqn (4.85). We included an ohmic resistance term in eqn (4.86) for the total polarisation of a corrosion system and noticed how, for zinc corrosion in a good electrolyte, we had to use a value of 300 Ω to indicate either resistance in the bulk solution, R_s, or resistance of a film on the electrode surface, R_f, or both. The limitations of dc measurements are obvious: if the ohmic resistance across the electrode and solution is large, the current density becomes very small for any given potential. Thus, the use of dc techniques in situations involving a large electrodic resistance requires very careful evaluation. This is the case in instances where a coating has been applied to a metal or the metal is passive, as in stainless steels, aluminium or titanium alloys. Impedance measurements are not subject to quite the same constraints and their use in corrosion science has recently been developed into a new technique called **electrochemical impedance spectroscopy (EIS)**.

Figures 4.2 and 4.7 represent the system as a corroding solid metal electrode adjacent to a solution containing varying concentrations of ions, known as the double layer. Let us first assume there is no resistance due to a film, R_f, and that diffusion effects are not important. There are two resistances we do need to consider. The first is the **charge transfer resistance**, which is the same as the polarisation resistance, R_p, resulting from the separation of charge across the solid interface to the outside edge of the double layer. The second is a resistance through the bulk solution, R_s. Inspection of Figs 4.2 and 4.7 also shows that charge has been separated and we can therefore give the double layer a capacitance value of C_{dl}. Thus, as a first approximation, a corroding electrode could be modelled using a combination of two electrical resistances and one capacitor, as in Fig. 4.20(a). This is called **equivalent circuit modelling** and this particular model is named after Randles, the first investigator in this area.

If a constant potential is applied, the system is dc only; the impedance has only a real component. The capacitance will act as an insulator and the system will behave as two resistances in series, total resistance $(R_s + R_p)$. Let us consider what will happen if we apply a sinusoidally varying potential across this system. The dc case represents zero frequency. When the frequency is very high, the impedance of the capacitor becomes very small and the charge transfer resistance is effectively short-circuited, making the total impedance of the system R_s.

We assume the component values, such as the electrolyte resistance, do not vary during the measurement process. The major difference in corrosion is that, because we are dealing with liquids rather than solids, the resistances and capacitances are leaky, they can vary as the ions move and may be affected by perturbations, albeit small, caused by the application of the alternating potential. Typical perturbations are usually much less than 50 mV, so this kind of representation has been considered very useful because electrochemical impedance techniques are quite rapid, impose only small perturbations to systems (compare the 2 V scan range of Fig. 4.19), and allow us to draw upon established ac circuit theory to characterise electrochemical systems in terms of their 'equivalent circuits'. Many workers, however, argue that equivalent circuits merely replace one poorly understood system with another. Some electrochemists believe that such equivalent circuits have no real meaning in terms of atomic models. Nevertheless, we shall see that it allows some useful progress to be made.

Measurements of electrochemical impedance are made by application of a sinusoidally varying current or potential and subsequent measurement of the response of the system at a range of frequency values from at or below the millihertz level to around 100 kHz. The amplitude of the response and its phase relationship can be measured with an oscilloscope or with phase-sensitive detectors such as lock-in amplifiers or correlators. Measurements are made over a wide range of frequencies, producing the electrochemical impedance spectrum. This would be a time-consuming experiment without the aid of computer-controlled instruments, since each measurement at a given frequency needs at least one cycle for data to be collected and this can result in long

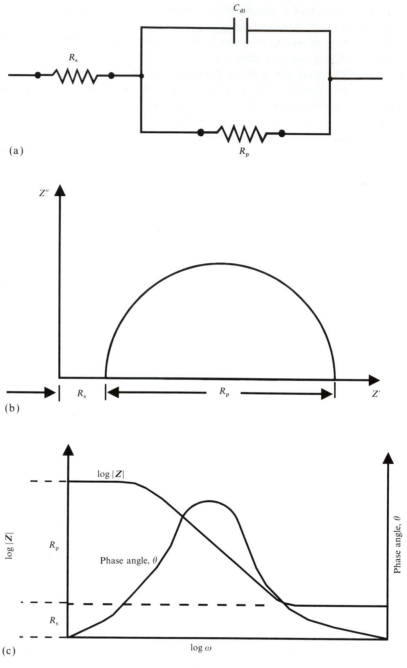

Fig. 4.20 Electrochemical impedance models of a corroding electrode. (a) Simple equivalent circuit for the system shown in Fig. 4.7(a). (b) A Nyquist plot of impedance data in which real (x-axis) and imaginary (y-axis) components of impedance are plotted. (c) A Bode plot of impedance data in which log $|Z|$ (left-hand y-axis) and phase angle, θ, (right-hand y-axis) are plotted against log ω (x-axis).

experiments (several hours) for impedance measurements at very low frequencies.

One approach which overcomes this problem is to perturb the sample with a signal consisting of a series of sine waves of different frequencies and make all the measurements simultaneously. **Fourier analysis** of the resulting current from the test cell gives Z for each frequency; it can break down any periodic waveform so that information is obtained across a range of frequencies for the chosen waveform (e.g. square wave). In practice, averaging over many cycles will eliminate the inevitable noise problems associated with small signal experiments, although in theory one cycle will contain the complete information needed for a full impedance spectrum.

A number of display formats may be used in helping the investigator to display and analyse impedance data and to determine the individual elements of the equivalent circuit. In the format known as the **Nyquist plot**, the frequency response in the complex plane is obtained by plotting the imaginary component of the impedance, Z'', against the real component, Z'. The Nyquist plot for mild steel in static aerated sea-water seems to result in a semicircle centred on the real axis (Fig. 4.20(b)). Each point on the plot represents a vector from the origin describing the magnitude (modulus) and phase angle of the impedance at that particular frequency. As the frequency decreases, the impedance increases and moves along the semicircle away from the origin. The high frequency intercept on the real axis gives the solution resistance, and the diameter of the semicircle gives the charge transfer resistance. The double layer capacitance can be determined from the angular frequency, ω, at maximum Z'' (top of the semicircle) by $R_p C_{dl} = \omega^{-1}$.

Closer inspection of the semicircle reveals that it is not as good as might be predicted by our simple theory. When the measured data *do* intercept the *x*-axis, a calculation of the centre of the semicircle shows that it frequently lies below the *x*-axis. Sometimes, the data obtained from the low frequency part of the scan do not curve around sufficiently to intercept the *x*-axis, in which case, extrapolation techniques are used, incorporated in the various software, to calculate the system parameters. One current theory [5] is that statistical analysis of the data can be used to make assessments of the type of corrosion occurring in the system. Thus, for example, the technique actually shows that the steel is corroding by means of the familiar pitting mechanism rather than by a uniform kind of general corrosion, a result which is not evident from the dc polarisation data.

A second format is the **Bode plot** (Fig. 4.20(c)). This consists of two graphs showing the magnitude of the impedance, $|Z|$, versus the frequency on a log-log scale, and phase angle, θ, against log frequency. The Bode plot is used by many workers [6, 7] because the overall impedance at any particular frequency is easily obtained and the phase angle spectrum can give information about changes in the electrode surface. [8]

Impedance measurements have been extensively used for the study of organic coatings such as paint. When the coating is new and at its most efficient, EIS techniques are not effective because the electrode is almost perfectly insulating, but as the film absorbs species, such as water molecules, ions and oxygen, gradual changes occur which can be monitored by impedance change. When the polymeric cross-linking begins to break down and the coatings start to degrade, then EIS can provide information about coating effectiveness and decay rates. [9]

EIS has been used to study the behaviour and growth of passive films. [10, 11] The usual reductionist approach treats the system by breaking it down into components:

$$Z_t = Z_{m/f} + Z_f + Z_{f/s} \qquad (4.89)$$

where Z_t is the total impedance, $Z_{m/f}$ is the metal/film interfacial impedance, Z_f is the film impedance and $Z_{f/s}$ is the film/solution interfacial impedance. Bode and Nyquist plots show features which can be ascribed to aspects of the behaviour of the components listed. Because the elements are in series, the largest will tend to predominate but since the terms are frequency dependent, any one element may dominate over different frequency ranges. Again, the approach in which the different aspects of the corrosion system are modelled by discrete electrical elements is obvious and has been much used.

Works giving a more detailed discussion of this extensive and rapidly developing discipline are listed in the bibliography.

4.14 REFERENCES

1. Trethewey K R, Sargeant D A, Marsh D J and Haines S 1994 New methods of quantitative analysis of localized corrosion using scanning electrochemical probes. In *Modelling aqueous corrosion*, edited by K R Trethewey and P R Roberge, Kluwer Academic, Dordrecht, The Netherlands, pp. 417–420.
2. McIntyre P and Mercer A D 1994 Corrosion testing and corrosion rates. In *Corrosion*, edited by L L Shreir, R A Jarman and G T Burstein, Butterworth-Heinemann, Oxford, pp. 19:1–132.
3. Mansfeld F 1986 Polarization resistance measurements — today's status. In *Electrochemical techniques for corrosion engineering*, edited by R Baboian, NACE International, Houston TX, pp. 67–71.
4. Barnartt S 1986 Electrochemical nature of corrosion. In *Electrochemical techniques for corrosion engineering*, edited by R Baboian, NACE International, Houston TX, pp. 1–12.
5. Roberge P R 1992 Analyzing electrochemical impedance corrosion measurements by the systematic permutation of data points. In *Computer modeling in corrosion*, edited by R S Munn, ASTM, Philadelphia PA, pp. 197–214.
6. Kendig M W, Mansfeld F and Tsai S 1983 Determination of the long-term corrosion behaviour of coated steel with ac impedance measurements, *Corrosion Science* 23: 317–29.
7. Mansfeld F 1981 Recording and analysis of ac impedance data for corrosion studies, *Corrosion* 37: 301–7.

8. Taylor S R, Cahen G L and Stoner G E 1989 Ion beam assisted deposition of thin carbonaceous films. III Barrier properties, *Journal of the Electrochemical Society* **136**: 929–35.

9. Scully J R 1989 Electrochemical impedance of organic coated steel: correlation of impedance parameters with long-term coating deterioration, *Journal of the Electrochemical Society* **136**: 979–90.

10. Armstrong R D and Edmondson K 1973 The impedance of metals in the passive and transpassive regions, *Electrochimica Acta* **18**: 937–43.

11. Chao C Y, Lin L F and MacDonald D D 1982 A point defect model for anodic passive films. III Impedance response, *Journal of the Electrochemical Society* **129**: 1874–9.

4.15 BIBLIOGRAPHY

Baboian R 1977 *Electrochemical techniques for corrosion*, NACE International, Houston TX.

Baboian R 1986 *Electrochemical techniques for corrosion engineering*, NACE International, Houston TX.

Baboian R and Dean S 1990 *Corrosion testing and evaluation*, ASTM, Philadelphia PA.

Bockris J O and Reddy A K N 1970 *Modern electrochemistry*, Plenum/Rosetta, New York.

Ives D J G and Janz G J 1961 *Reference electrodes*, Academic Press, New York.

Kaesche H 1985 *Metallic Corrosion*, 2nd edn. NACE International, Houston TX.

MacDonald J R 1987 *Impedance spectroscopy*, John Wiley, New York.

Pourbaix M 1974 *Atlas of electrochemical equilibria in aqueous solutions*, NACE International, Houston TX.

Rooyen D V 1994 The potentiostat and corrosion studies. In *Corrosion*, edited by L L Shreir, R A Jarman and G T Burstein, Butterworth-Heinemann, Oxford, pp. 19:133–53.

Shreir L L, Jarman R A and Burstein G T 1994 *Corrosion*, 3rd edn. Butterworth-Heinemann, Oxford.

Sykes J M 1990 25 years of progress in electrochemical methods, *British Corrosion Journal* **25**: 175–83.

Trethewey K R and Roberge P R 1994 *Modelling aqueous corrosion — from individual pits to system management*, Kluwer Academic, Dordrecht, The Netherlands.

Uhlig H H 1985 *Corrosion and corrosion control — an introduction to corrosion science and engineering*, 3rd edn. John Wiley, New York.

Verinck E D 1994 *Corrosion testing made easy: the basics*, NACE International, Houston TX.

5 DISSIMILAR METAL CORROSION

> Though ours is an age of high technology, the essence of what engineering is and what engineers do is not common knowledge.
>
> (Henry Petroski in *To Engineer is Human*)

Slightly over two hundred years ago, in 1791, Luigi Galvani made one of those chance discoveries that earned him a place in the pages of scientific history. An unexpected twitch in a frog leg muscle was followed by his discovery that an electrochemical potential results from the coupling of two dissimilar metals in an electrolyte. Since then, the name Galvani has been associated with the fundamental driving force of corrosion, although **galvanic corrosion** was formally demonstrated thirty years earlier in a naval context (see Chapter 1) and had actually been recognised by the Romans about 300 BC. In the museum at Paestum, south of Naples, an exhibit from the fourth century BC shows iron nails punched through copper washers for assembling the wooden sheets to make the hulls of ships. Just a century later the iron had been substituted, and the nails too were made of copper. [1]

Today **galvanic effect** is used in a broader context. On the macroscale, the galvanic effect might be the basis of modern battery technology, or the way in which sacrificial zinc anodes protect steel in cathodic protection (Section 16.3). On the microscale, however, it is non-homogeneous distributions of charge across the electrified solid/liquid interface that give rise to variations in the electrochemical potential over tiny areas within the whole. These, in turn, give rise to far more damaging localized corrosion mechanisms which are not indicated by the macrogalvanic effect. These are dealt with in Chapters 6, 7, 8 and 10.

Dissimilar metal corrosion is the more precise name given to corrosion that results when two different metals are coupled to form a basic wet corrosion cell, as defined in Section 4.2. It is also frequently called **bimetallic corrosion**, or **multimetallic corrosion** when there are more than two metals in the system.

As we have seen, the corrosion problems associated with the coupling of dissimilar metals are well known, as is the remedy of insulating them from each other.

Case 5.1: An aluminium alloy helicopter winch drum suffered pitting attack where it was in contact with the stainless steel winch wire (Fig. 5.1). A

Fig. 5.1 Aluminium alloy helicopter winch drums (Case 5.1). The drum on the left suffered attack where the aluminium contacted the stainless steel winch wire. A PVC coating (right) cured the problem.

flexible PVC coating, able to accommodate some chafing, insulated the aluminium from the stainless steel and cured the problem.

Even when this sound advice is carried out, problems still arise.

Case 5.2: Major restoration work was necessary on the Statue of Liberty. Built in 1886, its structure had suffered serious weakening by 1980, partly through dissimilar metal corrosion. In its construction, the copper skin of the statue was supported by a network of iron ribs. At intervals, the skin was riveted to a copper band called a saddle which, in turn, was supported by a rib (Fig. 5.2). The saddles were free to slip on the ribs to allow for contraction and expansion. In addition, the

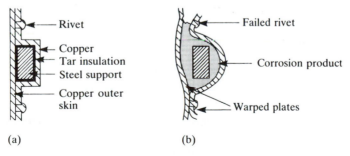

(a)   (b)

Fig. 5.2 Parts of the structure of the State of Liberty, which was given a major structural overhaul. Weakening of the iron support members had occurred because of bimetallic corrosion with the copper skin, despite the original use of insulated joints.

131

copper saddles were isolated from the ribs by a layer of tar, though the ribs and the skins were in contact. The tar deteriorated during the statue's hundred-year lifetime and, with the inevitable ingress of water, galvanic corrosion ensued. The build-up of corrosion products caused pull-out of saddle rivets, which held the copper to the supports, and resulted in serious deformation of the green patina-covered skin. [2]

Although the original design of the Statue of Liberty had guarded against dissimilar metal corrosion by insulating with tar, the case history is typical of what has probably been the most common problem of corrosion between dissimilar metals: the coupling of copper and copper-based alloys with iron and steel. The cheapness and availability of these engineering materials has inevitably resulted in their use in many situations in which for corrosion resistance they are quite incompatible.

Case 5.3:   The owner of a classic sports car spent many hours restoring his prize possession to showroom condition, but when renewing the plastic fasteners for his soft top, decided to fix them to the steel body with small brass screws because he believed they would not go rusty. Within weeks, rings of rust had broken through the paint coat immediately surrounding the fasteners.

Such a story is typical of the basic errors made, even with the best of intentions, through ignorance. On many occasions, the maintenance engineer is aware of the galvanic effect caused by dissimilar metal combinations, but faced with the need for a rapid repair and the non-availability of compatible materials, uses the wrong combination, only to remark with a shrug of the shoulders, 'Oh, that'll be all right.' Sometimes, dissimilar metal corrosion results from a quick, thoughtless action.

Case 5.4:   Steel stanchions around the steel upper deck of a warship were replaced with aluminium and correctly insulated to avoid dissimilar metal corrosion at the steel/aluminium joints. Unfortunately, this caused a different problem. It was soon decided that the aluminium needed to be in electrical contact with the steel after all, and earthing straps were attached from each stanchion to the deck. The straps were made of copper! Severe corrosion of the aluminium stanchions occurred in months.

This was a particularly bad mistake to make because copper has an especially severe effect on aluminium (see Case 12.7 and Section 15.4).

Case 5.5:   A hospital needed an emergency power supply in case of mains failure, so a diesel engine was provided. To ensure the diesel would cut in immediately the power failed, it was kept warm by means of a water heater with a cast aluminium casing (Fig. 5.3). After a few months, the aluminium casting was penetrated causing the water to leak out.

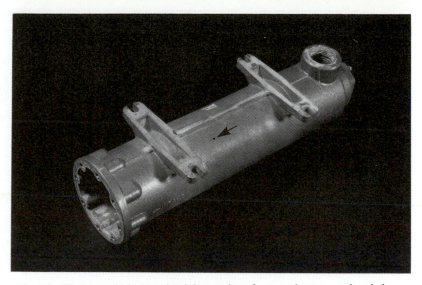

Fig. 5.3 The 8 mm thick, cast aluminium casing of a water heater on a hospital emergency diesel engine was penetrated (arrowed) when ions from a copper heating element deposited at the twelve o'clock position by convection and caused extremely rapid pitting (Case 5.5).

Inspection showed that a copper heating element had been used. Copper ions deposited at the twelve o'clock position by convection and caused extremely rapid pitting of the aluminium through a thickness of 8 mm. A stainless steel element would not have caused this damage.

Case 5.6: An aircraft handler experienced difficulty in operating a magnesium alloy fuel-pipe coupling (Fig. 5.4) during aircraft refuelling. The application of graphite grease proved to be a very short-term remedy, for the galvanic action between the graphite and the alloy caused seizure of the coupling soon after it next rained.

Failure to appreciate the problems posed by the noble and electrically conducting material, graphite, has been responsible for many cases of galvanic corrosion. Even in its amorphous form, carbon is a significant electrical conductor and is almost as likely to cause galvanic corrosion cells as graphite. Deposition on to metal surfaces of soot from smoke stacks is bad enough, but is often exacerbated because sulphur oxide emissions, which usually accompany the soot, cause acid rain. The resulting galvanic action is rapid and severe (see Case 12.8 and Fig. 12.2).

Numerous other case histories involving bimetallic corrosion will be found throughout this book, illustrating the frequency with which they occur in practice, and implying that dissimilar metals must never be coupled. With care, however, it *is* possible to use dissimilar metal combinations with relatively little adverse effect. Metals which are widely separated in the galvanic series

Fig. 5.4 A magnesium alloy aircraft fuel-pipe coupling; its threads seized up because of galvanic action after lubrication with graphite grease.

(discussed in the next section and illustrated in Fig. 5.6) should generally not be coupled — combinations with copper and copper alloys are always risky. However, the alliance of aluminium and steel can sometimes be made, providing the aluminium/steel area ratio is large.

Figure 5.5(a) shows how insulating plastic sleeves and washers enable the safe use of aluminium nuts and bolts to fasten steel plates to aluminium, but insulation is not always necessary. In the marine industry, the stability of ships has been improved by reducing top weight with aluminium alloy superstructures instead of steel. Lifeboats, fishing vessels, coastal patrol craft and even frigates have been constructed with a large proportion of aluminium in their superstructures. Normal welding techniques cannot be used to join aluminium alloy to steel so riveted joints were originally used, and corrosion protection was provided by suitable paint systems. This method has not proved entirely satisfactory and corrosion in the joint region has been observed. A more recent development has been to use an explosively bonded transition joint. Three sheets of metal — aluminium alloy (usually a 5000 series aluminium/magnesium alloy), commercial purity aluminium and steel — are joined by an explosion which creates millions of atmospheres pressure between the layers. The bonded sheets are then cut into bars of suitable dimensions. The bars are used in the transition between aluminium and steel structures by making two fillet welds: (1) of the steel in the joint to the steel of the deck and (2) the alloy in the joint to the aluminium alloy superstructure. Figure 5.5(b) illustrates the jointing method. To ensure maximum strength the transition joint is wider than the sheets being joined.

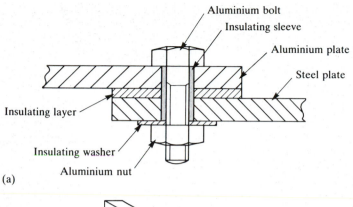

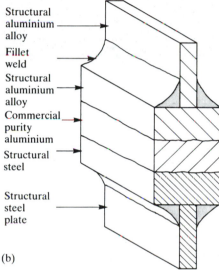

Fig. 5.5 (a) Aluminium nuts and bolts can be used to secure steel plates to aluminium if plastic insulating sleeves and washers are used. (b) The use of an explosively bonded transition joint for the fastening, by welding, of aluminium alloy to steel.

5.1 THE GALVANIC SERIES

The principles of dissimilar metal corrosion and the galvanic effect have already been discussed in some detail in Chapter 4 because they are fundamental to the understanding of all forms of corrosion. Experiment 3.3 described the establishment of an order of corrosion tendency, better known as a **galvanic series**. Such a series is of great practical value because it enables a rapid prediction of the corrosion resistance of a dissimilar metal couple. Table 4.1 is also a series, often called the **electrochemical series**. It compares the reduction potentials of the metals, but differs from the galvanic series in a number of ways:

The electrochemical series is an absolute quantitative series listing electro-chemical data for use in precise calculations; the galvanic series is a relative qualitative series listing an experimental order of nobility (or activity) of metals.

The electrochemical series necessarily lists data only for metal elements; the galvanic series contains both pure metals and alloys, a considerable practical advantage.

The electrochemical series is measured under standard conditions and is independent of other species in the environment; the galvanic series is measured under arbitrary (though specified) conditions of temperature, pressure and electrolyte.

Figure 5.6 shows the galvanic series of selected metals at 25°C in sea-water. The potentials listed are actually free corrosion potentials, and in general, it can be inferred that the greater the separation of any two metals, the greater will be the corrosion of the more active metal. Thus the considerable difference in activity between copper and iron or steel underlines the danger of using such a combination, not only in sea-water but in any aqueous medium. Very occasionally, such a combination may be justified on the grounds that the corrosion of the steel, rather than an expensive copper component, may be preferable, especially if the steel is of sufficient dimensions that failure is not possible. (The general rule is that large anodes may be tolerated with small cathodes, but small anodes are very dangerous with large cathodes. This is because the corrosion rate is determined by the current density, not by the current magnitude.)

The most anodic material listed in Fig. 5.6 is magnesium. This is a significant engineering material only when it is alloyed with small amounts of aluminium or zinc to improve the mechanical properties. It is extremely useful in the aircraft industry or for other applications requiring low density materials. The structures of many early helicopters were made of magnesium alloy and there are still many aircraft (particularly US military) which contain significant amounts of magnesium. Some very modern designs of racing cycle include magnesium alloy frames. But even in low-salt environments, magnesium alloys suffer severe attack, particularly when coupled to another metal such as steel. This has led to a general preference for substitution of magnesium alloys by aluminium alloys.

Case 5.7: Figure 5.7 is a magnesium alloy component from a helicopter which crashed into the sea and was immersed for three days before recovery. The component was secured to the main structure by steel bolts. The danger of using these materials in such an aggressive environment is obvious, even though they do not necessarily carry out their design function in immersed conditions!

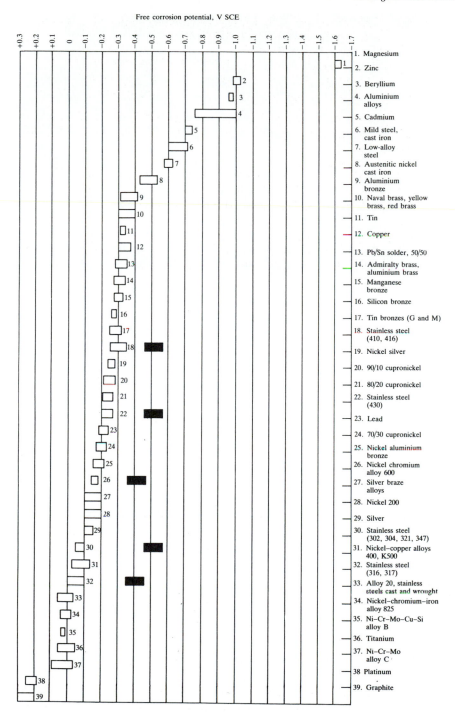

Free corrosion potential, V SCE

1. Magnesium
2. Zinc
3. Beryllium
4. Aluminium alloys
5. Cadmium
6. Mild steel, cast iron
7. Low-alloy steel
8. Austenitic nickel cast iron
9. Aluminium bronze
10. Naval brass, yellow brass, red brass
11. Tin
12. Copper
13. Pb/Sn solder, 50/50
14. Admiralty brass, aluminium brass
15. Manganese bronze
16. Silicon bronze
17. Tin bronzes (G and M)
18. Stainless steel (410, 416)
19. Nickel silver
20. 90/10 cupronickel
21. 80/20 cupronickel
22. Stainless steel (430)
23. Lead
24. 70/30 cupronickel
25. Nickel aluminium bronze
26. Nickel chromium alloy 600
27. Silver braze alloys
28. Nickel 200
29. Silver
30. Stainless steel (302, 304, 321, 347)
31. Nickel–copper alloys 400, K500
32. Stainless steel (316, 317)
33. Alloy 20, stainless steels cast and wrought
34. Nickel–chromium–iron alloy 825
35. Ni–Cr–Mo–Cu–Si alloy B
36. Titanium
37. Ni–Cr–Mo alloy C
38. Platinum
39. Graphite

Fig. 5.6   The galvanic series in sea-water. Black zones indicate active potential ranges.

137

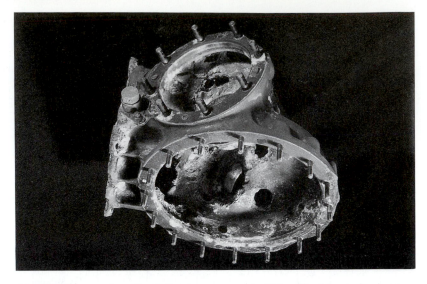

Fig. 5.7 A magnesium alloy helicopter component which suffered immersion in sea-water for three days. Corrosion is worst adjacent to steel bolts securing it to the main structure, where they are exposed to electrolyte.

The problem with graphite, the most noble material, and its coupling to one of the most active metals has already been mentioned in Case 5.6. The similar noble position of titanium, which is finding an ever increasing use in aggressive environments, indicates the need for care in the design of titanium components and structures.

Surprisingly, titanium is seldom a problem because it is a passivated material. This property will be discussed again in the next section, and in Sections 7.3, 15.6 and 16.6. The surface condition of passivated materials, which readily form stable oxide films in air, gives them an unexpected corrosion resistance and can sometimes lead to erroneous predictions about galvanic corrosion compatibility. For example, aluminium, a very active material, can be used in many applications with a reduced risk of galvanic corrosion because of its oxide layer (see Section 15.4).

Many other materials are similarly affected, notably stainless steels, titanium and nickel alloys. In some cases, an alloy which passivates only slowly can be used in the active condition instead of the preferred passive condition, and galvanic corrosion can ensue if it forms part of an incompatible couple. Particularly true of some stainless steels, this schizophrenic behaviour is clearly shown in Fig. 5.6, where the black zones indicate the active potential ranges (see Case 7.3).

The coupling of dissimilar metals can sometimes be used to advantage. Cathodic protection by sacrificial anodes is discussed at length in Chapter 16, but sacrificial wasters are also a good method of controlling the corrosion of long pipe runs (see Fig. 12.4). The wasters are anodic to the rest of the pipe and corrode preferentially. By locating them in easily accessible positions, they can be replaced quickly and cheaply.

Remember that, in certain circumstances, the order of metals in Fig. 5.6 may be significantly altered by a change in the environment. Some particular occurrences have been described by the term, cell reversal, the most important example of which is the reversal of zinc and iron in galvanised steel components at elevated temperatures in some potable water systems (see Section 14.10). Such instances have only served to emphasise the very complicated nature of metal/environment interactions, and the importance of supporting design of engineering components and structures with experimental testing of systems which accurately reproduce field conditions. The next section discusses methods for improving the understanding of galvanic effects and for predicting ways of achieving better corrosion control.

5.2   CORROSION QUANTIFICATION

Despite the usefulness of the galvanic series in predicting the relative tendencies of metals to corrode, we are again faced with the problem that it tells us nothing about the rate of corrosion. The true measure of corrosion rate is current density, but how do we analyse situations in which the system is multimetallic, of complex geometry, partly passivated and has varying concentrations, pH and velocities, etc.? In the first instance, it is necessary to examine corrosion behaviour by means of $E/\lg i$ plots. Elements of the mixed potential theory, originally devised by Evans, have already been described in Section 4.10 and Figs 4.15 and 4.16. Mixed potential theory and Evans diagrams have been much used. [3]

As a starting point, it is possible to make some statements about the effect of coupling dissimilar metals:

The galvanic series predicts that the most active metal will be the anode of the couple in a wet corrosion cell (as depicted in Fig. 4.2), and the least active will be the cathode. In combinations of more than two metals, there is usually only one anode, but geometry and potential distribution over the surfaces must be taken into account.

The corrosion rate of the most active metal is accelerated, whereas the others are retarded.

Despite this good start, to progress with a quantification of galvanic corrosion rate, we must consider a great many controlling factors, [3] listed in Table 5.1 and discussed throughout the book. Table 5.1 tells us at once how difficult it is to carry out reliable design calculations in situations containing multimetallic couples. Although $E/\lg i$ plots (or Evans diagrams), suitably amended to take into account the differences in areas of the metals present, form a firm theoretical foundation, the extrapolation of laboratory behaviour into large-scale engineering situations is extremely difficult to model. [4]

Table 5.1 Factors affecting galvanic corrosion

Factor	Comment
Alloy composition	Alloys used Major constituents Minor constituents Impurities
Electrode potential	Standard electrode potential of metal in solution Galvanic potential between metals
Reaction kinetics	Metal dissolution Oxygen reduction polarisation curve Hydrogen evolution polarisation curve
Protective film characteristics	Potential dependence pH dependence Solution dependence
Mass transport	Migration Diffusion Convection
Bulk solution environment	Temperature Flow rate Volume Height above surface
Bulk solution properties	Oxygen content pH Conductivity Corrosivity Pollutant level
Geometry	Area ratio Distances involved Surface shape Surface condition Number of galvanic cells
Type of join	Welded Fasteners Separated with external connection

The use of copper alloy propellers for steel-hull ships, for example, causes considerable problems because the galvanic effect from the couple is detrimental to the steel at the stern. Cathodic protection systems able to cope with the varying potential distributions are very hard to design (see Section 16.6). In industrial plant where there is a large throughput of sea-water or fresh water, the combination of different metal components in enclosed systems makes the prediction of corrosion extremely difficult because the mathematics required to calculate potential distributions over the surfaces is very complex. Conditions vary continuously within the system: flow rates change, turbulence may worsen as biological and mineral species slowly evolve in the system, temperature may fluctuate depending upon season or location, and the electrolyte composition itself may vary widely during operation. As turbulence increases, the degree of passivation of a surface may be lessened. As the biological content varies, the oxygen content of the electrolyte may change drastically, further affecting the degree of passivation of one of the metals. All these factors combine to make the corrosion rate at any point of a given system extremely difficult to quantify.

An example of such complexity is a condenser, illustrated schematically in Fig. 5.8. A large cylindrical drum contains a thousand tubes, all fitted at each end into a tube-plate or tube-sheet. The water enters at one end and leaves at the other through caps known as headers and there is a great quantity of large diameter pipework to deliver and remove the water. In the past, a sea-water condenser used in a ship or a power station might have a steel drum and headers, naval brass tube-plates and tubes. The steel would have been protected from the galvanic couple with the brass by sacrificial zinc anodes cast *in situ* in the headers. However, the life-limiting factor in such a design is the life of the condenser tubes, which suffer serious corrosion caused by turbulence at their inlet, where the flow changes from wide to narrow bore. Marine growth over a period of time also causes restrictions to the flow and eventual blockage, seriously reducing the efficiency of the plant.

Materials have been upgraded in several ways. Some power companies have

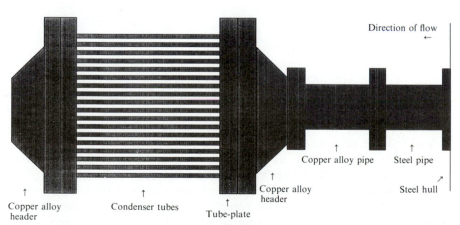

Fig. 5.8 Schematic layout of a sea-water condenser, illustrating the multimetallic combination of materials.

141

decided to adopt all-titanium designs, but titanium's very good corrosion resistance, particularly in high flow rates, is offset by its poor resistance to fouling and it has been necessary to chlorinate the feed-waters. This, in turn, causes different corrosion problems elsewhere in systems sensitive to chlorination. Other attempted solutions have used titanium tubes and tube-plates, with nickel aluminium bronze (NAB) headers and feed-pipes. In these designs considerable concern has been expressed over the galvanic effect between the supposedly noble titanium and the comparatively active NAB.

Determinations of the extent of the galvanic couple between the two metals are frought with problems, not least of which is the calculation of potential distribution over the structure. Many people use computer models applying standard mathematical techniques such as **finite element** [5] and **boundary element** modelling. [6, 7] In the second technique, a geometrical representation of the structure is made and a large number of points or **nodes** are defined. Then, using Laplace's theorem:

$$\nabla^2 \phi = 0 \tag{5.1}$$

approximate solutions are found using the previously defined boundaries and provision of suitable values for electrochemical parameters. Such modelling should never be used to provide a definitive answer, because there are uncertainties in the variables during operation, as outlined above. The metallurgical condition of the metal is critical. When made as large castings, NAB components frequently possess considerable passive films, which when eroded (perhaps in a very small region where severe turbulence or jet impingement occurs) may result in areas that are tiny but very active. Aggravated by galvanic coupling from a noble material such as titanium, these tiny areas may corrode severely.

Modelling is often carried out in the laboratory as well as on the computer. [4, 8] In cases where scaling is used, it is essential to take account of the Wagner parameter. [9, 10] The Wagner parameter is

$$\left[\rho \frac{di}{dE} \right]^{-1} = \sigma R_{\mathrm{p}} \tag{5.2}$$

where ρ = solution resistivity, dE/di is the slope of the polarisation curve at E_{couple}, i.e. the polarisation resistance R_{p}, and σ is the solution conductivity. The Wagner parameter has units of length, and when it takes large values, it signifies the potential does not change much over a surface. Conversely, when it takes small values, it implies the potential variation across a surface is great and likely to lead to large galvanic accelerations in corrosion.

It is with examples such as these that the limitations of science become painfully obvious. With the best models and the deepest understanding of the mechanisms, there can be no way of determining in advance the behaviour over an extended lifetime of the whole complex system with all its operating and environmental variables. The engineering approach is therefore to proceed with a design based upon previous experience, to define an envelope of operating ranges inside which the system performance is more reliable, to test at full-scale prior to installation, if possible, to monitor system performance during use, and to allow for inspection and maintenance throughout the system life. The engineer hopes that, with the inclusion of suitable safeguards, such as extra wall thickness in vulnerable areas or application of coatings in measured steps, a reasonable service life will result. Sometimes engineers are surprised by the longevity of a system, but other times they are embarrassed by a rapid failure. [11] There is only so much that can be done...

5.3 REFERENCES

1. Crolet J-L 1993 personal communication.
2. Chapin G 1984 Long-term deterioration requires Statue of Liberty rework, *Machine Design* **56**: 72–3.
3. Oldfield J W 1988 Electrochemical theory of galvanic corrosion. In *Galvanic corrosion*, edited by H Hack, ASTM, Philadelphia PA, pp. 5–22.
4. Astley D J 1988 Use of the microcomputer for calculation of the distribution of galvanic corrosion and cathodic protection in seawater systems. In *Galvanic corrosion*, edited by H Hack, ASTM, Philadelphia PA, pp. 53–78.
5. Fu J W 1988 Galvanic corrosion prediction and experiments assisted by numerical analysis. In *Galvanic corrosion*, edited by H Hack, ASTM, Philadelphia PA, pp. 79–95.
6. Adey R A and Niku S M 1988 Computer modelling of galvanic corrosion. In *Galvanic corrosion*, edited by H Hack, ASTM, Philadelphia PA, pp. 96–117.
7. Adey R A and Niku S M 1992 Computer modeling of corrosion using the boundary element method. In *Computer modeling in corrosion*, edited by R S Munn, ASTM, Philadelphia PA, pp. 248–64.
8. Scully J R and Hack H P 1988 Prediction of tube-tubesheet galvanic corrosion using finite element and Wagner number analyses. In *Galvanic corrosion*, edited by H Hack, ASTM, Philadelphia PA, pp. 136–57.
9. Wagner C 1951 *Journal of the Electrochemical Society* **98**: 116.
10. Hack H 1989 Scale modeling for corrosion studies, *Materials Performance* **28**(11): 72–7.
11. Petroski H 1992 *To engineer is human: the role of failure in successful design*, Vintage Books, New York.

5.4 BIBLIOGRAPHY

Baboian R, Cliver E B and Bellante E L 1990 *The Statue of Liberty restoration*, NACE International, Houston TX.
Hack H 1988, *Galvanic corrosion*, ASTM, Philadelphia PA.
Hack H 1993 *Galvanic corrosion test methods*, NACE International, Houston TX.
Munn R S 1992 *Computer modeling in corrosion*, ASTM, Philadelphia PA.
Ross R W and Tuthill A H 1990 Practical guide to using marine fasteners, *Materials Performance* **29**(4): 65–9.

6 SELECTIVE ATTACK

The present method of philosophising established by Sir Isaac Newton is to find out the laws of nature by experiments and observations.

(J Rowning, A *Compendious System of Natural Philosophy*, 1738)

The principles of galvanic corrosion discussed in the previous chapter are very important in the study of selective attack. Although galvanic corrosion normally applies to the coupling of dissimilar metals, the origins of very small-scale corrosion processes, occurring within a single piece of metal, are also galvanic.

Metals are rarely uniform in composition or structure, whether we consider them from a macroscopic or a microscopic viewpoint. Section 2.6 examined possible defects in the crystal structures of metals. Remember that a defect is considered to be any deviation from the perfect crystal lattice. The presence of defects can be beneficial or detrimental to the engineering properties of metals. For example, the movement of dislocations confers ductility, a useful property, but volume defects such as cracks reduce the failure stress (see Section 9.3). Metals often contain many undesirable volume defects created during production, yet even if these heterogeneities could be eliminated by careful quality control, the microscopic structure of most metals would still be non-uniform. One very significant defect is the grain boundary, resulting from the solidification process; other defects, such as dislocations or point defects, have a finite statistical probability of occurring at any temperature above absolute zero because of the thermodynamic energy of the metal atoms. An atom in the solid state attains its lowest thermodynamic energy only when it occupies a site within a perfect crystal lattice, thus any atom or group of atoms in a non-perfect lattice site will have a more positive free energy and be more liable to corrosive attack. Out of the total number of atoms in the lattice, only a small proportion occupy defect sites, so the corrosion processes described under the general heading of selective attack are localised and may be very penetrating. Considerable loss of strength often occurs and a dangerous condition can result, especially in pressurised or other stressed components.

Any corrosion which occurs at preferred sites on a metal surface, for whatever reason, can be described as selective attack. It is therefore possible to include in this category certain forms of corrosion described elsewhere in this book. Pitting corrosion (Section 7.2) is a form of selective attack, as is environment-sensitive cracking, a subject which occupies the whole of Chapter 10. These are discussed separately because of their importance and because other aspects of their mechanisms require special attention. Microbiologically influenced corrosion (MIC) has also been included in this chapter because it is a form of corrosion which is often localised and may be extremely damaging.

6.1 GRAIN BOUNDARY CORROSION

Most metals manufactured in bulk for general engineering purposes contain volume defects. Even in the case of a pure metal free from all production defects there can be selective corrosive attack at grain boundaries where, because of mismatch in crystal structure, atoms are less thermodynamically stable than those at perfect lattice sites, and have a greater tendency to corrode. This allows the examination of grain size and shape, a vital part of metallographic investigation.

If we wish to study the grain structure of a metal or alloy, the specimen must first be polished to obtain a scratch-free surface. In the polished state it is not possible to see the grain structure; the specimen is flat and, like a mirror, the illuminating light is reflected evenly from the surface, which exhibits no visible topography. If a mildly corrosive liquid is now applied to the surface, attack occurs more at the grain boundaries, whereas the bulk material remains largely unaffected. Some skill is required, otherwise the metal will corrode indiscriminately over its entire surface. The topographical detail which results from careful application of the corrodent corresponds to the grain structure and is readily visible because of the non-uniform reflection of the incident light. This process is known in the context of metallography as **etching**, but is actually a form of controlled grain boundary corrosion on the surface.

You may well have seen these effects without realising the corrosion significance. Many zinc-coated or brass articles have large grain structures which are very noticeable after being exposed to the atmosphere for a period of time. (See the large specimen in Fig. 6.4.) Brass door handles in particular soon show beautiful and intricate patterns after the touch of hundreds of sweat-moistened hands has caused non-uniform surface corrosion. Experiment 3.9 examines this further.

Figure 6.1 is a photograph of the microstructure of a sample of mild steel containing 0.15% carbon. The specimen was prepared by the method described above and etched with a solution of 2% nitric acid in ethanol. The specimen, magnified 500 times, shows two distinct regions: a light-coloured area of ferrite, the solid solution of carbon in body-centred cubic iron, and a striped grey area of pearlite, a fine mixture of ferrite and iron carbide. The grain boundaries are clearly visible, and so are manganese sulphide inclusions, which appear as dark specks randomly scattered through the material. The corrosion of this material is further discussed in Sections 7.2 and 15.1. Without grain boundary corrosion, the examination of such an important engineering material would not be

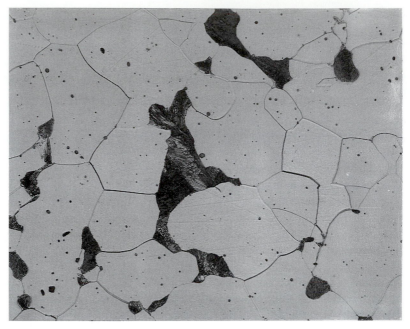

Fig. 6.1 Micrograph of a fully annealed 0.15% carbon steel, etched with 2% nitric acid in ethanol (× 500).

possible and engineers' understanding of the properties of steels, indeed all metals, would be seriously deficient. Not all corrosion is unwanted!

6.2   INTERGRANULAR CORROSION

Intergranular corrosion occurs when a grain boundary area is attacked because of the presence of precipitates in these regions. Grain boundaries are often the preferred sites for the precipitation and segregation processes observed in many alloys (Sections 2.5 and 2.6). Segregates and precipitates are distinguished only by their genesis; from a corrosion point of view they may be considered as physically distinct from the remainder of the material with their own thermodynamic energy. These intruders into the metal structure are of two types:

Intermetallics: also known as intermediate constituents, intermetallics are species formed from metal atoms and having identifiable chemical formulae. They can be either anodic or cathodic to the metal.

Compounds: compounds are formed between metals and the non-metallic elements, hydrogen, carbon, silicon, nitrogen and oxygen. Iron carbide and manganese sulphide, two important constituents of steel, are both cathodic to ferrite.

In principle, any metal in which intermetallics or compounds are present at grain boundaries will be susceptible to intergranular corrosion. It has been most often reported for austenitic stainless steels, but it can also occur in ferritic and two-phase stainless steels, as well as in nickel-base corrosion-resistant alloys. For a discussion of the composition and properties of these alloys, see Chapter 15.

Aluminium alloys may suffer severe intergranular corrosion. In the high strength aluminium alloys used for aircraft, it is the control of precipitates both at the grain boundaries and within the grains that determines the material strength. Two common precipitates, $CuAl_2$ and $FeAl_3$, are both cathodic, but Mg_5Al_8 and $MgZn_3$ are anodic to the surrounding metal. The presence of these constituents, whether anodic or cathodic, creates minute localized galvanic cells when an electrolyte is present. Corrosion can then occur such that, if the precipitates are anodic, they dissolve and leave a porous material, or if they are cathodic, the surrounding metal is attacked. In either case the material may be seriously weakened. The precipitation process can also cause depletion of alloying elements, particularly near the grain boundary regions, which in turn has a deleterious effect on corrosion behaviour.

Intergranular corrosion has also been observed in some zinc die-cast alloys, and also in lead. But the most significant problem is found in the austenitic group of stainless steels, often called **weld decay** because of its frequent occurrence in association with the welding of the materials. The remainder of the discussion will be a case study of weld decay, but the principles apply equally well to any alloy system in which precipitation occurs at grain boundaries.

Figure 6.2 shows the simplified diagram of the solid solubility of carbon in an Fe–18Cr–8Ni (type 304) alloy. When the carbon content is less than about 0.03% only the γ-phase is stable. For compositions with greater than 0.03% carbon the equilibrium stable phases are γ and a mixed carbide, believed to have the formula $(FeCr)_{23}C_6$ and called chromium carbide. The proportions of carbide actually obtained depend upon the rate of cooling; fast cooling by water or oil quench from >1000°C suppresses carbide formation. If the material is subsequently reheated, especially within the range 600–850°C, there is a considerable likelihood that carbide precipitation will occur at the grain boundaries. The material is then said to be **sensitised**, a dangerous condition from a corrosion point of view. Below 600°C the rate of diffusion of chromium is too slow for carbide precipitation to occur. The presence of chromium (>12%) in a steel significantly improves the corrosion resistance, hence the term 'stainless' steels. However, the precipitation of chromium carbides causes a depletion of chromium to below 12% in the metal adjacent to the precipitates; this metal is thus no longer 'stainless'. Compared to the rest of the grain, the chromium-depleted region is very anodic and severe attack occurs adjacent to the grain boundary if the metal comes into contact with electrolytes. In extreme cases whole grains become detached from the material, which is considerably weakened (Fig. 6.3).

The problems of using such materials are obvious. Even if the alloy is in an

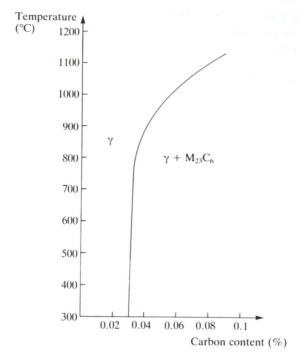

Fig. 6.2 Simplified solid solubility diagram of carbon in an Fe–18Cr–8Ni (type 304) alloy.

unsensitised condition after manufacture with no carbides present, there is always a danger that subsequent use may involve heating that causes sensitisation. The welding of austenitic stainless steels is one example which in the past has led to many serious failures. The problem was so serious that other alloys have now been developed in which there is far lower probability of obtaining grain boundary precipitates. Even so, type 304 stainless steel was still being specified for boiling water reactors, despite knowledge of its susceptibility to weld decay. [1] Furthermore, sensitisation and the resulting danger of weld decay can occur at 300–320°C if chromium carbide nuclei pre-exist in grain boundary regions.

Stabilised stainless steel describes an alloy not susceptible to intergranular corrosion. It is possible to stabilise an austenitic stainless steel, such as the Fe–18Cr–8Ni example used above, by the addition of a small amount of titanium or niobium (known as columbium in the United States). These elements in sufficient amounts form carbides in preference to those of chromium and as a result the grain boundary areas are not chromium-depleted. It is usual to add about 5–10 times as much titanium or niobium as the amount of carbon present. This is to ensure no chromium carbides are obtained.

Correct training or supervision of welders can assist in guarding against weld decay. For example, prior to welding it is essential the work is not wiped clean

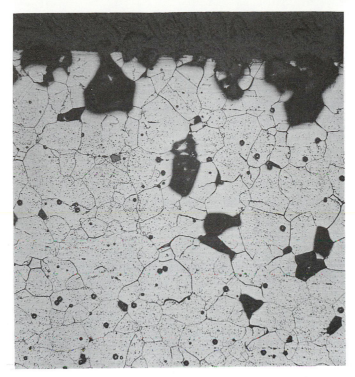

Fig. 6.3 Optical micrograph of the cross-section through a typical sample prepared in Expt 6.1 showing severe intergranular corrosion of an unstabilised 304 stainless steel. Large dark areas in the material and at the top edge are holes remaining where complete grains have become detached.

with an oily rag because this may lead to carbon pick-up by the metal when it is heated.

In principle, there are three ways of reducing the susceptibility to intergranular corrosion of a metal such as a 304 stainless steel:

Use a low carbon steel, i.e. less than 0.03%, so that the carbides are not stable (e.g. type 304L).

Apply post-weld heat treatment to dissolve the precipitate.

Add titanium or niobium to form carbides preferentially (for example, type 321).

Case 6.1: While attempting to fold the tail pylon of a helicopter, the locking assembly could not be disengaged because of a fractured unstabilised stainless steel pin. Examination showed the most likely cause of failure was intergranular cracking initiated by a network of grain boundary precipitates. Subsequent heat treatment of the material redissolved the precipitates and it was recommended that heat treatment be applied to all similar pins in service or in store.

Experiment 6.1

For this experiment you will need specimens of unstabilized AISI type 304 and stabilized type 321 stainless steels. If they can be obtained in the form of tensile test specimens, so much the better. Take two specimens of each material; heat treat in a furnace at 1,050°C for 3 hours then water quench. This is to ensure the material contains only one phase. Next reheat the specimens to 750°C for an hour and furnace cool to about 450°C. To save time, the specimens may be water quenched when they have cooled to this temperature. The purpose of the second heat treatment is to simulate the effect of a welding process on the alloys.

Prepare a solution of 25 g of hydrated copper (II) sulphate in 200 ml of water. Add **CAREFULLY** 25 g of sulphuric acid and make up to 250 ml. Reflux the four specimens in this solution for 72 hours making sure they do not lie across each other; wastage because of crevice corrosion (Section 7.1) will occur at points of contact. After refluxing, use one of each type to examine the microstructures. Section, mount and polish each one, then electrolytically etch using a solution of oxalic acid, 10 g in 100 ml of water. Make the specimen the anode of a corrosion cell by connecting it to the positive terminal of a dc power supply set to deliver 7.5 V. A suitable piece of stainless steel may be used as a cathode. The etching process takes approximately 1 minute. Examination under a microscope should show that the unstabilised specimen is severely attacked, whereas the stabilised sample is very much less affected. Figure 6.3 shows a cross-section of a typical specimen of unstabilised material.

Now perform a tensile test on the two remaining specimens. You should observe that the unstabilised stainless steel has a much reduced tensile strength compared to the stabilised specimen. Indeed, it is quite common for the unstabilised specimen to break in the hand, in which case you will be able to perform the experiment without a tensile testing machine!

This experiment is based upon a standard method of testing devised by Hatfield [2] to predict the susceptibility of alloys to intergranular corrosion. Several variations exist, as well as other test procedures, and today the methods of testing have been much refined. Hatfield's original idea remains the simplest and is an effective way of demonstrating the phenomenon of intergranular corrosion. It is a severe test and exaggerates the effect observed *in vivo*. The work of Streicher [3] contains a detailed review of the methods of testing.

6.3   SELECTIVE LEACHING

Selective leaching is the net removal of one element from an alloy and is thus often referred to as **dealloying** or **demetallification**. The whole of an exposed surface may be attacked, leaving the overall geometry unchanged, yet the removal of a large proportion of one of the alloying elements leaves a porous material with virtually no mechanical strength. Sometimes the effect is very localised, in which case perforation may occur. As with the other types of corrosion discussed in this chapter, the prime cause of dealloying is the galvanic effect between the different elements or phases which compose the alloy, although factors such as differential aeration and temperature are also important.

In the past, many problems have been reported [4] with brasses that corrode by loss of zinc in a process known as **dezincification**. Components designed for use in sea-water and fresh water have failed by this form of corrosion in such applications as condensers, valves, taps and pipework, as well as screws, nuts and bolts. The problem is not confined to brasses; other reports [4] of this form of corrosion have included the loss of nickel, aluminium and tin from copper alloys in processes known respectively as **denickelification**, **dealuminification** and **destannification**. This subject is discussed further in Section 15.5, but because the overwhelming majority of reports have concerned dezincification, this section is restricted to the dezincification of brasses. The principles are the same for dealloying in general.

The compositions of a number of important brasses are listed in Table 15.5. There are two main categories of brass: single-phase and two-phase. The most common single-phase brasses are the 70/30 brasses; the two-phase brasses contain the so-called 60/40 brasses. Because of their poor cold-working properties, 60/40 alloys are restricted to castings; 70/30 brasses are used where malleability is important, e.g. in tubes.

The problem of dezincification was quickly overcome in the case of the single-phase, lower zinc brasses by the discovery that the addition of very small amounts of arsenic, usually about 0.05%, virtually eliminated this form of corrosion. The 70/30 alloy with the arsenic addition became known as Admiralty brass (inhibited) and today, most commercial production of 70/30 brass contains arsenic for this purpose. Uninhibited 70/30 brass should never be specified for uses involving immersion in either fresh water or sea-water.

Unfortunately, the addition of arsenic does not significantly affect the susceptibility of the 60/40 alloys and the problem of dezincification has remained. Efforts have been made to find other elemental additions which might alleviate the corrosion. The addition of 1% tin to Muntz metal, an alloy with 60 Cu–40 Zn, gives another two-phase material known as Naval brass, widely claimed to have greatly reduced susceptibility to dezincification, though there is little evidence to support it. Rogers [5] states that Naval brass does not provide improved resistance and there is no reliable method of inhibiting the dezincification of two-phase brasses.

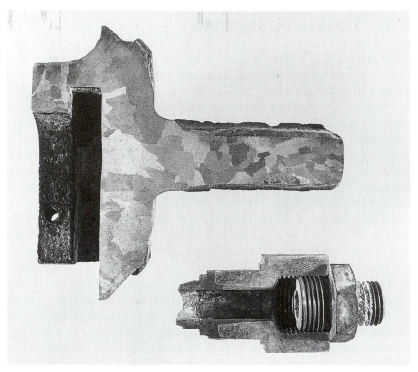

Fig. 6.4 Two brass components which failed because of dezincification. The smaller component is described in Case 6.2 and the larger component in Case 6.3.

Case 6.2:   The smaller component in Fig. 6.4 is a 60/40 brass casting used as a connector in a water supply. The dark inner areas are dezincified and the left extremity has fractured because of the embrittlement caused by the high porosity.

Case 6.3:   The larger of the two components in Fig. 6.4 is part of a cast aluminium bronze sea-water valve, originally wasted by erosion but not dealloyed. The component was recovered by depositing 60/40 brass along the shank and around the closing surface. The reclaimed areas have suffered dezincification and some has been broken off. Note the beautiful grain structure exhibited by this specimen. Alloys formed by the addition of lead (about 4%) to 60/40 brass to improve the machinability are also susceptible to dezincification.

Case 6.4:   Figure 6.5 shows a leaded brass bolt, 12 mm in diameter, dezincified to a depth of 3 mm over several years of exposure to sea-water. The fractures in Fig. 6.5(a) are a direct result of the embrittlement of the material and they are clearly visible. In Fig. 6.5(b) the extent to which the attack has penetrated is evident from the dark and beautifully layered structure.

(a)   (b)

Fig. 6.5   (a) A dezincified leaded brass bolt, 12 mm diameter. (b) The corrosion penetrated to a depth of 3 mm, causing severe porosity and embrittlement, and a remarkable layered microstructure within the corrosion product (Case 6.4).

The fact that such corrosion can take a long time to develop in practice makes accelerated laboratory simulation a desirable technique. This can be achieved using a method described below.

Experiment 6.2

Set up a three-electrode cell, as described in Section 4.11, using a specimen of leaded 60/40 brass as a working electrode, a carbon rod auxiliary electrode, an SCE reference electrode and a solution of 3.5% sodium chloride as electrolyte. Set the potentiostat to apply an anodic potential of -120 mV SCE ($E_{corr} = -240$ mV SCE) and leave the specimen for a convenient length of time. The effect will be observable after one or two days, but several weeks will give better results. After the chosen exposure time, mount the specimen longitudinally to examine the penetration of the dezincified layer. Figure 6.6(a) shows an example of the microstructure you should obtain. The features of the specimen prepared in this way are very similar to those of the corroded brass bolt in Fig. 6.5. Several distinct regions can be identified and for clarity these are illustrated in Fig. 6.6(b).

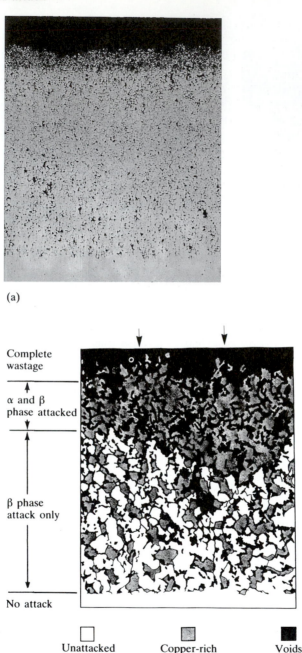

(a)

Complete
wastage

α and β
phase attacked

β phase
attack only

No attack

☐ Unattacked material   ▨ Copper-rich areas   ■ Voids

(b)

Fig. 6.6 (a) Microstructure of a laboratory-prepared dezincified specimen of 60/40 leaded brass ($\times$ 200), ammonia/hydrogen peroxide etch. (b) Schematic illustration of the dezincified layers of the specimen shown in (a).

In the 60/40 brasses there are two phases, α and β. The α phase contains much less zinc than the β phase. In leaded brasses the lead atoms are totally insoluble in the copper–zinc lattices and the lead exists as tiny globules dispersed at random throughout the solid. We use this material in the experiment because it gives good examples of this form of corrosion. The presence of the lead can be ignored for this discussion, although it does influence the corrosion rate. The E_{corr} values of the α and β phases are respectively -230 mV and -285 mV SCE; on this basis we would predict from the galvanic effect that the β phase would be more anodic compared to the α phase and that the β phase would be attacked first. The attack of the electrolyte on the β grains is interesting because it seems they behave as if they were themselves simply galvanic couples of zinc and copper; the zinc is dissolved quickly, whereas the copper is much less affected. Dezincified specimens produced in Expt 6.2 and illustrated in Fig. 6.6 show the following features:

1. A band that appears pink to the naked eye and contains a mixture of unattacked α grains and pink copper-rich grains which were initially β.
2. A band in which the β phase is missing and the α phase is dezincified.
3. Complete wastage of the material from the outside surface.

One explanation for the observed corrosion proposes the following sequence of events. In the early stages, the α is relatively unaffected. The zinc leaching from the β leaves behind an increasing number of pores through which the electrolyte is able to penetrate much deeper into the material. A corrosion front begins to progress into the material. Once the band has been established (step 1), two more processes occur. The copper in the dezincified β grains begins to corrode and soon the complete grains, which were originally β, have disappeared. At the same time the α grains are leached of zinc. This yields the band in step 2. Eventually the dezincified α is dissolved completely and wastage occurs (step 3).

There has been much speculation about the exact nature of the mechanism of dezincification. Evidence has been produced that only zinc is leached from brass, while it has also been claimed that both zinc and copper are dissolved. Both mechanisms are possible, but the factor which determines the type of behaviour is the corrosion potential. [6] This would certainly seem to apply to the experiment described above. Measurements on the specimens obtained from similar experiments have shown that porosity of up to 40% is normal in the dezincified layers of two-phase brasses. Furthermore, the free corrosion potential of the specimen becomes more anodic when the solution is de-aerated by bubbling argon into the flask; indeed, the E_{corr} of Muntz metal changes from -235 mV SCE to -380 mV SCE when the solution is deoxygenated. It is reasonable to assume that this simulates the differential-aeration cell formed once it has become porous. There is a potential difference between the cathodic areas on the outside of the specimen, where oxygen is readily available, and the corroding interface deep within the metal, where there can be little oxygen; this potential difference constitutes a major driving force. [7]

6.4   MICROBIOLOGICALLY INDUCED CORROSION

Biofouling is a macroscale problem. On ship hulls it involves weed and barnacle growth and is more a problem of speed loss and poor fuel economy than corrosion. In 1982 biofouling in heat exchangers was estimated to cost between £300 million and £500 million per annum in the United Kingdom. [8] The kind of effect embraced by **microbiologically induced corrosion (MIC)** is quite different and in some cases serious and rapid corrosion problems have been experienced. In 1968, the contribution of MIC to biodeterioration of materials as a whole was a large proportion of the overall $1 billion per annum in the United States, [9] while in 1972 buried pipelines were incurring expenditure of between $0.5 billion and $2 billion due to microbial corrosion. There is a common relationship between biofouling and MIC. Once macrospecies such as weed, barnacles and mussels become established, they often create micro-habitats in which various bacterial species can proliferate beneath the biofilm layer.

Micro-organisms are commonly subdivided into fungi, algae and diatoms, and bacteria. Fungi typically metabolise hydrocarbons and, when associated with corrosion, they are filamentous and yeast-like, branching and weaving themselves into a tangled mass. This catalyses the formation of differential-aeration and concentration cells which then lead to pitting and other localised attack. Algae and diatoms are chlorophyll-containing plants found in sea-water and fresh water. Their role in corrosion is similar to that of fungi. Bacteria are unicellular organisms with a wide range of behaviour, some of which will be described below.

An early example of serious corrosion attributed to fungal growth was in the fuel tanks of military aircraft such as the Comet. Here, the fungus *Cladosporium resinae* thrived on the nutrient combination of water, fuel and dissolved oxygen and large mats of fungal growth appeared at the water/fuel interface. [10] It was shown that the microbe excreted organic acids which lowered the pH and caused serious pitting of the aluminium fuel tank plates (Fig. 6.7). In other cases, the fungal mats were drawn into fuel pipes, where serious blockages occurred. Today, the problem is regarded as under control because of better design of fuel tanks and extensive use of biocides as additives to the fuel; indeed, the use of biocides remains the main method of corrosion control for all forms of MIC.

In recent years there has been some debate about whether the corrosion attributed to microbes actually constitutes a specific form of corrosion. It is said that the microbes simply create changes in the environment by secreting chemicals as by-products of their metabolism. This causes other standard forms of corrosion to occur, so the microbes are just a secondary factor in the corrosion mechanism. Indeed, it can often be difficult to unambiguously associate microbes with observed corrosion. Nevertheless, MIC is now regarded by many corrosion scientists and engineers as a form of corrosion in its own right.

Fig. 6.7 Microbiologically induced corrosion of the internal surface of an aluminium alloy Comet aircraft fuel tank.

Microbes are not usually associated with general corrosion, but with localized attack. This results from the small size of the microbe compared to microdefects in materials. Bacteria are typically 0.2–5 μm wide and 1–10 μm long, although some filaments may be several hundred micrometres in length. Surface defects in steel products can easily be 100 times as large, thus micro-organisms can be found deep inside metal defects, where they can quickly develop the optimum conditions they need to flourish. These include a suitable temperature, pH, water and available nutrient. In general, micro-organisms are either **aerobic** (requiring the presence of oxygen) or **anaerobic** (requiring the absence of oxygen). Some micro-organisms are able to survive in a wide range of oxygen concentrations.

The decision about whether a species of microbe is responsible for a given example of corrosion always results from a series of careful investigations.

Case 6.5: When a 76 mm diameter carbon steel pipe was sectioned, it was found that serious pitting occurred at the six o'clock position. Despite favourable temperatures and the presence of some water in the pipe, it was established that the chemical flowing through the pipe was more dense than water. This meant the corrosion would have been expected at the top of the pipe, rather than at the bottom. And the chemical was not a nutrient, so MIC was discounted.

Desulfovibrio and *Desulfotomaculum* are **sulphate-reducing bacteria (SRB)**, by far the most troublesome micro-organisms associated with corrosion of iron and steel. Rates of SRB-induced corrosion as high as 2 inches per year have been

157

reported. [11] They are anaerobic bacteria which reduce sulphate ions to sulphide ions. They can be found in virtually all industrial aqueous processes, including cooling water systems, paper-making systems and petroleum production and refining. The pH range most suited to growth of SRB is 5–10. When present in biofilms, SRB are known as **sessile micro-organisms**. These are particularly resistant to biocides, probably because the biofilm contains extracellular polymeric material as a protective layer.

Sulphides are highly corrosive to most metals. In the oil and gas industries, for example, the problems of sour gas (i.e. containing H_2S) have been extreme. In one group of US oil wells, 77% of all failures were attributed to MIC. Sulphide appears as dissolved or gaseous H_2S, HS^- ions and S^{2-} ions or metal sulphides. Thus, MIC caused by SRB is frequently diagnosed when a site is excavated and a smell of hydrogen sulphide, combined with the presence of black-stained soil, is discovered. The sulphide film on the metal is soft and loose. When removed, a bright metal surface appears, but quickly develops a brown rust film. [12]

Biological slimes are commonly present in the water phases of industrial plant and it is here that species such as *Pseudomonas* and *Falvobacterium* metabolise large amounts of organic materials. Conditions at the base of these slimes are ideal for the growth of SRB and thus even though the bulk material may not support SRB, localised conditions within the bulk create ideal locations for SRB to flourish.

SRB have been implicated in hydrogen pick-up by metals. [11] In this mechanism, the micro-organisms interfere with the association of hydrogen atoms which would otherwise form molecular hydrogen on cathodic surfaces. The hydrogen atoms are absorbed into the metal with consequent increase in the likelihood of other hydrogen-induced problems (see Section 10.3). There is also some evidence for depolarisation of the cathodic reaction.

There are many examples of microbes being attracted to welds and heat-affected zones, where they then participate in localized corrosion. The reason for this remains the subject of speculation, but there is good evidence that microbes are able to identify anodic sites and colonize them. Microbes have been observed to 'build' deposits and tubercles which are needed for survival in a given environment. Some are hard shells offering protection from hot demineralized water. *Thiobacillus* and *Desulfovibrio* do not appear to require the same protection and inhabited aggregations of loose fine particles.

Microbes have been shown to participate in dealloying. In a case of denickelification, sulphate-reducing and sulphate-oxidising bacteria, *Pseudomonas*, was implicated in an alloy 400 heat exchanger tube. [13] Another bacterium, *Bacillus* sp., was found to have formed a deposit on a 304 stainless steel component. Underneath the deposit was a pit which had initiated a stress corrosion crack. It was concluded that the bacterium thrived at 75–90°C and accumulated chloride ions, depositing them at the crack tips which developed beneath the deposit.

Gallio, and to a lesser extent *Sphaerotilus*, are iron bacteria frequently associated with other micro-organisms. They result in serious corrosion of iron and steel components and structures. *Gallionella* has also been held responsible for chloride/low oxygen attack of stainless steels, as the following case history illustrates.

Case 6.6: A chemical plant in the tropics used a large quantity of 304 stainless steel pipes. After hydrotesting with local water for several weeks, the 5 mm thick pipes began to leak in hundreds of places in the six o'clock position. Large quantities of slime containing *Gallionella* were found in the pipes, together with high concentrations of chloride ions. This, together with the low oxygen levels beneath the slime, was enough to cause this very rapid penetration of the pipes. [14]

Despite the fact that copper acts as a toxin to many macrofouling marine organisms, copper and copper alloy components are still affected by MIC.

Case 6.7: Growth of SRB in stagnant water in a marine pump caused severe corrosion of a gunmetal impeller only weeks after the pump was switched off. The component had previously given two years trouble-free service in aerated water. [14]

Micro-organisms colonize and form gelatinous films on all engineering materials exposed in natural marine environments, including cathodically protected surfaces. This is of considerable importance in situations in which the presence of a calcareous film, generated by cathodic protection, is regarded as providing protection against corrosion. [15] The calcareous deposit supplements the protection afforded by the impressed cathodic current, and much interest focuses on how micro-organisms influence its growth and consequently the level of protection.

The collapse of concrete sewers has been attributed to both aerobic and anaerobic bacteria. SRBs are thought to start the mechanism by producing hydrogen sulphide in the anaerobic sewage. The gas accumulates at the top of the pipe and is then converted into sulphuric acid by *Thiobacillus*. The pH drops, leading to concrete decay by attack of steel reinforcement and by spalling. The spalling mechanisms are similar to those discussed more fully in Sections 14.7 and 16.5.

Control of MIC is very much tailored to the situation. For pipelines, use of suitable backfills may be advantageous. Alternatively, protective coatings can be used, but some fungi are capable of penetrating certain kinds of elastomeric coatings. Replacement of metal with polymeric materials is always to be considered. In flow-through systems and tanks, biocides represent the most important control measure, and in potable water much of the problem has been alleviated by the use of chlorination. But remember that chlorination can induce other forms of corrosion. Where water can be considered an impurity, then obviously careful control over its access to the system will control the amount of growth that can occur. Finally, as always, good housekeeping and maintenance is usually a most effective preventative action.

6.5 REFERENCES

1. MacDonald D D, Begley J A, Bockris J O, Kruger J, Mansfeld F B, Rhodes P R and Staehle R W 1981 Aqueous corrosion problems in energy systems, *Materials Science and Engineering* **50**: 19–42.
2. Hatfield W H 1933 *Journal of the Iron and Steel Institute* **127**: 380–383.
3. Streicher M A 1978 Theory and application of evaluation tests for detecting susceptibility to intergranular attack in stainless steels and related alloys. In *Intergranular corrosion of stainless alloys*, edited by R F Steigerwald, ASTM, Philadelphia PA, pp. 3–84.
4. Heidersbach R H 1968 Clarification of the mechanism of the dealloying phenomenon, *Corrosion Quarterly Report* **Feb**: 38–43.
5. Rogers T H 1968 *Marine corrosion*, Newnes, London, p. 120.
6. Trethewey K R and Pinwill I 1987 The dezincification of free-machining brasses in seawater, *Surface and Coatings Technology* **30**: 289–307.
7. Trethewey K R and Murphy P W 1988 New aspects of the dezincification of brasses, *Industrial Corrosion* **6**(7): 10–15.
8. Pritchard A M 1979 Heat exchanger fouling in British industry, *Fouling Prevention Research Digest* **1**(3): iv.
9. Knox J 1968 The role of government in international co-operative research. In *Biodeterioration of materials*, edited by A H Walters and J J Elphick, Elsevier, London.
10. Parberry D G 1968 The role of *Cladosporium resinae* in the corrosion of aluminium alloys, *International Biodeterioration Bulletin* **4**(2): 79.
11. Hardy J A and Brown J L 1987 Sulphate-reducing bacteria — their contribution to the corrosion process, *Industrial Corrosion* **5**(2): 8–10.
12. Tiller A K 1982 Aspects of microbial corrosion. In *Corrosion processes*, edited by R N Parkins, Applied Science Publishers, Barking UK, pp. 115–59.
13. Stoecker J 1984 Guide for the investigation of microbiologically induced corrosion, *Materials Performance* **23**(8): 48.
14. Stott J F D 1988 Assessment and control of microbially-induced corrosion, *Metals and Materials* **4**: 224–9.
15. Little B J and Wagner P A 1993 Interrelationship between marine biofouling and cathodic protection, *Materials Performance* **32**(9): 16–20.

6.6 BIBLIOGRAPHY

Anon 1990 *Microbiologically influenced corrosion and biofouling in oilfield equipment*, NACE International, Houston TX.

Dexter S C 1986 *Biologically induced corrosion*, NACE International, Houston TX.

Freedman A J 1993 Microbiologically-influenced corrosion: a state-of-the-art technology. Paper 635 in *Corrosion 93 plenary and keynote lectures*, edited by R D Gundry, NACE International, Houston TX.

Gooch T G 1984 *Corrosion and marine growth on offshore structures*, Ellis Horwood, Chichester UK.

Isaacs H, Bertocci U, Kruger J and Smialozska S 1990 *Advances in localized corrosion*, NACE International, Houston TX.

Kobrin G 1993 *A practical manual on microbiologically influenced corrosion*, NACE International, Houston TX.

Staehle R W, Brown B F, Kruger J and Agrawal A 1974 *Localized corrosion*, NACE International, Houston TX.

Wilkes J F and Shifler D A 1993 Historical perspectives of intergranular corrosion of boiler metal by water. Paper 44 in *Corrosion 93, New Orleans LA*, NACE International, Houston TX.

7 CREVICE AND PITTING CORROSION: CONCENTRATION CELLS

It is the natural habit of a metal to corrode unless prevented by human endeavour.

(T H Rogers, *Marine Corrosion*)

In the corrosion cells of Chapters 5 and 6, the corrosion results from changes in the nature and composition of metals exposed to electrolytes of *homogeneous composition*. In contrast, **concentration cells** exist when metals of uniform nature and composition are exposed to aqueous environments in which the components are *heterogeneous*. Experiments 3.6 and 3.7 showed how changes in oxygen or electrolyte concentration give rise to changes in corrosion potentials. The two forms of corrosion described in this chapter, **crevice** and **pitting corrosion**, result directly from such changes in composition within an electrolyte, although they are initiated differently.

It is commonly believed that stainless steel has excellent resistance to corrosion and that its selection will provide an infallible solution to corrosion problems. While it is certainly true that stainless steels do perform well in most situations, they are a very poor choice for some applications and must be selected with caution (see Section 15.2). Experiment 3.15 highlights a common problem. Many failures of stainless steel components have occurred because of corrosion in crevices or shielded areas, where small volumes of stagnant electrolyte can become much more aggressive than in the bulk solution. Figures 7.1 and 7.2 are good examples in which the designer thought that fabrication from stainless steel (type 304, 18Cr–8Ni) would provide good service.

Case 7.1: Figure 7.1 shows a thermometer pocket from a sea-water-cooled steam condenser. Pitting and serious wastage occurred along the stem on the downstream side where there was a layer of stagnant water.

Case 7.2: The stainless steel universal joint in Fig. 7.2 was badly pitted in the region over which the rubber grommet fitted. The grommet provided a stagnant environment in which the low level of dissolved oxygen created the concentration cell with the bulk electrolyte.

Case 7.3: In the renovation of a building, a large number of 316 stainless steel panels were used as architectural features to improve the look of the

Fig. 7.1 Stainless steel thermometer pocket from a sea-water-cooled steam condenser. A stagnant layer of sea-water formed on the downstream side of the stem when the pocket was inserted into the cooling water flowing in the condenser. The difference in oxygen concentration between the stagnant and flowing water initiated a crevice attack upon the steel (Case 7.1).

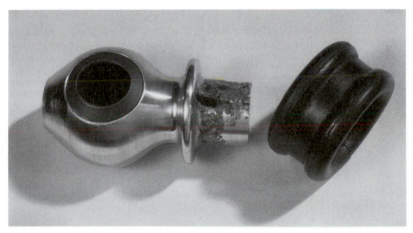

Fig. 7.2 Stainless steel universal joint showing crevice corrosion and pitting in the region where the rubber grommet fitted (Case 7.2).

balconies on the outside of each apartment. The panels of sheet material each had a series of neat circular holes stamped into them before they were mounted vertically between the concrete balustrade and the balcony floors. Even during installation, the panels were beginning to show trails of rust, emanating from the edges of the punched holes and running vertically downward across the sheet. Panels were not so badly affected where they had been mounted in locations exposed to more severe weather.

163

There is good reason for considering crevice and pitting corrosion together. Although they may be treated as different forms of corrosion many aspects of their mechanisms and many of their application are similar. The distinction between pitting and local wastage becomes academic when the corrosion is severe.

The problems are not restricted to stainless steels: many other alloy systems suffer from the localised corrosion described in this chapter.

Case 7.4: The pitting of copper pipes for freshwater systems is rare, but can occur when incorrect fabrication techniques are employed. Figure 7.3 shows a section of 25 mm diameter copper piping which failed because a layer of organic lubricant had been left after the pipe was formed. The pipe was annealed and the high temperature formed a carbon film along the inside of the pipe. Any subsequent break in the carbon film resulted in a very active pitting site which penetrated the copper in a short time. The pits are characterised by the formation of tubercles, small scabs of corrosion product over the hole (see Section 7.2).

Despite the well-known problems of crevice corrosion, engineers still fail to appreciate the importance of correct design and materials selection for adequate corrosion resistance.

Case 7.5: Denting of steam generator tubes at tube support plate intersections arises from corrosion of the carbon steel support plate in the crevice

Fig. 7.3 Pits (arrowed) penetrated the wall of a copper water-pipe because of the presence of a film of carbon along the bore. The pit on the left is beneath a tubercle of corrosion product; the pit on the right is clearly visible because the tubercle has been removed (Case 7.4).

between the plate and the tube. The corrosion product has a greater volume than the metal from which it forms and rapidly fills the crevice. Continued corrosion results in the tube being squeezed and eventually causes permanent deformation of both the tube (denting) and the support plate. By 1977 denting of Inconel 600 tubing had become a major issue. [1] With about 60 steam generators affected, the cost of rectifying the known problems was estimated to be $6 billion. [2]

7.1 THE MECHANISM OF CREVICE CORROSION

In the past, crevice corrosion was used to describe only the attack upon oxide-passivated alloys by aggressive ions such as chloride in crevices or other shielded areas of a metal surface. Attack in similar circumstances upon non-passivated metals was called **differential-aeration corrosion**. Current practice has tended to neglect such distinctions, and consequently crevice corrosion has been given many other names. Differential-aeration and concentration cell corrosion are terms which arise because of aspects of the mechanism of corrosion in crevices. Less common alternatives arise from the specific situations in which crevice corrosion is found, for example, deposit, fissure, gasket, interface, poultice, water-line and wedge corrosion. A good general definition of crevice corrosion is

> **The attack which occurs because part of a metal surface is in a shielded or restricted environment, compared to the rest of the metal which is exposed to a large volume of electrolyte.**

Thus, crevice corrosion is very much associated with the geometry of structures such as riveted plates, welded structures and threaded components; contact of metal with non-metallic solids such as plastics, rubber and glass; or deposits of sand, dirt or corrosion products. The common factor is the initial geometrical arrangement of the solid elements of the system, arising from the way in which the solid components of the system have been assembled or through the surface accumulation of solid debris. This leads to a heterogeneous distribution of species dissolved in the electrolyte, which in turn provides the conditions for localised attack.

Considerable research has been devoted to the electrochemistry within crevices, mainly because of advances in technology which enable accurate measurements to be made within the extremely small confines of a crevice; it is thought that typical widths are 25–100 μm.

A widely accepted general mechanism for crevice corrosion of stainless steels has been proposed by Fontana and Greene (Fig. 7.4). The steps involved are as follows:

1. Initially, the electrolyte is assumed to have uniform composition. Corrosion occurs slowly over the whole of the exposed metal surface, both inside and

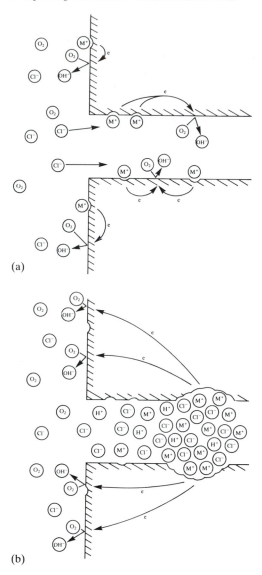

Fig. 7.4 The Fontana–Greene mechanism of crevice corrosion. (a) Initial conditions: corrosion occurs over the whole of the metal surface. (b) Final conditions: metal dissolution occurs only inside the crevice where acidity increases, concentration of chloride ion increases, and the reaction becomes self-sustaining.

outside the crevice. The normal anode and cathode processes occur, as described in Section 4.2. Under such conditions, the generation of positive metal ions is counterbalanced electrostatically by the creation of negative hydroxyl ions (Fig. 7.4(a)). See the discussion on the principle of electroneutrality (Section 2.2). Some association of positive and negative ions is assumed.

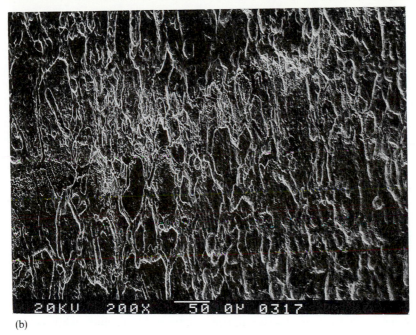

(b)

Fig. 7.9 Typical example of crevice corrosion resulting from a specimen used in Expt 7.1. (a) The PTFE collar, having been moved up to expose the corrosion in the crevice, appears white at the top. Three corrosion 'scars' are visible, the centre scar is the most severe. (b) Scanning electron micrograph of part of the centre scar. The surface topography shows directionality associated with the elongation of grains due to the manufacturing process of the bar.

At 2 Hz, the above experiment run for 10 hours produces 72,000 data points, a very large amount of data, and you may need further software to help with display and analysis. To display as a graph, you can take an average of every 10 successive points. This will not distort the appearance of the data and a program such as Microsoft Excel can be used to plot the result. More detailed analysis techniques are discussed elsewhere. [25] The record of electrochemical noise you measure will have many of the *features* of Fig. 7.10, although it will probably not look the same. The information contained in the noise record seems mysterious. Why does a freely corroding specimen show such a wide variation in behaviour? How is this behaviour related to the corrosion occurring on the specimen?

For many years the freely corroding metal surface was treated as a system in which the oxidation and reduction processes were deterministic; corrosion parameters obeyed precise mathematical equations which allowed their prediction in new situations. Using Tafel, Stern–Geary, Butler–Volmer equations and other relationships, seemingly useful values of well-established electrochemical parameters have been evaluated. In particular, rest or free

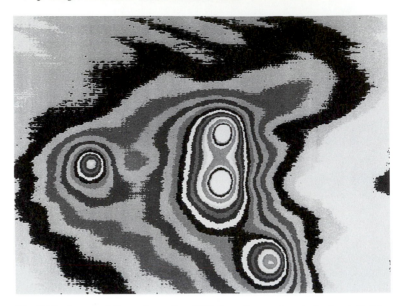

2. The consumption of dissolved oxygen results in the diffusion of more oxygen from electrolyte surfaces exposed to the atmosphere. Oxygen is more readily replaced at metal surfaces in the bulk electrolyte than at those within the crevice. Within the confined space of the crevice this lack of oxygen impedes the cathodic process and generation of negative hydroxyl ions is diminished.

3. The production of excess positive ions in the gap causes negative ions from the bulk electrolyte to diffuse into the crevice to maintain the potential energy at a minimum. In the presence of chlorides, it is likely that complex ions are formed between chloride, metal ions and water molecules. These are thought to undergo hydrolysis (reaction with water), giving the corrosion product, and more importantly, hydrogen ions which reduce the pH. This can be described by the simplified equation:

$$M^{z+} + H_2O \rightarrow MOH^{(z-1)+} + H^+ \tag{7.1}$$

The equation describes a general hydrolysis reaction, in which the role of the electrolyte anion (chloride in this case) is important but too complicated to be described here. Presence of chloride is well known to be conducive to the development of low pH because of its extremely low tendency to associate with hydrogen ions in water. Hydrogen chloride (HCl) dissociates completely in water. Additionally, stainless steels, which rely upon the protection of passive films, are notoriously unstable in chloride environments, and in the active crevice the very species needed to maintain passivity, oxygen, is denied access. Titanium, another passivating metal, has excellent resistance to crevice corrosion because its oxide film is particularly unreactive towards chloride ions.

4. The increase of hydrogen ion concentration accelerates the metal dissolution process, which in turn exacerbates the problem; so does the increase of anion (chloride) concentration within the crevice. An important feature of active crevice corrosion cells is that they are **autocatalytic**, once started they are self-sustaining. This situation is shown in Fig. 7.4(b). The metal within the crevice is corroding rapidly while that outside is cathodically protected.

Interestingly, the corrosion of the iron in stainless steels is not the most damaging process. The dissolution and subsequent hydrolysis of chromium is thought to lead to the most significant fall in pH, summarised in its simplest form by

$$Cr^{3+} + 3H_2O \rightarrow Cr(OH)_3 + 3H^+ \tag{7.2}$$

There is little doubt that the pH of an electrolyte within an active crevice can become extremely acidic; two reported instances of pH change within a crevice in a titanium alloy are from 8.3 (bulk) to 2.3 (crevice), and from 6 (bulk) to 1 (crevice). Of the two concentration cells which act in the mechanism — oxygen concentration and ion concentration — the ion concentration is thought to have the most significant effect upon the level of attack because of its effect upon local pH.

Limitations of space preclude the description of other models of crevice corrosion proposed since the Fontana–Greene model, which remains the most widely accepted to date. Other mechanisms include those of Oldfield and Sutton, [4] Bernhardsson, [5] Alkire, [6] Fu and Chan, [7] Pickering, [8] Watson and Postlethwaite [9, 10] and Sharland. [11]

7.2 PITTING CORROSION

Pitting is localised corrosion which selectively attacks areas of a metal surface where there is

A surface scratch or mechanically induced break in an otherwise protective film

An emerging dislocation or slip step caused by applied or residual tensile stresses

A compositional heterogeneity such as an inclusion, segregate or precipitate

The observation of corrosion pits as a result of crevice corrosion can sometimes cause confusion about the difference between the two forms of corrosion.

Pitting is distinguishable from crevice corrosion in the initiation phase.

Whereas crevice corrosion is initiated by differential concentration of oxygen or ions in the electrolyte, pitting corrosion is initiated (on plane surfaces) by metallurgical factors alone. Once initiated, the pit takes on very similar geometrical characteristics to those of a crevice and the propagation electrochemistries of pitting and crevice corrosion converge. [12] This will be discussed again in relation to Expt 7.1.

Much research has now been carried out into the detailed mechanisms of pitting, in particular the initiation of pits. [13–15] According to Burstein, [16] pit nucleation in stainless steel is viewed as a microscopically violent process which is **unstable** because pit propagation may not be achieved. Nucleation current dies continuously and most pit initiation events terminate. If the pit survives nucleation then the pit growth is called **metastable** because continued survival depends upon maintenance of an effective barrier to diffusion provided by a perforated cover of corrosion product over the pit mouth. If the cover is lost, but the current density is not sufficiently great, this stage dies too. If the pit survives then **stable** pitting occurs. This process is also diffusion-controlled; the diffusion barrier depends upon the pit depth. These are the three identifiable phases to pitting of stainless steel.

There has also been much recent research into surface modification leading to improved resistance to pitting corrosion, particularly for aluminium alloys and stainless steels. The procedure, known as the Ce–Mo process, consists of immersion in boiling 0.01 M cerium (III) nitrate for 2 hours, followed by immersion in boiling 0.05 M cerium (III) chloride for 2 hours and polarisation at +500 mV SCE in 0.1 M sodium molybdate for 2 hours at room temperature. [17,

18] Aluminium alloys 6061 and 6013 subjected to this treatment have been called 'stainless aluminum' and are reported to be extremely resistant to both general and localised corrosion. Extension of the technique to both the 2- and 7-series aluminium alloys is expected. The latest work on stainless steels using an alternating voltage (AV) treatment for surface modification [19] has produced encouraging results, but it is too early to comment on the general applicability. The mechanism for improved performance is not known, but rare earth atoms are believed to be incorporated into the surface passive films, improving their properties.

7.3   DIFFERENTIAL-AERATION CORROSION

The mechanism of pitting of carbon steel was first described by Evans, [20] and for many years represented a significant advance in the understanding of pitting. Today, the features are better classified under the heading of differential-aeration corrosion.

It is well known that, if a sheet of clean mild steel is exposed to rain, within a few days it will be rusting rapidly with the *rust* occurring as hard deposits, scabs or tubercles, in localised areas where water droplets have remained longest. If the *rust* is subsequently removed with a wire brush the surface will be found to be pitted in the areas previously covered by corrosion products. At this stage, we are using italics because *rust* is commonly understood to mean *the brown corrosion product formed on corroded iron or steel surfaces*. As we shall see below, this corrosion product is actually a mixture of chemical species and has a more precise definition. Figure 7.5 shows that the initiation of a pit is preceded by general corrosion over the whole of the wetted surface, probably as a result of simple grain boundary effects. The consumption of oxygen by the normal cathode reaction in neutral solution causes an oxygen concentration gradient within the electrolyte. Obviously, the wetted area adjacent to the air/electrolyte interface receives more oxygen by diffusion than the area at the centre of the drop, which is at a greater distance from the oxygen supply. This concentration gradient anodically polarises the central region, which actively dissolves:

$$Fe \rightarrow Fe^{2+} + 2e^- \tag{4.6}$$

The hydroxyl ions generated in the cathode region diffuse inwards and react with the iron ions diffusing outwards, causing the deposition of insoluble corrosion product around the depression or pit. This further retards the diffusion of oxygen, accelerates the anodic process in the centre of the drop and causes the reaction to be autocatalytic (Fig. 7.5(b)).

At this point, it is worth noting exactly how much we have simplified in eqn (4.6). The actual steps involved are believed to be as follows:

$$Fe + H_2O \rightarrow Fe(H_2O)_{(ads)} \tag{7.3}$$

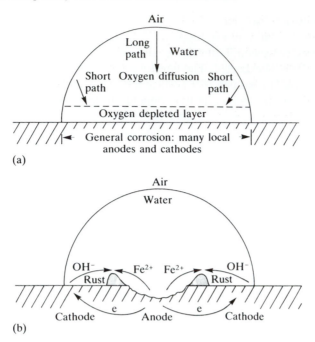

Fig. 7.5 The mechanism of pitting because of differential aeration beneath a water droplet. (a) General corrosion over the whole of the wetted metal surface depletes the oxygen levels in the adjacent electrolyte. (b) Longer diffusion path for oxygen to reach the central area makes this the anode. Metal dissolution occurs in the centre of the droplet and reaction of metal ions with hydroxyl ions formed at the edge generates a ring of rust around the corrosion pit.

$$Fe(H_2O)_{(ads)} \rightarrow Fe(OH^-)_{(ads)} + H^+ \tag{7.4}$$

$$Fe(OH^-)_{(ads)} \rightarrow Fe(OH)_{(ads)} + e^- \tag{7.5}$$

$$Fe(OH)_{(ads)} \rightarrow Fe(OH)^+ + e^- \tag{7.6}$$

$$Fe(OH)^+ + H^+ \rightarrow Fe^{2+} + H_2O \tag{7.7}$$

In the above expressions, *ads* represents *adsorbed* and implies that reaction occurs in the solid phase at the solid/liquid interface. The summation of the five equations leads directly to eqn (4.6). The example is not intended to intimidate students, but is merely a reminder of the complexity of reactions which may otherwise appear very simple, especially in a process as common as rusting.

It was mentioned in Section 2.2 that iron has two valency states; it can lose either two or three electrons. In aqueous solution, numerous iron ions can exist, some oxidised to iron (II) and others to iron (III). The oxidation state can be disguised by reaction with negative hydroxyl ions, and for clarity distinguishing labels will be given in the equations which follow. It is also probable that neutral water molecules are combined with the complex iron ions, but these will not be shown as they have no direct bearing on the chemistry.

Evans' simple explanation for the pitting of iron has been extended by Wranglen [21] and well summarised by Shreir [22] to better explain pit formation on carbon steel.

First, a hydrolysis reaction occurs, similar to that in the mechanism of crevice corrosion, in which acidity is increased:

$$Fe^{2+}(iron(II)) + H_2O \rightarrow FeOH^+(iron(II)) + H^+ \tag{7.8}$$

The formation of iron (III) ions is an oxidation reaction facilitated by the presence of oxygen. Even when the iron is combined in the $Fe(OH)^+$ ion, it can still be oxidised to the iron (III) state:

$$2Fe^{2+}(iron(II)) + \frac{1}{2}O_2 + 2H^+ \rightarrow 2Fe^{3+}(iron(III)) + H_2O \tag{7.9}$$

or

$$Fe(OH)^{2+}(iron(II)) + H_2O \rightarrow Fe(OH)_2^+(iron(III)) + H^+ \tag{7.10}$$

More hydrolysis reactions are possible, in which the solution is further acidified:

$$Fe(OH)^{2+} + H_2O \rightarrow Fe(OH)_2^+ + H^+ \tag{7.11}$$

and

$$Fe^{3+} + H_2O \rightarrow Fe(OH)^{2+} + H^+ \tag{7.12}$$

All iron ions in eqns (7.11) and (7.12) are iron (III) ions. The two major corrosion products, magnetite and rust, are respectively denoted by the formulae Fe_3O_4 and $FeO(OH)$, and they are formed from the complex ionic species:

$$2Fe(OH)^{2+} + Fe^{2+} + 2H_2O \rightarrow Fe_3O_4(magnetite) + 6H^+ \tag{7.13}$$

and

$$Fe(OH)_2^+ + OH^- \rightarrow FeO(OH)(rust) + H_2O \tag{7.14}$$

Rust is now a specific chemical species. At cathodic sites outside the pit, along with the usual oxygen reduction reaction, eqn (4.22), the rust is reduced to magnetite:

$$3FeO(OH) + e^- \rightarrow Fe_3O_4 + H_2O + OH^- \tag{7.15}$$

The corrosion products 'grow' over the pit and its immediate surroundings, forming a scab or tubercle and isolating the environment within the pit from the bulk electrolyte. It is thought that the autocatalytic process is assisted by an increased concentration of chloride ions within the pit.

In Section 6.1 it was described how corrosion can be initiated at various inclusions and precipitates. In steels, it is common to find manganese sulphide inclusions which are strongly cathodic to uncombined metal. In stainless steels, the areas immediately adjacent to these inclusions make ideal pit initiation sites. The increased acidity described in eqns (7.11), (7.12) and (7.13) above is thought to attack the manganese sulphide:

$$MnS + 2H^+ \rightarrow H_2S + Mn^{2+} \tag{7.16}$$

This reaction produces S^{2-} and HS^- ions, which themselves promote more rapid dissolution of the iron because they decrease the activation polarisation. A further consequence of the acidity increase is the reduction of hydrogen ions by electrons generated in the oxidation reactions. Hydrogen evolution causes occasional breaking of the crust of corrosion product.

Most hot-rolled steel is produced with a layer of mill scale, a layer of oxide which protects the metal while the scale remains intact but which promotes corrosion at breaks in the film. Cold-rolled steels are produced with a bright surface without mill scale, so they are usually more susceptible to rusting.

The corrosion of iron and steel has been described in some detail because they are very common engineering materials. Pitting mechanisms may differ in their details, for example aluminium or copper alloys [22] are different from steel, but remember that events within such micro-environments have broadly similar features.

7.4   ELECTROCHEMICAL TECHNIQUES

Polarisation curves of passivating metals

The most significant differences between pitting and crevice corrosion lie in their mechanisms of initiation. The previous two sections emphasised the similarities in the propagation mechanisms. Recent developments in electrochemical techniques have helped to distinguish these two very similar forms of corrosion, while offering methods of assessing the relative performance of alloys under similar conditions. Polarisation curves are particularly valuable in quantifying the behaviour of materials under these conditions. This section explains how they are used and what information can be obtained.

The solid line in Fig. 7.6(a) is a schematic $E/\lg i$ plot for a metal in a mild environment, such as an austenitic stainless steel in dilute sulphuric acid. The dashed lines should be ignored for the moment. AB represents **cathodic** behaviour, while BG is the **active zone**. The metal is not passivated at its free corrosion potential, B, and Tafel straight lines, AC and DC, could be drawn for the reduction and oxidation reactions of the normal metal dissolution, eqn (2.7).

At potentials more positive than B, corrosion rate increases, and reaches a maximum at the **passivation potential**, G, which is often given the symbol E_{pp}. The transition from active dissolution occurs as a solid species becomes more thermodynamically stable than the metal ion. A protective film begins to form

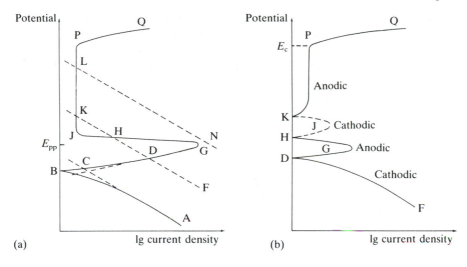

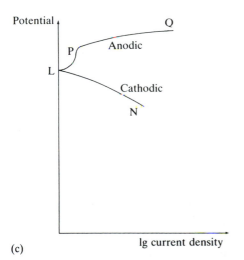

Fig. 7.6 Schematic potentiodynamic polarisation behaviour of passivating metals. (a) general polarisation plot showing three possible theoretical cathode processes; (b) polarisation curve showing unstable passivity; and (c) polarisation curve showing stable passivity.

and causes a sudden drop in corrosion current density along limb GJ. Along limb JP, the *passive zone*, the current density is maintained at a low and steady level, until breakdown of the protective film begins at P. At P the likelihood of pitting is greatest, consequently potential E_c, often called the **critical pitting** or **breakdown** potential, is a useful parameter in assessing pitting properties of materials. [23] E_c is not an absolute parameter; it varies according to both metallurgical and environmental conditions. At potentials more positive than P, the current density begins to rise as more and more pits propagate.

173

When current density greatly increases because of the onset of new anodic processes, rather than because of pitting, this region of the curve is known as the **transpassive** region.

Some passivating metals have $E/\lg i$ plots which, at first sight, seem to bear no relation to Fig. 7.6(a). For example, considerable instability may be noticed in certain potential ranges, as in Fig. 7.6(b). These differences are caused by variation in the dominant cathodic process. In Fig. 7.6(a) three possibilities are shown as dashed lines: CA, KF and LN. A different type of polarisation plot will be measured, depending upon which cathodic process predominates. Remember, there are three usual cathodic processes: hydrogen evolution, oxygen reduction and metal cation reduction. And the experimental plot is always the sum of all anodic and cathodic currents. In Fig. 7.6(a) the observed cathodic line, BA, results from the addition of the anodic and cathodic currents in that potential range. The cathodic process alone is represented by CA.

If a new cathodic process is represented not by line CA but by dashed line KF, it can be seen that KF intersects the anodic curve in three places: D, H and K. Each point of intersection represents equal anodic and cathodic current densities, so each is an E_{corr} value. The measured polarisation plot is shown in Fig. 7.6(b). Between K and H the cathodic current is greater than the anodic current, while between H and D the reverse is true. Thus, in potential range KD, the material is unstable and alternates between cathodic and anodic behaviour. The use of a material in such conditions of instability is usually considered to be unwise.

A third type of behaviour is observed when the cathodic process is represented by dashed line LN in Fig. 7.6(a); then the measured $E/\lg i$ plot will be as shown in Fig. 7.6(c). E_{corr} may be quite noble and the metal is completely passivated in range PL. Such a situation is regarded as exhibiting resistance to localised corrosion because the metal is well protected by its passive film.

The behaviour we have just described as a thought experiment is illustrated as real experimental curves for the polarisation behaviour of 304 stainless steel in 5% sulphuric acid at room temperature (Fig. 7.7). Three curves are shown: (a) with an aerated electrolyte (b) with a deaerated electrolyte and (c) with an intermediate oxygen concentration.

Cyclic pitting scans and pitting potentials

Further data relating to both crevice and pitting behaviour can be obtained. [24] The anodic polarisation scan described in Fig. 7.6(a) is continued to Q, but instead of terminating there, the potential is now reduced at the same stepping rate until E_{corr} is reached once more. The resulting graph is often described as a **corrosion hysteresis loop**, but is more accurately called a **cyclic pitting scan**. In general, there are two types of behaviour:

The metal may be slow to repassivate
The pits which have initiated at P continue to propagate, but no new pits are

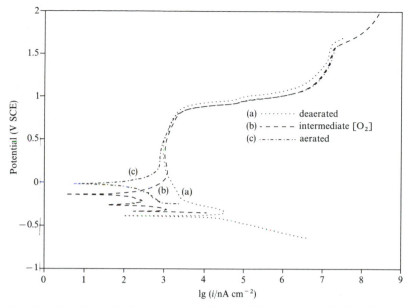

Fig. 7.7 Experimental polarisation curves for type 304 stainless steel in 5% sulphuric acid, ambient temperature, scan rate 0.1 mV s^{-1}: (a) de-aerated, (b) intermediate oxygen concentration, (c) aerated.

formed. The corrosion current density remains high until a sufficiently negative potential has been reached, whereupon repassivation occurs once more. This will lead to a large area in the loop, and usually indicates poor resistance to pitting because if breaks in the passive film occur they do not repair themselves easily. Figure 7.8(a) shows such a pitting scan for type 304 stainless steel in aerated sea-water at 25°C. Although reasonably stable at potentials more positive than E_{corr}, -0.260 V SCE, breakdown occurs at $+0.300$ V SCE. Once the material has passed this potential, attack is severe, even though the potential may be made more negative again. Only if the potential can be brought down to below -0.160 V SCE does the attack cease. A polarisation scan carried out in acid with a high concentration of chloride ions, conditions likely to exist within a crevice, shows the material to be very active at all potentials more positive than E_{corr}. This is shown as the dashed line in Fig. 7.8(a). The graph illustrates the unsuitability of type 304 stainless steel in a marine environment, say, where there is a danger of crevice conditions.

The metal repassivates easily
The pits which initiated at P and propagated along PQ in Fig. 7.6 are quickly filmed and both initiation and propagation cease. If this is the case, then the reverse scan will retrace a very similar path, returning to the low level of current density of the passive state. The loop, if there is one, has a small area. Such materials would be expected to show good pitting resistance, since the reforming of surface protection eliminates the local active sites. Figure 7.8(b) shows the

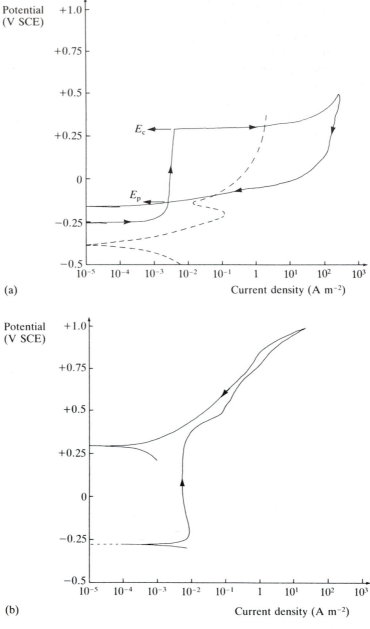

(a)

(b)

Fig. 7.8 (a) A cyclic polarisation (pitting hysteresis) curve for type 304 stainless steel in aerated sea-water at 25°C, scan rate 0.1 mV s^{-1}. The broken line shows a similar plot using aerated sea-water, 4.5 M sodium chloride and HCl, adjusted to pH 1.2. (b) Cyclic polarisation curve for Hastelloy C-276 in aerated sea-water at 25°C.

pitting scan for Hastelloy C276 in aerated sea-water at 25°C. It shows no hysteresis; the material is efficiently passivated from the moment of its immersion. During the reverse scan period, smaller current densities are recorded for the same values of potential. The final E_{corr} at +0.285 V SCE is much more noble than the starting E_{corr} at −0.280 V SCE. Such a material is recommended for use in a marine environment.

Pourbaix and his colleagues [24] defined a **protection potential**, E_p, corresponding to the potential at which the reverse scan intersected the forward scan and completed the hysteresis loop. They proposed that, at potentials more negative than this, pit initiation could not occur and existing pits would repassivate. Such a concept has proved useful because it is considered that the area of the hysteresis loop is related to the amount of pit propagation that occurs during the potential sweep. Furthermore, since a growing pit can be considered as a special case of an active crevice, it has been proposed that if a series of alloys is tested under similar conditions, relative susceptibilities to crevice corrosion can be determined by comparison of loop sizes. Alloys having the smallest loops are the least susceptible. Indeed, to a good approximation, it is often sufficient to measure $E_c - E_p$ instead of loop area. Wilde [12] showed that such an analysis on a series of stainless steels tested in sea-water correlated well with similar specimens with simulated crevices, subjected to four years of exposure tests. However, Wilde showed that care should be exercised in the interpretation of such data. He showed that a plane sample of material with no crevice gave a very narrow hysteresis loop, and a very small value of $E_c - E_p$; it showed no signs of pitting corrosion. However, when a crevice was introduced into the same material, a large loop was obtained. The specimen showed considerable crevice corrosion but no pitting. It was judged that a growing crevice and the associated anodic micro-environment cathodically protect the rest of the metal surface. The conclusions reached were that E_c measurements relate only to pitting in the absence of crevices and that for stainless steels in sea-water, crevice corrosion is more important than pitting, especially since it is difficult to imagine real structures with no crevices whatsoever. Although, once formed, pits and crevices are thought to propagate by similar mechanisms, Wilde concluded that the initiation of pits was primarily a potential-controlled passivity breakdown by adsorption of chloride ions at specific sites. In a crevice, however, concentration effects were important and initiation occurred when differential aeration caused passivity breakdown.

Electrochemical noise measurement and crevice corrosion

A technique which is attracting much attention at present is the measurement and analysis of electrochemical noise, apparently random behaviour emanating from corroding systems. The following experiment illustrates the electrochemical noise associated with crevice corrosion of a 304 stainless steel.

Experiment 7.1

Cut a cylindrical specimen about 100 mm in length and 6 mm in diameter from an AISI Type 304 stainless steel bar. Machine one end (which is to be exposed to electrolyte) so as to produce a rounded or bullet shape with no sharp edges. Manufacture a tight-fitting PTFE collar and push it onto the shank of the specimen about 60 mm from the round end, leaving the blunt end free for an electrical connection. PTFE tubing of internal diameter 6 mm can be obtained which is then simply cut to length. Immerse the bullet, rounded end down, into the electrolyte, 0.01 M aqueous iron (III) chloride acidified to pH 2 with aqueous hydrogen chloride then heated to 60°C, so the electrolyte half covers the collar, thus allowing a crevice at the bottom of the collar. Connect the reference electrode to one input of an A/D conversion board and the specimen to the other. Measure and display the potential noise fluctuations by comparison with an SCE or SSC reference electrode, taking samples at 2 Hz. Data can be digitized using any appropriate A/D converter, e.g. a PCI-20377W low power multi-function board from Intelligent Instrumentation, Tucson, Arizona. Data is stored on a PC using, for example, the Visual Designer software package supplied by the same company. Allow the experiment to continue for 8–10 hours. Stop the measurement, remove the sample, remove the collar and examine the corrosive attack. You should obtain a specimen exhibiting crevice corrosion similar to that shown in Fig. 7.9.

(a)

(b)

Fig. 7.9 Typical example of crevice corrosion resulting from a specimen used in Expt 7.1. (a) The PTFE collar, having been moved up to expose the corrosion in the crevice, appears white at the top. Three corrosion 'scars' are visible, the centre scar is the most severe. (b) Scanning electron micrograph of part of the centre scar. The surface topography shows directionality associated with the elongation of grains due to the manufacturing process of the bar.

At 2 Hz, the above experiment run for 10 hours produces 72,000 data points, a very large amount of data, and you may need further software to help with display and analysis. To display as a graph, you can take an average of every 10 successive points. This will not distort the appearance of the data and a program such as Microsoft Excel can be used to plot the result. More detailed analysis techniques are discussed elsewhere. [25] The record of electrochemical noise you measure will have many of the *features* of Fig. 7.10, although it will probably not look the same. The information contained in the noise record seems mysterious. Why does a freely corroding specimen show such a wide variation in behaviour? How is this behaviour related to the corrosion occurring on the specimen?

For many years the freely corroding metal surface was treated as a system in which the oxidation and reduction processes were deterministic; corrosion parameters obeyed precise mathematical equations which allowed their prediction in new situations. Using Tafel, Stern–Geary, Butler–Volmer equations and other relationships, seemingly useful values of well-established electrochemical parameters have been evaluated. In particular, rest or free

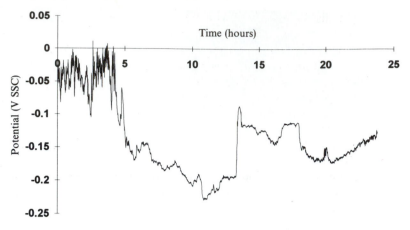

Fig. 7.10 Time series showing the electrochemical noise generated by a sample in Expt 7.1. For display purposes, each data point represents an average of 10 successive points.

corrosion potentials, E_{corr}, have been identified as extremely useful indicators of corrosion behaviour of a system supposedly at equilibrium, while corrosion rate has been identified with free corrosion current density, i_{corr}. Treated in this way, it was hoped that extrapolations about behaviour in slightly different situations could be made and predictions about the lifetime of large chemical plant placed on a 'scientific' footing. In the course of such laboratory investigations, systems which 'misbehaved' by generating data completely at odds with that predicted by the deterministic theories were ascribed to artefacts, the source of which were often considered to be instrumentation effects or induced from external sources. In this sense noise became synonymous with garbage. The results were discarded as unreliable and the experimenters turned to other systems, with which they were more comfortable.

Realising that systems were behaving in a random fashion at the atomic level, experimenters have accounted for this by making statistics an intrinsic part of the theory. By saying that a very large number of atoms were involved in the corrosion, the statistical probability of a process occurring with a given rate placed a measure of reliability on the corrosion rate measurement. In the light of continuing corrosion problems in every aspect of modern industry, this assumption is plainly not valid in all cases. Indeed, although it might be true on a sufficiently large scale, in which metal dissolution is general across the entire surface, it need not be the case for Expt 7.1, or another highly localized corrosion process such as environment-sensitive cracking in which extremely localised cracks initiate and propagate by processes involving a comparatively small number of atoms. The panoply of metallurgical factors contributing to microscopic variations — composition, grain structure, dislocations, segregates, precipitates, inclusions, porosity, surface condition — also contributes to the non-deterministic way in which corrosion occurs. In addition, the extremely variable nature of the material/environment partners in a corroding system dictates that dissolution events must contain a highly random element not amenable to analysis by the statistical approach outlined above.

It is now well recognized that the kind of behaviour in which potential and current vary in apparently useless ways is actually a record of *real* behaviour — passive films exhibiting the first signs of breakdown, pits being initiated, cracks propagating — the very processes in which corrosion scientists and engineers are *most* interested. But they require interpretation in order to provide additional information for evaluation of corrosion situations. Far from being useless, electrochemical noise is now well established as extremely *useful*. [26]

Electrochemical noise measurements are represented by data records in which current or potential variations are plotted as a function of time, **time series**. Cursory inspection shows them to be comprised of many apparently unrelated short sequences in which stability may be followed by sudden large step changes or other periods of varying oscillatory character, as is the case in Fig. 7.10. The task of interpreting such behaviour as recognisable events in the history of the system has been enigmatic and workers have scanned the literature (mostly electronic signal processing) looking for new tools with which to examine the data. As if this were not difficult enough, there is much argument at present about the best way to *measure* the noise.

At the extremes of a spectrum of behaviour, data is described as either **deterministic** or **stochastic**. Spanning these extremes is a range of behaviour composed of varying degrees of *both* deterministic *and* stochastic character; recently its interpretation has preoccupied many corrosion scientists. The new theories of chaos and fractals have been used, and we have described the representation of corroded surfaces by time series [27] and their analysis for fractal dimension. [28, 29] Since surfaces can be represented by time series, the same analysis techniques are valid when applied to electrochemical noise data.

Compared to a laboratory scientist, an engineer is often better off. For as the structures become larger, the statistical effects increase and a better 'average' behaviour is obtained. In the laboratory, experiments with samples of 1 cm^2 have poor reproducibility. This is especially true when a passive film is present and breakdown is required before a corrosion process can propagate. The initiation phase has always been a problem in corrosion fatigue and stress-corrosion cracking; the best engineering solution is to consider the initiation time as zero. Although this is a safe choice, it is seldom cost-effective and is certainly not a solution with which engineers should be satisfied. It is much better to determine exactly the initiation phase of the mechanism and then to determine the lifetime.

In industry the reliable interpretation of electrochemical noise is particularly valuable since it promises new types of on-line monitoring devices. Systems can be observed in their natural condition, i.e. without perturbation, which has always been considered to introduce artificiality into the evaluation. A suitable device mounted in a chemical system could give an early warning of an undesirable level of corrosion, as long as it was accompanied by a reliable analysis algorithm.

Our analysis of the system described in Expt 7.1 used new software based on fractals, the full details of which are beyond the scope of this book. However, we concluded the following about the data in Fig. 7.10.

In the early part of the series, there was a considerable amount of potential variation (Burstein [16] writes of 'microscopically violent' events), but the specimen occupied a free corrosion potential domain in the range −0.1 to −0.05 V SSC. After about 4 hours, activity was observed which lasted for about an hour and was followed by a period of comparatively little potential noise as the corrosion potential fell to −0.15 to −0.25 V SSC. Thereafter, the corrosion potential did not return to the original domain more positive than -0.1 V SSC.

Figure 7.11 shows the polarisation curve for the system, together with a model of the corrosion processes. E_{corr} occurred at the mixed potential resulting from the two dominant processes of iron dissolution and oxygen consumption (the hydrogen evolution kinetic line was at much more negative potentials). In the early stages, the electrolyte was so aggressive as to cause immediate local breakdown of the passive film over parts of the exposed surface outside the crevice. Although identifiable pits were not necessarily formed at this stage, very active sites were created over short times, in agreement with present models. [16, 30] In the circumstances of an enhanced dissolution rate, we would expect the exchange current density for the Fe → Fe^{3+} redox process (marked as i_0 in Fig. 7.11) to increase. Inspection shows that a hundred fold increase (i.e. a change of +2 on the x-axis) results in a negative potential shift of the free corrosion potential by about 80 mV. Successive local site activation and repassivation over short times was represented by the comparatively rapid oscillations of E_{corr}

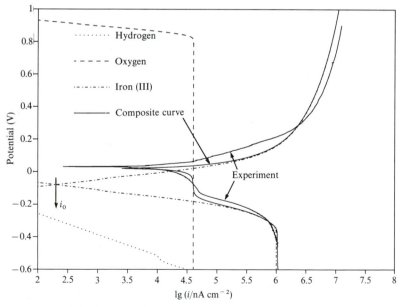

Fig. 7.11 Polarisation curve for the system in Expt 7.1 and the analysis of the curve using the Simpler electrochemical analysis software.

in the range 0.0 to −0.1 V SSC. As time proceeded, metastable pits may well have begun to form. This was not evident from the noise series, but another analysis method based upon determinations of fractal dimension did indicate a separate phase at times between 2.5 and 4 hours. At some time into the experiment, crevice corrosion became established as the dominant process and the pitting outside the crevice subsided in favour of cathodic reaction sites to support the anodic activity inside the crevice. The generally higher levels of net metal dissolution and the elimination of the oxygen cathodic process from the reaction scheme would be expected to depress E_{corr} to even more negative potentials. This was indeed the case; inspection of Fig. 7.10 indicates values in the range −0.1 to −0.25 V SSC, although the noise trace contained far less of the oscillatory character exhibited in the early phase because there was less variation in the level of activity.

Scanning electrochemical probes

There are many demonstrations that show variations in electrochemical potential across metal surfaces in aqueous solution, variations which lead to localised attack. Early work [31] involved macroscopic bimetallic couples of zinc and steel. Equipotential and ion flux lines were determined in the adjacent electrolyte by means of reference electrodes in combination with Luggin capillaries, moved laboriously by hand across the surface of interest. Since the early work, progress has been made both in the miniaturizing of electrodes and in the development of automatic scanning and data handling tools. Ultramicro-electrodes have been used in many electrochemical applications — particularly advantageous for localised corrosion studies [32, 33] — while numerous researchers have constructed dedicated apparatus to measure localised corrosion by means of manual, or more recently, automated scanning probes. [34–38] Various names have been adopted for the many variations used, but the name which best fits the general technique is **scanning electrochemical microscopy (SECM)**. [39] In this, a dc coupled scanning tip generates signals from currents carried by redox processes at tip and substrate. The signals are amplified then analysed by computer to generate a variety of information and visual displays.

Figures 7.12 and 7.13 show a commercial quality instrument known as SRET Scanning Reference Electrode Technique. [40] It consists of a conventional three-electrode corrosion cell with reference (saturated calomel electrode) and auxiliary electrodes connected to a potentiostat controlling the dc polarization of the specimen, the working electrode. A cylindrical specimen (in the form of either sealed tube or bar) immersed in the electrolyte is made the working electrode and is rotated at speeds of 5–250 rpm by a computer-controlled indirect drive stepper-motor. A magnetic triggering device mounted on the motor shaft provides synchronised pulses for the data collection program and allows the precise identification of the position of the probe above the surface of the specimen.

Fig. 7.12 A modern instrument for the Scanning Reference Electrode Technique, known as SRET.

The probe consists of a pair of platinum electrodes made from wire of diameter 0.2 mm. The electrodes have electrochemically sharpened tips of approximate radius 1 μm and, apart from the tips, have as little platinum exposed to the electrolyte as possible; protruding platinum is coated with an insulating layer once the body of the probe has been fabricated. One tip is a few millimetres closer to the specimen surface than the other. The front probe samples the electric field created by the ion flux close (10–20 μm) to the surface of the specimen, while the rear probe samples the noise in the bulk electrolyte. The output from the platinum electrodes is taken to an ac-coupled differential amplifier before being digitised for computer analysis and display.

This technique has been shown to be powerful in many different applications. [41] Figure 7.14 is a map of a 3 mm $\times$ 2 mm area of the surface of an 11% Cr steel, polarised to 0 V SCE and suffering from pitting attack in sea-water. The lines are potential contours indicating the areas of extreme anodic activity over pits. Using a calibration technique, localised current densities can be measured. In an experiment with a 304 stainless steel, polarised to 0.1 V SCE in sea-water, an average corrosion rate over the surface of 10 mm y^{-1} (high) was found; immediately over a pit the actual corrosion rate was found to be >250 mA cm^{-2} (2,933 mm y^{-1}). [42]

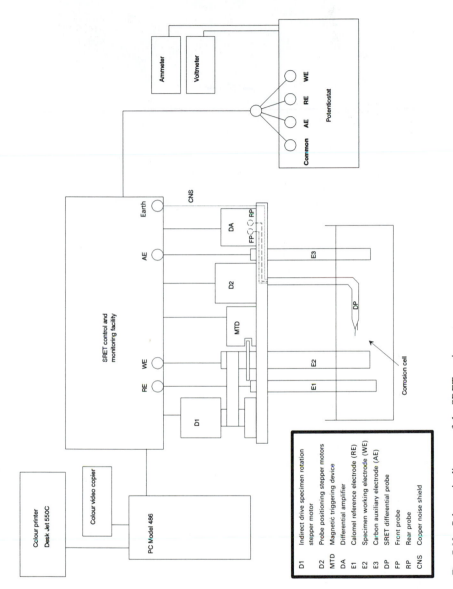

D1   Indirect drive specimen rotation
      stepper motor
D2   Probe positioning stepper motors
MTD   Magnetic triggering device
DA   Differential amplifier
E1   Calomel reference electrode (RE)
E2   Specimen working electrode (WE)
E3   Carbon auxiliary electrode (AE)
DP   SRET differential probe
FP   Front probe
RP   Rear probe
CNS   Copper noise shield

Fig. 7.13 Schematic diagram of the SRET equipment.

185

Fig. 7.14 A map of the surface of an 11% chromium steel, polarised to 0 V SCE and suffering from pitting attack in sea-water. The dimensions of the area shown are 3 mm × 2 mm and the lines are potential contours indicating the areas of extreme anodic activity over pits. This is actually obtained in colour with user-defined potential scales, in this case 10 mV between contour lines.

7.5 MATERIAL SUSCEPTIBILITY

It has been stated that the most damaging environment giving rise to pitting and crevice corrosion is one of high chloride ion content. In sea-water, the resistance of alloys to these forms of corrosion has been ranked [24, 43] and is given in Table 7.1. It can be seen that stainless steels fare badly in such an environment, in accordance with the discussion so far. It has been well documented that additions of chromium, nickel and molybdenum are especially advantageous in combating such corrosion. Thus, for a typical 18% Cr, 8% Ni austenitic stainless steel, it is considered that the addition of 2–3% molybdenum in the 316, or 3–4% in the 317, causes an improvement in the resistance. Only in some of the more highly alloyed materials, not shown in the table but discussed in Chapter 15, does the performance of stainless alloys in crevice conditions reach a satisfactory standard. However, it can be seen from the table that high nickel alloys have been found to give good resistance, together with titanium and cupronickels.

The information given in Table 7.1 should be regarded cautiously. In a comprehensive review of crevice corrosion, France [44] states that the absence of a standard procedure for comparative testing of alloys in environments which lead to crevice and pitting corrosion has created a situation where there is much conflicting evidence about material susceptibilities. Flow rate, for example, is a

Table 7.1 The relative crevice corrosion resistance of metals and alloys in quiet seawater

Metal or alloy	Resistance
Hastelloy C276 Titanium	Inert
Cupronickel (70/30 + 0.5% Fe) Cupronickel (90/10 + 1.5% Fe) Bronze Brass	Good
Austenitic cast iron Cast iron Carbon steel	Moderate
Incoloy 825 Carpenter 20 Copper	Low
316 stainless steel Ni–Cr alloys 304 stainless steel 400 series stainless steels	Poor: pit initiation at crevices

factor of considerable importance; titanium and Hastelloy C276 show excellent resistance over a considerable range of flow rate, while stainless steels benefit from a rapid flow rate which allows oxygen to repassivate the material. Indeed, in some chloride-free environments there is evidence to suggest that 316 does not have superior resistance to 304. [45] Thus, the order given in Table 7.1 is likely to be considerably altered in a different environment. France stresses the importance of carrying out evaluation testing in an environment which simulates as closely as possible the real conditions (see Case 15.5).

7.6 REFERENCES

1. Jones R L 1993 Corrosion experience in US light water reactors — a NACE 50th anniversary perspective. Paper 168 in *Corrosion 93, New Orleans LA*, NACE International, Houston TX.
2. MacDonald D D, Begley J A, Bockris J O, Kruger J, Mansfeld F B, Rhodes P R and Staehle R W 1981 Aqueous corrosion problems in energy systems, *Materials Science and Engineering* **50**: 19–42.
3. Fontana M G and Greene N D 1967 *Corrosion Engineering*, 2nd edn. McGraw-Hill, New York, pp. 39–44.
4. Oldfield J W and Sutton W H 1978 Crevice corrosion of stainless steels. 1 A mathematical model, *British Corrosion Journal* **13**: 13.
5. Bernhardsson S, Eriksson L, Opelstrup J, Puigdomenech I and Wallin T 1981 Crevice corrosion of stainless steels, calculations of concentration and pH changes.

In *Proceedings of 8th international congress on metallic corrosion, Mainz, Germany*, pp. 193–8.

6. Hebert K and Alkire R 1983 Dissolved metal species mechanism for initiation of crevice corrosion of aluminium. II Mathematical model, *Journal of the Electrochemical Society* **130**: 1007–14.

7. Fu J W and Chan S 1984 *Corrosion* **40**: 540.

8. Pickering H W 1989 The significance of the local electrode potential within pits, crevices and cracks, *Corrosion Science* **29**: 325.

9. Watson M K and Postlethwaite J 1990 Numerical simulation of crevice corrosion of stainless steels and nickel alloys in chloride solution, *Corrosion* **46**: 522–30.

10. Watson M K 1991 Numerical simulation of crevice corrosion: the effect of the crevice gap profile, *Corrosion Science* **32**: 1253–62.

11. Sharland S M 1992 A mathematical model of the initiation of crevice corrosion in metals, *Corrosion Science* **33**: 183–202.

12. Wilde B E 1974 On pitting and protection potentials: their use and possible misuses for predicting localized corrosion resistance of stainless alloys in halide media. In *Localized corrosion*, edited by R W Staehle, B F Brown, J Kruger and A Agrawal, NACE International, Houston TX, pp. 342–52.

13. Williams D E, Westcott C and Fleischmann M 1985 Stochastic models of pitting corrosion of stainless steels. I Modelling of the initiation and growth of pits at constant potential, *Journal of the Electrochemical Society* **132**: 1796.

14. Isaacs H S 1989 The localized breakdown and repair of passive surfaces during pitting, *Corrosion Science* **29**: 313.

15. Baker M A and Castle J E 1993 The initiation of pitting corrosion at MnS inclusions, *Corrosion Science* **34**(4): 667–82.

16. Burstein G T, Pistorius P C and Mattin S P 1993 The nucleation and growth of corrosion pits on stainless steel, *Corrosion Science* **35**: 57–62.

17. Mansfeld F, Wang Y and Shih H 1991 Development of stainless aluminium, *Journal of the Electrochemical Society* **138**: L74.

18. Mansfeld F, Wang Y and Shih H 1992 The Ce-Mo process for the development of a stainless aluminium, *Electrochimica Acta* **37**: 2277.

19. Kwiatkowski L and Mansfeld F 1993 Surface modification of stainless steel by an alternating voltage process, *Journal of the Electrochemical Society* **140**(3): L39.

20. Evans U R 1961 *The corrosion and oxidation of metals*, Edward Arnold, London, p. 127.

21. Wranglen G 1985 *An introduction to corrosion and protection of metals*, Chapman and Hall, London, p. 24.

22. Shreir L L 1994 Localized Corrosion. In *Corrosion*, edited by L L Shreir, R A Jarman and G T Burstein, Butterworth-Heinemann, Oxford, pp. 1:181–3.

23. Greene N D 1962 *Corrosion* **18**: 136–42.

24. Pourbaix M, Klimzack-Mathieu L, Merterns C, Meunier J, Vanleugen-Haghe C, Munck L D, Laureys J, Neelemans L and Warzee M 1963 *Corrosion Science* **3**: 239–59.

25. Trethewey K R, Marsh D J, Sargeant D A, Burvill J P and Roberge P R 1994 An indicator of crevice corrosion initiation based upon electrochemical noise analysis. In *Proceedings of international symposium on materials performance, maintenance and plant life assessment*, CIM, Toronto Canada pp. 47–56.

26. Searson P C and Dawson J L 1988 Analysis of electrochemical noise generated by corroding electrodes under open-circuit conditions, *Journal of the Electrochemical Society* **135**: 1908–15.

27. Trethewey K R, Keenan J S, Sargeant D A, Haines S and Roberge P R 1993 The

fractality of corroding metal surfaces. In *Proceedings of 12th international corrosion symposium, Houston TX*, NACE International, Houston TX.

28. Trethewey K R and Roberge P R 1994 Towards improved quantitative characterization of corroding surfaces using fractal models. In *Modelling aqueous corrosion*, edited by K R Trethewey and P R Roberge, Kluwer Academic, Dordrecht, The Netherlands, pp. 443–64.

29. Keenan J S, Trethewey K R, Sargeant D A, Haines S and Roberge P R 1993 Towards better quantitative models for the corrosion of aluminium using fractal geometry. In *Proceedings of 32nd annual conference of metallurgists: light metals, Quebec City, Canada*, The Metallurgical Society of CIM, Toronto, Ontario.

30. Baroux B and Gorse D 1994 The respective effects of passive films and non-metallic inclusions on the pitting resistance of stainless steels — consequences on the pre-pitting noise and the anodic current transients. In *Modelling aqueous corrosion*, edited by K R Trethewey and P R Roberge, Kluwer Academic, Dordrecht, The Netherlands, pp. 161–82.

31. Evans U R 1940 Report on corrosion research at Cambridge University interrupted by the outbreak of war, *Journal of the Iron and Steel Institute* **141**: 219.

32. Engstrom R C, Meany T, Tople R and Wightman R 1987 Spatiotemporal description of the diffusion layer with a microprobe electrode, *Analytical Chemistry* **59**: 2005.

33. Luo J L, Lu Y C and Ives M B 1992 Use of microelectrodes to determine local conditions within active pits, *Materials Performance* **31**: 44.

34. Gainer L J and Wallwork G R 1979 An apparatus for the examination of localized corrosion behaviour, *Corrosion* **35**: 61-67.

35. Bhansali K J and Hepworth M T 1974 The corrodescope, its description and application to the study of pitting phenomena, *Journal of Physics: Part E* **7**: 681–5.

36. Sargeant D A, Hainse J G C and Bates S 1989 Microcomputer controlled scanning reference electrode technique apparatus developed to study pitting corrosion of gas turbine materials, *Materials Science and Technology* **5**: 487–92.

37. Cottis R A and Holt D 1987 A semi-automatic device for the rapid measurement of pit depths and position, *Corrosion Science* **27**: 103–106.

38. O'Halloran R J, Williams L F G and Lloyd C P 1984 A microprocessor based isopotential contouring system for monitoring surface corrosion, *Corrosion* **40**: 344–9.

39. Bard A J, Fan F-R F, Kwak J and Lev O 1989 Scanning electrochemical microscopy: introduction and principles, *Analytical Chemistry* **61**: 132–8.

40. Isaacs H S and Brijesh V 1981 Scanning reference electrode techniques in localized corrosion. In *Electrochemical corrosion testing*, edited by F Mansfeld and U Bertocci, ASTM, Philadelphia PA, pp. 3 33.

41. Trethewey K R, Sargeant D A, Marsh D J and Haines S 1994 New methods of quantitative analysis of localized corrosion using scanning electrochemical probes. In *Modelling aqueous corrosion*, edited by K R Trethewey and P R, Roberge Kluwer Academic, Dordrecht, The Netherlands, pp. 417–42.

42. Trethewey K R, Marsh D J and Sargeant D A 1994 Quantitative measurements of localized corrosion using the scanning reference electrode technique. Paper 317 in *Corrosion 94, Baltimore MD*, NACE International, Houston TX.

43. Fontana M G and Greene N D 1967 *Corrosion engineering*, 2nd edn. McGraw-Hill, New York, p. 269.

44. France W D 1972 Crevice corrosion of metals. In *Localized corrosion — cause of metal failure*, edited by M Henthorne, ASTM, Philadelphia PA, pp. 164–200.

45. Sedriks A J 1979 *Corrosion of stainless steels*, John Wiley, New York, p. 93.

7.7 BIBLIOGRAPHY

Baboian R 1986 *Electrochemical techniques for corrosion engineering*, NACE International, Houston TX.

Friedrich H, Killian H, Knornschild G and Kaesche H 1994 Mechanism of stress corrosion cracking and corrosion fatigue of precipitation hardening aluminium alloys. In *Modelling aqueous corrosion*, edited by K R Trethewey and P R Roberge, Kluwer Academic, Dordrecht, The Netherlands, pp. 239–60.

Henthorne M 1972 *Localized corrosion — cause of metal failure*, ASTM, Philadelphia PA.

Isaacs H, Bertocci U, Kruger J and Smialozska S 1990 *Advances in localized corrosion*, NACE International, Houston TX.

Mansfeld F B and Bertocci U 1981 *Electrochemical corrosion testing*, ASTM, Philadelphia PA.

Staehle R W, Brown B F, Kruger J and Agrawal A 1974 *Localized corrosion*, NACE International, Houston TX.

Szklarska-Smialowska Z 1986 *Pitting corrosion of metals*, NACE International, Houston TX.

Trethewey K R and Roberge P R 1994 *Modelling aqueous corrosion — from individual pits to system management*, Kluwer Academic, Dordrecht, The Netherlands.

Turnbull A 1994 Mathematical modelling of localized corrosion. In *Modelling aqueous corrosion*, edited by K R Trethewey and P R Roberge, Kluwer Academic, Dordrecht, The Netherlands, pp. 29–64.

8   FLOW-INDUCED CORROSION

0 ruin'd piece of nature! This great world shall so wear out to nought.
(Shakespeare: *King Lear*)

Many engineering systems operate with electrolytes flowing either through or around them. When corrosion occurs because of predominantly enhanced electrochemical processes, it is termed **flow-induced corrosion**. Many people still call it **erosion corrosion**, a poetic and self-explanatory name. Although electrochemical processes do occur, mechanical effects predominate, such as wear, abrasion and scouring, and the term, erosion corrosion, is a better description in such cases. Soft metals are particularly vulnerable to this form of attack, for example, copper, brass, pure aluminium and lead. However, most metals are susceptible to flow-induced corrosion in particular flow situations.

We discuss corrosion associated with laminar and turbulent flow, damage caused by impingement, and cavitation, a special form of erosion corrosion of materials in very rapidly moving environments.

8.1   FLUID FLOW

Laminar flow occurs when a fluid flows across a metal surface as a series of parallel layers, each moving at a different velocity. The slowest layer is adjacent to the metal surface, where frictional forces and molecular collisions at surface irregularities are greatest. Layer velocity rises to a maximum at some distance into the bulk fluid. This effect can have several consequences, some of which may actually be beneficial.

Equilibrium cannot be established
We saw in Section 4.6 that an equilibrium is established at metal surfaces in static systems in which the cathodic and anodic processes are occurring at the same rate. The ionic distribution in the immediate vicinity of the surface is called the double layer. In principle, when the ions from the corroded metal are removed from the system by a flowing electrolyte, equilibrium cannot be established and an increased rate of

dissolution would be expected. In practice, this is rarely the cause of accelerated corrosion. In neutral environments, systems susceptible to general corrosion suffer accelerated corrosion due to the enhanced transport of oxygen.

Oxygen is replenished

In other systems, the replenishment of oxygen can be a mitigating factor. Differential-aeration cells, a very common cause of attack, are minimised and overall corrosion resistance is improved in cases where the relative motion is not too great. Stainless steels usually have improved corrosion resistance in electrolytes flowing above a given minimum velocity because the replenishment of oxygen maintains the protective oxide films, and flow prevents the development of critical pit chemistry.

Aggressive ions are replenished

A detrimental effect of increased flow rates is the replenishment of aggressive ions such as chloride or sulphide. Conversely, a high flow rate can be an advantage in situations where a steady concentration of added inhibitor is important in controlling a corrosion process (see Section 13.3).

Protective layers may be scoured

If any solid particles are present in the fluid, protective layers may be scoured away and corrosion enhanced. On the other hand, the flow may be sufficient to prevent deposition of silt or dirt which might otherwise cause differential-aeration cells in the crevices beneath.

Such a combination of factors makes the effect of increased laminar flow rate somewhat unpredictable. Perhaps the most significant effect occurs when the flow rate increases such that the frictional forces between the laminar layers cause changes in the direction of flow, a situation known as **turbulence**. Flow towards and away from the surface becomes very complex in such conditions and is dependent upon many factors. Some of them are geometrical:

A sudden change in the bore diameter or direction of a pipe
A badly fitting gasket or joint, which introduces a discontinuity in the otherwise smooth metal surface
A crevice, which allows liquid flow outside the main body of fluid
The presence of a corrosion product or other deposit, which may disturb the laminar flow

Figures 12.10 to 12.12 show typical examples of the first three cases and Section 12.5 further discusses turbulence in tanks and pipe systems.

Fig. 8.1 Copper alloy flange showing erosion by impingement (arrowed) at badly fitting gasket.

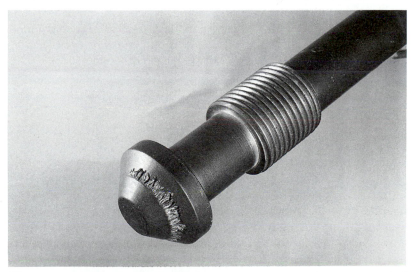

Fig. 8.2 Valve spindle showing erosion caused by valve being only half-open instead of fully open or closed.

Case 8.1: Figure 8.1 illustrates a copper alloy flange suffering from flow-induced corrosion caused by a badly fitting gasket.

Case 8.2: Figure 8.2 shows a valve spindle from a hydraulic system which should have been fully closed or fully opened. It was left slightly open and the constriction of flow caused turbulence and flow-induced corrosion. The damage shown occurred in only 48 hours.

193

The classic example of flow-induced corrosion caused by turbulence is damage to a heat exchanger tube, damage which extends a few centimetres from the inlet. Here, turbulence is created by flow from the (large diameter) header into the (small diameter) tube. Once inside the tube, laminar flow can develop when the liquid has passed the critical first few centimetres. That is why the corrosion is usually found only at the inlet ends.

Flow-induced corrosion is easily recognisable because it can create some strange and beautiful effects, often characterised by grooves, rounded holes or gullies.

Case 8.3: The upper of the two specimens in Fig. 8.3 is an example of the flow-induced corrosion found at the end of a brass condenser tube. Wet steam flowed through the tube, but the corrosion was caused in a restricted pocket where the sea-water was turbulent.

The peculiar effects produced by erosion corrosion have been attributed partly to a time dependence. On smooth surfaces, the corrosion rate is slow, but accelerates as roughening proceeds. When a certain depth of surface roughness has been achieved, a layer of water adheres to the surface or becomes trapped inside the pits to reduce the eroding effects of subsequent liquid flow. Consequently, the rate is observed to decelerate after a maximum rate has been achieved. Undercutting of the metal on the downstream side of the pit is often a characteristic feature of flow-induced corrosion.

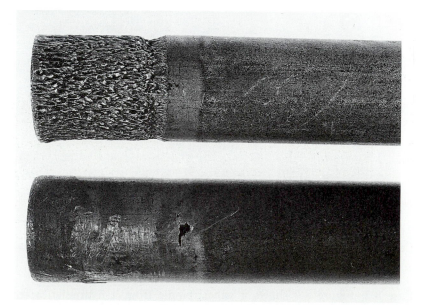

Fig. 8.3 (top) Admiralty brass condenser tube exhibiting the characteristic well-rounded erosion pits. (bottom) Admiralty brass condenser tube pierced by impingement of wet steam.

Impingement may be thought of as a rather more extreme form of severe turbulence effects. Molecules of fluid impinge directly upon the metal because the bulk flow has a large component perpendicular to the metal surface. In the earlier flow examples the large component was parallel to the metal surface. Impingement causes more mechanical wear than increased electrochemical activity. It has been a serious problem in several important applications. The destructive effects of high velocity impingement of rain upon the skins of aircraft caused considerable problems when the first supersonic jet fighters were introduced. Considerable testing was necessary to evaluate the erosion resistance óf new materials to the high impact velocities of wet steam in high speed turbine generators. The impingement corrosion which occurs when a liquid is forced to change direction may be quite different from the frequently beautiful effects produced by erosion corrosion.

Case 8.4: The lower specimen in Fig. 8.3 shows an Admiralty brass (see Table 15.5) condenser tube which has been pierced by the direct impact of wet steam upon the metal surface.

Case 8.5: An aluminium alloy exhaust manifold of a marine diesel engine (Fig. 8.4) carried hot gases overboard. Coolant water also left the boat at the same point after joining the exhaust through a pipe with a smaller bore. When it joined the exhaust pipe, the stream of water entered at an angle, impinged upon the opposite wall and penetrated the metal to a depth of 6 mm in about two years.

Sometimes, resistance to flow-induced corrosion can be achieved in surprising ways.

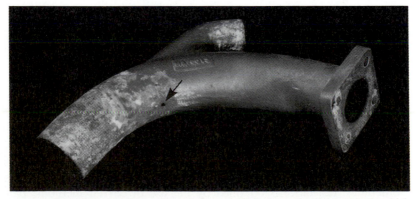

Fig. 8.4 Part of the aluminium alloy exhaust manifold of a marine diesel engine in which hot gases exited overboard (left) from the engine (right). Coolant water also exited the boat at the same point (left) after joining the exhaust by means of a slightly narrower pipe (top). When it joined the exhaust pipe, the water entered at an angle, impinged upon the opposite wall and penetrated the metal to a depth of 6 mm (arrowed) in about two years.

195

Case 8.6: In the Shell 250,000 dwt vessel, *Mitra*, commissioned in 1969, failures occurred among the 6,000 aluminium brass tubes in the condenser within the first year of service. According to common practice, soft iron sacrificial anodes were fitted to the water-box and plastic inserts fitted to the tube inlets, but within five months, the anodes corroded away and attack still occurred beyond the extent of the inserts at water velocities of only 2 m s^{-1}. The problem was completely eliminated by the addition of 1 ppm of ferrous sulphate for 1 hour per day. The presence of Fe^{2+} ions had a beneficial effect on the copper alloy surface films, improving the alloy resistance to flow-induced corrosion. The treatment was then successfully applied to all other vessels experiencing similar problems.

8.2   CAVITATION

Cavitation is a particular form of erosion corrosion caused by the formation and collapse of bubbles of vapour on metal surfaces. This form of corrosion tends to be associated with components being driven at high velocity through a fluid, unlike pipes or tanks, where the fluid flow occurs across stationary metal surfaces. Propellers, impellers and hydraulic turbine gear are the most common places to encounter corrosion by cavitation.

When the fluid flow over a metal surface becomes sufficiently great, very localised reductions in hydrodynamic pressure cause the fluid to vaporise and bubbles to nucleate on the metal surface. The same mechanical effect that reduces the pressure also creates pressure increases which cause the bubbles to collapse with considerable force. If these forces exceed the elastic limit of the metal (see Section 9.2), its surface deforms, the protective films are broken and corrosion ensues. The surface roughening provides better nucleation sites for new bubble formation and the corrosion process is aggravated.

Cavitation has also been observed on wet liners of diesel engines, where vibration resulting from the piston motion is the primary cause of the pressure variations within the flowing coolant. Figure 8.5 is one such liner; it is described in Case 12.19. The cavitation damage is clearly visible on the upper and lower parts of the specimen. In addition, the central part of the liner shows erosion damage where the water was pumped through the specially machined channels. Crevice corrosion is also evident around the seat of the fuel injector nozzle in the centre of the photograph.

An important method of controlling corrosion by cavitation is to use very smooth and well-machined components which offer fewer nucleation sites. Sometimes, resistance is obtained by the use of barrier methods, particularly varieties of rubber coatings which absorb the shock waves, though in severe environments they are less effective. Cathodic protection has been reported as beneficial because evolution of hydrogen can protect against the damaging forces. However, the large currents necessary to produce sufficient protective

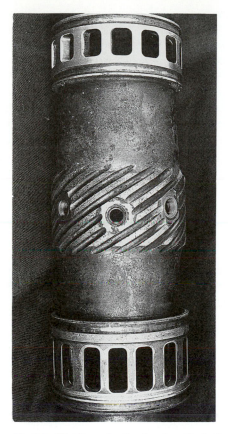

Fig. 8.5 Wet liner of a diesel engine. Cavitation damage caused by engine vibration is at the top and bottom. Erosion damage is visible in the grooves at the middle of the liner. Crevice corrosion has occurred around the fuel-injector seat in the centre hole.

hydrogen make this a rather inefficient method of corrosion control, and the danger of hydrogen embrittlement cannot be neglected (Section 10.3).

Selection of the correct material is equally important in the control of cavitation damage. Stainless steel is widely regarded as the material with the greatest resistance to cavitation, although Stellite, a Co–Cr–W–Fe–C alloy, is known to have a much greater resistance to wear effects than a 304 stainless steel in a severe environment. Indeed Ultimet, a newer alloy with a cobalt base (Co–26Cr–9Ni–5Mo–2W–3Fe) possesses a combination of excellent wear and corrosion resistance.

8.3 BIBLIOGRAPHY

Kennelley K J, Hausler R H and Silverman D C 1992 *Flow-induced corrosion: fundamental studies and industry experience,* NACE International, Houston TX.

197

Levy A V 1979 *Corrosion-erosion of coal conversion materials,* NACE International, Houston TX.

Levy A V 1982, *Corrosion-erosion wear of materials in emerging fossil energy systems,* NACE International, Houston TX.

Levy A V 1991 *Corrosion-erosion wear at elevated temperatures,* NACE International, Houston TX.

9   ENABLING THEORY FOR ENVIRONMENT-SENSITIVE CRACKING

Ut tensio sic vis. (Robert Hooke, 1660)

Engineering structures are frequently required to withstand the action of forces which tend to deform them. When forces act upon these structures in an environment which may cause corrosion of the metal, new types of corrosion may be observed which can be broadly grouped under the heading of **environment-sensitive cracking**. Before describing the nature of these corrosion processes it is necessary to achieve an understanding of the effects of forces upon those materials in the absence of a corrosive environment.

9.1   ELASTICITY: STRESS AND STRAIN

Robert Hooke's discovery that elastic properties could be described in the simple relationship:

$$\frac{\textbf{load}}{\textbf{extension}} = \textbf{constant} \tag{9.1}$$

was to have far-reaching consequences for the science of materials subjected to the action of forces. Provided that the load is not too great, most materials will extend and compress uniformly in the direction of the load, in accordance with Hooke's Law. The behaviour is termed **elastic** if it is both linear and reversible. To examine more precisely the effect of forces upon materials, some fundamental parameters must be defined.

When a force acts upon a material so as to deform it, we use the term, **stress**. The stress in the material is the ratio of the force to the area of the material which resists the force, i.e.

$$\textbf{stress} = \frac{\textbf{force}}{\textbf{area}} \tag{9.2}$$

Although the unit of measurement would seem to be $N\ m^{-2}$, the SI unit is called the **pascal**, symbol Pa, such that $1\ Pa = 1\ N\ m^{-2}$.

199

(a)

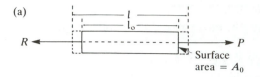
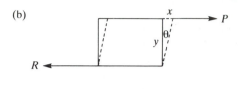

(b)

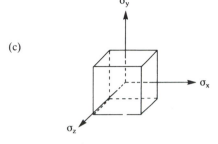

(c)

Fig. 9.1 Three simple stress situations: (a) tensile stress (uniaxial); (b) shear stress; and (c) triaxial stress.

The material is said to be in a state of **strain** when it has suffered a deformation as a result of a stress. Strain is defined as follows:

$$\text{strain} = \frac{\text{deformation}}{\text{dimension of material}} \qquad (9.3)$$

Strain is a dimensionless quantity, but is often expressed as a percentage of the original dimension.

The deformation which results from the application of a stress depends upon the way in which the stress is applied. In Fig. 9.1(a) force P is applied to a material of cross-sectional area A_0. The material is fixed in space and resists the applied force with a reaction R which is equal, opposite and colinear. The stress, σ, is thus given by

$$\sigma = \frac{P}{A_0} \qquad (9.4)$$

This type of stress is known as **tensile** stress. It is also called **uniaxial** stress because it acts along only one of the principal axes.

The tensile strain is uniform in the direction of the applied stress and is usually given the symbol ϵ such that:

$$\epsilon = \frac{(l - l_0)}{l_0} \qquad (9.5)$$

where l_0 is the original length and l is the length under load. Although the material extends elastically in the direction of the applied force, its width and its thickness contract (not shown in Fig. 9.1(a)). As a good approximation, it can be said that the volume of material remains constant. The reduction in width and thickness causes a reduction in the area of cross-section and although the load may have remained constant, the *true* stress has increased. It is normal in such situations to call the stress, as defined by eqn (9.4), **engineering stress** in order to distinguish it from the **true stress**. Similarly, the tensile strain calculated on the basis of eqn (9.5) is better called the **engineering tensile strain**.

Fig. 9.1(b) shows another kind of stress, **shear stress**. The material is fixed in space at the bottom, and the force is applied parallel to the top face. The reaction is equal and opposite but *not* colinear. The shear strain is non-uniform in the direction of the shear stress, τ. Shear strain is measured in terms of the angle of deformation, θ radians, such that for small angles:

$$\theta \text{ (radians)} \approx \tan \theta = x/y \tag{9.6}$$

The most general case is depicted in Fig. 9.1(c) which shows an element of material subjected to a **triaxial stress state**. In the figure, σ_x, σ_y, and σ_z are three principal stresses. If they are equal and if they act so as to compress the material, rather than to extend it, the situation models a body immersed in a fluid. And the stress is a **hydrostatic stress**. If one principal stress is zero the stress state is called **plane stress**, a situation found in the loading of thin sheets which do not develop a tensile stress through their thickness.

When a material is subjected to a uniaxial tensile stress which increases from zero, the tensile strain increases uniformly and the material initially behaves in an elastic manner. For most materials, we can write this relationship:

$$\frac{\text{stress}}{\text{strain}} = \text{constant} \tag{9.7}$$

For tensile stress and strain, the constant is called **Young's modulus** and is given the symbol E, not to be confused with potential. Because strain is a dimensionless quantity, the units of E are the same as for stress. Young's modulus is an important material property used in engineering design.

The linear elastic behaviour of materials ceases when stresses are increased beyond the elastic limit. From this point onward, the behaviour of the material is of even greater importance to the engineer, and this will be discussed next.

9.2 THE TENSILE TEST

A useful way of monitoring the behaviour of materials under the action of tensile stresses is by means of the **tensile test**. A specimen of defined dimensions (it may have either cylindrical or rectangular cross-section) is extended uniformly at a predetermined rate between the parting crossheads of a loading machine and the values of the load on the specimen at any given extension are measured. It is usual for this data to be produced on a chart recorder in graphical form as a load/displacement curve. However, it is instructive to consider the graph of engineering stress against engineering strain; there is no difference in shape, but the axes are changed. Load is divided by a constant quantity, the original cross-sectional area, while extension is converted into strain by eqn (9.5).

A schematic curve which illustrates some important features of tensile test results is shown in Fig. 9.2. During the initial period, the relationship between the two parameters is linear (eqn. (9.7)) and elastic, i.e. if the test were stopped and the stress removed, the material would return to its original dimensions. At point A the material reaches the limit of linear elastic behaviour. As the increments of extension take the specimen beyond A, smaller increments of load are measured. This situation continues until B, where the material begins to **neck**; it narrows considerably at one point along its length, rather than along the whole of its length. The stress on the specimen at the point of necking, B, is called the **tensile strength**.

Further strain leads to actual decreases in engineering stress until the specimen breaks at C. The two broken pieces can be matched together to determine the plastic elongation of the specimen at the point of fracture. The greater this elongation, the greater is the property known as **ductility**. If the fracture surfaces of a ductile material are examined, they will exhibit much tearing. The specimen has undergone a **ductile fracture**. Fig. 9.3 (right) shows a typical example.

In some cases, there is little or no permanent deformation of a material before fracture occurs. The surfaces are typically bright and shiny with almost none of the macroscopic tearing associated with ductile fracture. The specimen in Fig. 9.3 (left) is said to have fractured in a **brittle** manner. **Brittle failure** is the

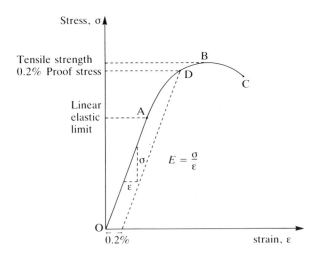

Fig. 9.2 Typical stress/strain curve for a tensile test specimen.

Fig. 9.3 Fracture surfaces exhibiting brittle fracture (left) and ductile fracture (right).

scourge of engineers, because components and structures give no warning of impending failure, as they do if the metal deforms plastically.

It is often difficult to determine the point at which plastic deformation (dislocation movement, which occurs to a very small degree from the first application of a load) begins to predominate over elastic behaviour (atomic bond stretching). Consequently, engineers have defined another parameter known as the **proof stress**. For example, a 0.2% proof stress will result in a permanent deformation of 0.2%. If, in Fig. 9.2, the specimen were strained from O to D and then the stress were removed, the specimen would contract to release the elastic part of the extension, but the permanent 0.2% elongation would remain. For the sake of clarity, the curve shown in the figure has been exaggerated; real curves may look rather different from this, with, for example, D much closer to A and a much steeper slope, E.

9.3 STRESS CONCENTRATION

Let us now consider the effects of a tensile stress upon the bonding between the atoms of a metal. In a perfect crystal structure, the stress is absorbed uniformly between all the bonds concerned, but when a defect in the structure is present, the stress distribution cannot be uniform. Some bonds will therefore be under greater stress than others (Fig. 9.4(a)). The atomic structure is shown as a simple cubic arrangement for simplicity and arrows indicate how, when a tensile stress σ_{app} is applied to the metal, the stress is concentrated in the bonds which are adjacent to the defect. Thus bonds 1 and 2 are subjected to greater stresses than

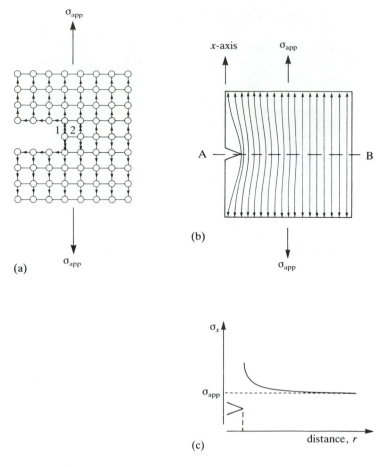

Fig. 9.4 (a) Effect of an applied stress, σ_{app}, upon the atomic bonds in a crystal lattice containing a crack. (b) Stress concentration at a crack tip. (c) Variation of effective stress in the x-direction, σ_x, with distance r ahead of the crack tip.

σ_{app}. As distance is increased away from the defect and into the material, the stress diminishes until an average level is reached which is the same as the external stress. For small values of σ_{app} the bonds are merely stretched, but for large σ_{app} the stress upon bond 1 may be larger than the bond strength, in which case the bond will break. Not only is the full effect of the stress concentration now transferred to bond 2, but the concentration is greater than before and the material is very likely to undergo total failure by a domino effect of bond rupture through the material.

The action of a stress on the atomic bonds within a material may be likened to lines of magnetic force (except that stress lines act only through solids). Figure 9.4(a) is sometimes drawn like Fig. 9.4(b) to show how the concentration of stress lines is greatest at the crack tip. Consider line AB through the material, and define the x-axis as the direction of the applied stress. It has been shown that, to a fair approximation, the effective stress in the *x*-direction at points along AB is similar to Fig. 9.4(c).

Of the defects described in Section 2.6, any which can locally reduce either the number or the strength of bonds have the potential to concentrate stress in this manner. Not surprisingly, they are referred to as **stress raisers**. They may be contained within the body of the material, in which case they may have been caused by the manufacturing process. Or they may be on the exterior, perhaps caused mechanically or by corrosive means.

9.4 LINEAR ELASTIC FRACTURE MECHANICS

In earlier times, the most obvious criterion by which engineering structures could be designed was the tensile strength, the greatest load which could be applied to a structure without causing immediate failure (see Fig. 9.2). As engineering knowledge increased, this idea was superseded by the elastic limit. Even so, it was not good enough to design structures so they deformed; a safety factor was divided into the chosen proof stress to produce a design stress and the magnitude of the safety factor was largely dependent upon the scale of disaster which would result from a failure. For example, a component upon which human lives were dependent might have a safety factor of 10 or more, so that the maximum stress which the designer anticipated being applied to the component was still only one-tenth of the stress which *might* lead to failure. Even so, it was found by experience that catastrophic failure occurred unexpectedly. A better method of predicting failure was necessary and, as a result, the subject of linear elastic fracture mechanics, often simply called **fracture mechanics**, was devised.

The failure of a metal component occurs as a result of the presence of defects, whether they be present in the original material or introduced during the lifetime of the component. The likelihood of catastrophic failure increases with increasing defect size. The aim of fracture mechanics is to predict whether or not a given defect size will lead to catastrophic failure. In Chapter 2, a number of different defects were described, any one of which may contribute to the failure process. For simplicity, we shall refer to defects as **cracks**. In any failure, there are two distinct phases to be considered. First, there is an **initiation phase**. Cracks may be present in the original material as a result of the manufacturing process, or they can be created mechanically or by corrosive mechanisms. Once initiated, cracks may **propagate** mechanically or by corrosive means. Obviously, this book is concerned with the corrosive aspects but consideration of the mechanical mechanism is also essential.

A very important material property is **fracture toughness**, a measure of the resistance to crack growth. If a material is able to sustain the presence of a crack and perhaps allow the growth of the crack to occur slowly, then it will probably have desirable engineering properties. Brittle materials, on the other hand, have little resistance to crack growth once a critical tensile stress is applied. Above

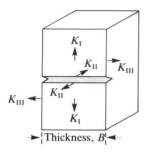

Fig. 9.5   The three modes of crack growth. I, crack opening mode; II, in-plane sliding mode; and III, antiplane shear mode.

this stress, a crack propagates very rapidly to give a **catastrophic failure**. Some brittle materials such as concrete are used only in compression because their resistance to tensile stress is so low.

It is an unfortunate fact that most materials become more brittle as their strength is increased, a property which works against the designer and enforces compromise. The engineer who seeks to use the strongest materials must have a very good predictive capability for typical crack growth rates under different loading conditions. Such information can be gleaned from a study of fracture mechanics. A determination of fracture toughness, for example, can be made so that it is a conservative parameter and represents the worst possible case which a material must withstand. This can be done in the following way.

Figure 9.5 shows three modes by which crack growth can occur. A crack in a material is simulated by a sharp-edged notch cut across the whole of the thickness of the specimen.

Mode I is considered to be the most potentially damaging and is most frequently described because it leads to the most conservative value of fracture toughness. It is the crack opening mode, in which crack surfaces move directly apart. Therefore, mode I relates precisely to a material under the action of a tensile stress.

Mode II is called the in-plane sliding mode and corresponds to crack surfaces sliding over one another in a direction perpendicular to the leading edge of the crack.

Mode III is the antiplane shearing mode, a tearing mode in which crack surfaces slide over one another in a direction parallel to the leading edge of the crack.

Modes II and III are usually less damaging situations than mode I and will not be discussed further. Henceforth, all terms with the subscript I relate to mode I crack opening.

Consider a specimen similar to that shown in Fig. 9.5 and subjected to a constant applied mode I tensile stress, σ_{app}. We wish to study the effect of increasing the size of a material defect (represented by the machined notch) upon the stress immediately ahead of the crack tip. Initially, the stress concentration for small cracks is not sufficient to cause significant growth. As crack length increases, so does the stress concentration at the crack tip until the effective stress becomes so great that the crack begins to grow at a very fast rate and failure ensues. It is possible to determine a critical crack length below which the material can be considered stable. Just as a simple tensile test determines the tensile strength of a material, a similar test can also determine the fracture toughness for failure under a mode I crack opening regime.

In Fig. 9.5, the thickness of the material is represented by B, and the crack extends throughout the whole of the thickness. This is called a **through-thickness crack**. Using the schematic graph of Fig. 9.4(c) to represent the variation of effective stress with distance in an infinite body containing a through-thickness crack of length 2*a*, it can be shown that the effective stress at a distance *r* ahead of the crack tip, when *r* is very small, is given by

$$\sigma = \frac{K}{(2\pi r)^{1/2}} \tag{9.8}$$

K is the **stress intensity factor** and is given by

$$K = \sigma_{app}(\pi a)^{1/2} \tag{9.9}$$

For mode I crack opening, K is replaced by K_I. When the crack is of critical size such that failure results, K_I is minimised and becomes the useful material property, fracture toughness, denoted K_{Ic}.

$$K_{Ic} = \sigma_{app}(\pi a_{crit})^{1/2} \tag{9.10}$$

It would appear that the determination of fracture toughness for engineering materials is a straightforward exercise. Unfortunately, the situation is not quite so simple; different values are obtained with specimens of the same material but different dimensions. This is mostly because stress concentration is strongly dependent upon the dimensions of the specimen, and in particular, the thickness. If a thin specimen is used (B is small) the material will be found to be quite tough. Such a result might be thought to be favourable. But if a thicker specimen were to be tested a much lower critical stress intensity factor, K_c, would be measured. The engineer must know the least possible value of K_c for a material in order to avoid the possibility of catastrophic failure. Figure 9.6 is a graph showing the variation of critical stress intensity factor with thickness of a test material. The lowest value of K_c is the fracture toughness, K_{Ic}. Fracture toughness is a major design parameter in modern engineering.

If the fracture surface of a broken specimen exhibits shear lips (Fig. 9.7 (right)), it is usually associated with ductility and relatively high toughness value. The specimen is not thick enough to make a valid measurement of K_{Ic}. The shear lips almost disappear in specimens which have the correct thickness for valid K_{Ic} measurement (Fig. 9.7 (left)). In thin specimens that exhibit shear

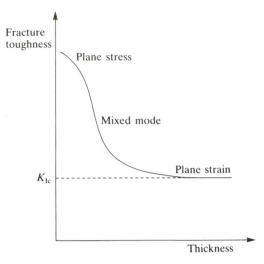

Fig. 9.6 Variation of fracture toughness with thickness.

Fig. 9.7 Fracture surfaces: (left) a titanium alloy specimen showing almost no evidence of shear lips (fracture in plane strain, specimen thick enough for valid test); and (right) an aluminium specimen exhibiting shear lips (fracture in plane stress, specimen too thin for valid K_{Ic} test).

lips, the condition is known as **plane stress**; the condition causing brittle fracture in thicker specimens is called **plane strain**. Intermediate cases, which show elements of both, are known as **mixed mode fractures**.

The procedures for performing mechanical tests are laid down in the references and must be carefully adhered to if meaningful data are to be obtained. Experimentally, K_{Ic} is evaluated by a series of careful time-consuming

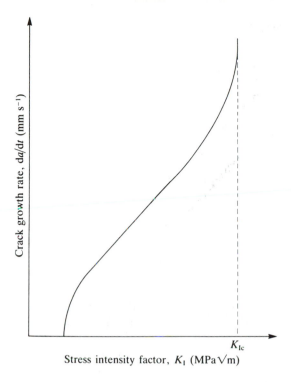

Fig. 9.8 Variation of crack growth rate with stress intensity factor.

experiments. A number of identical notched specimens are stressed under mode I conditions and measurements of crack growth rates are made. When the variation of stress intensity factor is plotted against the crack growth rate, a graph similar to Fig. 9.8 is obtained. For convenience, log scales are used for crack growth rates. Units of stress intensity factor (and fracture toughness) are MPa $m^{1/2}$. Such graphs are very important in the analysis of corrosion mechanisms for environment-sensitive cracking.

9.5   FATIGUE

The discussions so far have assumed that applied stress is constant in magnitude and direction. In reality, this may not be so. Engineering structures and components may be subjected to stresses which continuously rise and fall in magnitude, and may also change direction. If a metal component fails as a result of such a condition, it is referred to as a **fatigue failure**. The potential problems caused by fatigue were known over a century ago when the British engineer, Sir William Fairbairn, found that a steel girder would support a static load of 120 kN indefinitely, yet it failed under a cyclic load of only 30 kN

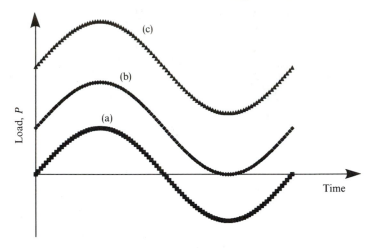

Fig. 9.9   Three common load cycles: (a) $R = -1$, (b) $R = 0$, (c) $R > 0$.

applied 3 million times. It was only after a series of Comet airliner crashes in the 1950s that the severity and the unpredictability of fatigue failure were brought to the attention of the public.

Fatigue is produced by stress cycling, a process which introduces many new variables not present in static stress situations. Figure 9.9 illustrates three different load cycles to which a material may be subjected. Figure 9.9(a) represents a sinusoidal load variation which oscillates with equal magnitude about a zero mean. Figure 9.9(b) shows a similar load cycle which is never negative in sense, but increases from zero and returns to zero. Figure 9.9(c) shows the load always positive. It is possible to describe each situation by considering the ratio of the minimum and maximum applied loads. P_{min}/P_{max} is a parameter denoted by R. In the first case, $R = -1$, in the second, $R = 0$; and in the third, $R > 0$. As well as the amplitude of the stress cycle, the frequency (measured in cycles per second, Hz) is also important. It should be emphasised that in real engineering situations, loading patterns may be most irregular and quite unlike the idealised cycles used for laboratory simulations. In such situations the **mean stress**, indicated in Fig. 9.9, is a useful parameter. The parameters described above must be carefully defined for all fatigue tests.

One common type of fatigue test applies a particular stress cycle to a material and observes the number of applied cycles necessary to cause failure. The stress amplitude, S, is plotted against the logarithm (base 10) of the cycles to failure, N. Such graphs are called **S–N curves**, and a typical example is shown in Fig. 9.10.

The upper curve demonstrates the existence of a stress level below which the material will never fail by fatigue. Such behaviour is typical of iron and ferrous alloys, but many non-ferrous alloys do not possess such a property, as evidenced by the lower curve. Structures made of these metals must be designed with fatigue firmly in mind.

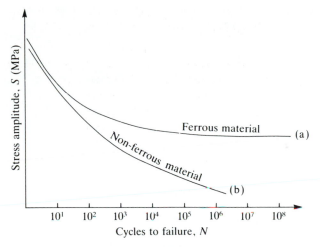

Fig. 9.10 Typical *S–N* curves for fatigue tests in air: (a) a ferrous alloy and (b) a non-ferrous alloy.

As fatigue is a mechanical not a corrosion process, it has little respect for the microstructure. A fatigue crack, such as shown in Fig. 9.11(a), literally tears its path across the material. Crack propagation deviates here and there at places of local weakness, but its path is primarily perpendicular to the direction of the maximum load vectors. When the material breaks to expose a fracture surface (Fig. 9.11(b)) it is often possible to see **fatigue striations**, markings in the material which run perpendicular to the direction of crack propagation. They actually indicate the amount of crack growth during each positive phase of the load cycle, and a measure of the average distance across each striation gives a measure of the crack propagation rate. In Fig. 9.11(b), the average width of a striation is 1.5 μm (note the 5 μm scale marker at the bottom); the frequency used was 0.25 Hz. The specimen was part of a corrosion fatigue programme.

The development of methods of analysing and predicting failure using fracture mechanics has proved to be a major advance. The theory described in Sections 9.3 and 9.4 applies equally well to fatigue and modification of the experimental technique to include stress cycling is straightforward. Data obtained from fatigue testing can be plotted on a graph similar to Fig. 9.8, except that stress intensity factor is replaced by the difference between the maximum and minimum stress intensities encountered in each cycle, i.e.

$$\Delta K = K_{max} - K_{min} \tag{9.11}$$

ΔK is plotted against the crack growth rates measured in terms of the increment of crack growth per cycle, *da/dn*. Observations of fatigue crack growth rates in the intermediate range of stress amplitudes are often found to behave in accordance with the equation:

$$\frac{da}{dN} = C\Delta K^m \tag{9.12}$$

(a)

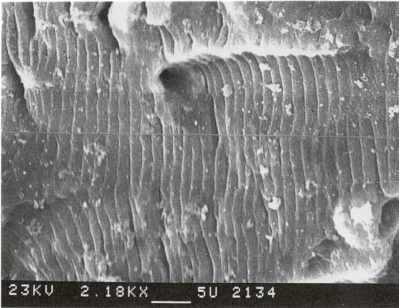

(b)

Fig. 9.11 (a) Optical micrograph of a fatigue crack in a specimen made from type 2024 aluminium extruded bar. The area shown measures approximately 2 mm × 1.5 mm. The direction of crack propagation is from left to right, with K_I mode load vectors operating along the axis from top to bottom. (b) Scanning electron micrograph of fatigue striations in the fracture surface of a specimen similar to (a). Crack propagation is from left to right, with the load vectors in and out of the plane of the paper.

where C is an experimental constant. Over a wide range of materials, it has been found that the average value of m is between 2 and 3. Equation (9.12) is frequently referred to as the **Paris Law**.

9.6 BIBLIOGRAPHY

Broek D 1986 *Elementary engineering fracture mechanics,* 4th edn. Martinus Nijhoff, Dordrecht, The Netherlands.
Colangelo V J and Heiser F A 1974 *Analysis of metallurgical failures,* John Wiley, New York.
Fuchs H O and Stephens R I 1980 *Metal fatigue in engineering,* John Wiley, New York.
Knott J F 1973 *Fundamentals of fracture mechanics,* Butterworths, London.
Knott J F and Withey P A 1993 *Fracture mechanics – worked examples,* 2nd edn. The Institute of Materials, London.
Nair S V, Tien J K, Bates R C and Buck O, 1988 *Fracture mechanics: microstructure and mechanisms,* ASM International, Ohio.
Unterweiser P M 1979 *Case histories in failure analysis,* ASM International, Ohio.

10 ENVIRONMENT-SENSITIVE CRACKING

The image of stress corrosion I see
Is that of a huge unwanted tree,
Against whose trunk we chop and chop,
But which outgrows the chips that drop;

And from each gash made in its bark
A new branch grows to make more dark
The shade of ignorance around its base,
Where scientists toil with puzzled face.

(S P Rideout, *On Stress Corrosion)**

In the late nineteenth century, British rule in India was at its strongest. Infantry, artillery and cavalry were all present in strength and well supported with arms and ammunition. But during the wettest months of the monsoon season, military activity was diminished and ammunition was stored in the stables until the dry weather returned. Then a problem with cartridges became apparent; many were found to be useless because of considerable cracking of the brass. It was not until 1921 that the problem was satisfactorily explained, when, in a classic investigative study, Moore, Beckinsale and Mallinson [1] found that residual stresses caused cracking in 70/30 cartridge brass in the presence of specific chemical agents. It was concluded that residual stresses from the manufacturing process, combined with ammonia from horse urine, had been the agents responsible for the cracking of the cartridges in India. A new form of corrosion had been identified, which because of the circumstances in which it was found, was first called season cracking. Later it was called stress-corrosion cracking.

Case 10.1:   Figure 10.1 shows a 70/30 brass fitting to a high pressure air bottle. When withdrawn from stores after a long period on the shelf, the fitting was found to be cracked from end to end. Inspection of the remainder of the stock showed that 80% were suffering from the same problem. The source of the aggressive species responsible for the corrosion was never identified, but was probably airborne ammoniacal vapours. The remaining 20% were given a stress-relieving heat treatment and did not crack thereafter (see also Case 12.6).

---

*Published in *Proceedings of the conference on fundamental aspects of stress corrosion cracking*, 1969. Reprinted with permission from NACE International, Houston, Texas.)

Fig. 10.1 Stress-corrosion cracking of 70/30 brass. A fitting for a high pressure air bottle which had cracked after a period in store in an environment containing an unidentified aggressive species. Application of a stress-relieving heat treatment to other uncracked specimens resulted in no further failures.

During the long period which intervened between the observation and explanation of season cracking, a new and seemingly unrelated problem was regularly occurring in riveted steam boilers. The age of steam power had arrived but the materials technology was still deficient in important areas. Boiler explosions were common, not just caused by poor operation and design but many of them a direct result of corrosion. In the final 20 years of the nineteenth century, 600 lives were lost in the United Kingdom alone and fatal explosions were still occurring until quite recently. In one case, a man lost his life, sitting with a 'for sale' notice on top of his boiler under excessive steam pressure, as he demonstrated to passers-by his conviction in its safety. [2] In another, no less than five men were sitting upon a boiler in a tug as it waited to enter a dock in Cardiff. The boiler exploded and was propelled, like a rocket, out of the boat for 230 m where it landed on an unfortunate sixth individual on another ship.

Although new methods were being introduced to control the corrosion of boiler tubes, it was found that the presence of caustics (alkalis), usually concentrated in crevices around rivets, combined with the considerable fabrication stresses around rivet holes to cause cracking of the steel boiler shells and tube-plates with consequent catastrophic failure. [3]

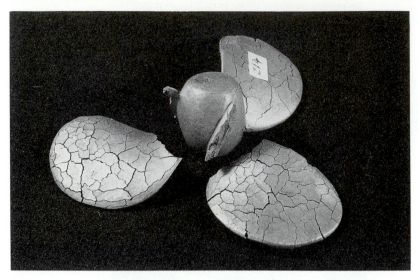

Fig. 10.2 A small casting of Al–4%Si (120 mm diameter) suffering from acute stress-corrosion cracking after 20 years in a damp drawer.

The caustic embrittlement of boilers and the season cracking of brass are linked by an insidious combination; a *particular metal/environment pair* in the presence of *tensile stress*. They were the forerunners of a problem which has blighted the technological society for decades, for with the development of each new alloy has come the seemingly unpredictable risk of catastrophic failure. In the early 1900s, mild steel was found to crack in nitrates as well as caustics, while some cast aluminium–silicon alloys were found to crack simply in moist atmospheres, such as the small casting made as a test piece for a practical examination and left for 20 years in a drawer in a damp garage (Fig. 10.2). In the 1930s, magnesium alloys were also found to be susceptible to cracking in moist atmospheres, and stainless steels suffered badly from cracking in aerated chloride-containing environments.

Case 10.2:   The pressuriser of a PWR nuclear reactor is maintained regularly. To allow access to the working parts, handholes of 304L stainless steel are present (Fig. 10.3(a)). These are in the form of bosses, 250 mm outside diameter with a 150 mm concentric hole sealed by welding a diaphragm plate into it. This is then covered by a backing plate, 40 mm thick, seamed onto eight 25 mm, 8 tpi (teeth per inch) studs. During routine dye penetrant examination of the root run of the steel weld of the plate, multibranched intergranular cracks typical of stress-corrosion were found in an area between two weldments where grinding had been used unsuccessfully to remove a welding defect (Fig. 10.3(b)). The stud holes had originally (wrongly) been tapped to only 7 tpi. It was necessary to fit blank studs and seal them in with welding. New holes

(a)

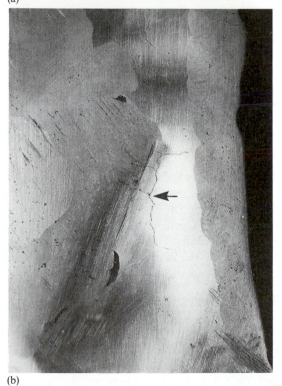

(b)

Fig. 10.3 (a) Handhole for a pressuriser of a PWR (Case 10.2), diaphragm plate at centre. Blank studs of the filled 7 tpi holes are just visible between the present stud holes. (b) Multibranched stress corrosion cracks (arrowed) in the region between the fillet weld of the diaphragm plate (right) and the filled stud hole (left). The micrograph is taken at just above the nine o'clock position of (a). The weld defect is visible at the bottom, left of centre.

were then drilled and tapped to 8 tpi. The combination of residual stresses from welding and grinding, with a tiny chloride contamination (sweaty hands?) had caused the cracks in only two years.

In the 1950s and 1960s martensitic steels for aerospace applications were discovered to be vulnerable to stress-corrosion cracking, and even the very corrosion-resistant titanium alloys could be cracked in environments containing methanol. The problem seemed worse with every new application, and the poem at the beginning of this chapter captures well the cloud of despair which would often descend upon those who worked in the field of stress-corrosion cracking. The efforts made by both pure and applied researchers to solve the problems have been considerable. Those stress-corrosion failures which result in fatalities or injuries usually receive detailed investigation, yet many others are still not identified as such. When, in 1975, a detailed survey of corrosion failures in the chemical industry was carried out, it was discovered that more than one-third were caused by environment-sensitive cracking. [4] In 24 years of annual NACE corrosion conferences between 1968 and 1992, some 27% of the 962 papers presented were on the subject of stress-corrosion cracking. This statistic highlights recent concern over this form of corrosion. [5]

Meanwhile, the failures continue.

Case 10.3: At Scaramanga in Athens, Greece, in 1985, a floating concrete dry dock was used as a refit dock for cargo ships. Suddenly the floor buckled and rose like a balloon before bursting. Water flooded in and destroyed the dock, which cost $60 million. The failure was attributed to stress corrosion cracking of steel tendons used as anchor piles to fix the dock to the seabed.

This chapter uses **environment-sensitive cracking** to describe all failures in which stress combines with particular metal/electrolyte pairs to cause cracking. There is an important subdivision in which *cyclic* or *periodic* stress is differentiated from *static* stress. Cyclic stress is called **corrosion fatigue**. Traditional texts have further separated cases of **hydrogen embrittlement**, but most current workers believe this is so closely allied to stress-corrosion as to be an integral part of mechanistic discussions.

10.1 STRESS-CORROSION CRACKING

Stress-corrosion cracking (SCC) is a term given to the intergranular or transgranular cracking of a metal by the conjoint action of a static tensile stress and a specific environment. This form of corrosion is very common throughout industry. Despite decades of intensive research, workers are only now beginning to understand the processes involved and control measures are still frequently unsuccessful.

In boiling water reactors, intergranular SCC of stainless steel (type 304) piping is said to be the major corrosion problem, while in pressurised water reactors the same material has been found cracked in such applications as boric acid lines and spent fuel liners. Stress-corrosion failure of turbine blades made from stainless steel (type 403) was said to run at 4% per annum. [6] In the chemical industry, SCC of stainless steel owing to chloride leaching from thermal insulation continues to be a problem, despite the fact that its cause is well known. In 1973, one instance alone of the failure of a stainless steel component caused the loss of $1 million. [7] Similar problems continually plague the oil industry, where pipes in deep, high pressure wells require the use of high strength steels. These are known to be susceptible to SCC, particularly in the presence of hydrogen sulphide. An inhibitor has been used consistently in an attempt to alleviate corrosion problems in such situations, yet failures in the presence of the inhibitor were still being reported 10 years after the inhibitor had been proved ineffective. [8]

A repeated feature of stress-corrosion cracking is its unexpectedness. Often, a material, chosen for its corrosion resistance in a given environment, is found to fail at a stress level well below its normal fracture stress. Rarely is there any obvious evidence of an impending failure, and because it can occur in components which are apparently unstressed, it is even more of a surprise when components are found to be defective.

Case 10.4: Stainless steel pipe was stored close to the sea while waiting to be used in a construction programme in the Middle East. The high daytime temperature and low night-time temperature, together with the salt environment, caused a massive salt build-up and resulted in SCC of the pipe, even before it was installed in the plant.

Problems with pipes and tubes are very common because of the hoop stresses which result from the fabrication processes. Stress-relieving heat treatments are thus a vital part of the control of SCC. Experiment 3.16 shows in minutes how the residual stresses in a cold-drawn brass tube combine with the aggressive mercury (I) ion to destroy the integrity of the tube. An identical tube which has been annealed is unaffected. Such a test has frequently been used to show the presence of residual stress in metal components; it is *not* a test for susceptibility to stress-corrosion cracking, though as a quick and simple demonstration it is ideal. As will be seen later, the testing for SCC susceptibility is rather more involved.

It is now generally agreed that there is no single mechanism for stress-corrosion cracking. [9] Any combination of a number of significant factors may contribute to a given SCC failure. Rather than list a large number of specific instances, this chapter will focus on the main elements involved in general types of environment-sensitive cracking. In particular, it is a good idea to refer frequently to the **stress-corrosion spectrum** proposed by Parkins, [10] which lists the important alloy/environment combinations as well as the essential features of the mechanism involved. This important information source is

Table 10.1 The stress-corrosion spectrum (after Parkins [10])

	INTERGRANULAR CORROSION	
Corrosion dominated failure: solution requirements highly specific	Carbon steels in NO_3^- solns	Intergranular fracture along pre-existing paths
	Al–Zn–Mg alloys in Cl^- solns	
	Cu–Zn alloys in NH_3 solns	
	Fe–Cr–Ni steels in Cl^- solns	Transgranular fracture along strain-generated paths
	Mg–Al alloys in CrO_4^{2-} and Cl^- solns	
	Cu–Zn alloys in NH_3 solns	
	Ti alloys in methanol	
Stress dominated failure: solution requirements less specific	High strength steels in Cl^- solns	Mixed crack paths by adsorption, decohesion or fracture of brittle phase
	BRITTLE FRACTURE	

shown as Table 10.1. Information about specific cases will be found in the bibliography.

The main features of stress-corrosion cracking have been listed by Brown [11]:

Environment-sensitive cracking is the synergism of stress and corrosion: the absence of either eliminates the problem. The stress can be applied directly during the operational lifetime; alternatively, it can be present in the component as a result of the fabrication or installation process.

In general, it is found that alloys are more susceptible than pure metals, though there are some well-documented exceptions, notably copper.

Cracking of a particular metal is observed only for relatively few chemical species in the environment, and these need not be present in large concentrations.

In the absence of stress, the alloy is usually inert to the same species in the environment which would otherwise lead to cracking.

Even when a material is particularly ductile, stress-corrosion cracks have the appearance of a brittle fracture.

It is usually possible to determine a threshold stress below which SCC does not occur.

It is now also clear that there are certain potential ranges within which SCC is likely or unlikely, and that corrosion potential is an important consideration.

10.2 ENVIRONMENT-SENSITIVE CRACKING: INITIATION

The failure of metal components can be conveniently divided into the **initiation phase**, during which a stress raiser is formed, and a second phase of **crack propagation** leading to failure.

Question: *Stress-corrosion cracking is observed in metal/environment pairs where, in the absence of stress, corrosion would not be a serious problem. So, how does stress initiate SCC?*

In the first instance, some form of attack upon very localised anodes on the metal surface must occur, the result of which is best described as a pit. The action of tensile stresses upon materials can have many effects, some of which were described in Chapter 9. The most fundamental possibility is that the application of a tensile stress to a crystal lattice, which is otherwise in equilibrium, results in a raising of the thermodynamic energy of the atomic bonds. If this effect is localised at the surface, anodes will be formed, even though the material is being stressed within its elastic limit. Such arguments can be used only in the cases of SCC which occur when the stresses are well below the yield strength and there is no evidence of significant structural defects in the original material. As we shall see, this is unlikely.

Once the stress exceeds the yield stress of the material, plastic deformation is obtained; the crystal structure suffers bond breaking and reforming; its shape is permanently altered. The mechanism for this can be thought of simply as the creation and motion of defects, usually dislocations, through the crystal structure. The movement of dislocations will be halted when they reach either the metal surface or a grain boundary. Dislocation movement can be prevented in a number of other ways, but these are the most significant in stress-corrosion mechanisms. The pile-up of dislocations at grain boundaries results in their anodic polarisation because of the increased irregularities in the crystal structure. This has no effect upon the initiation phase if it is within the material, but is most important in the propagation stage. At the surface, a local blemish occurs on an otherwise 'smooth' surface; this is known as a **slip step** and is the site where the material is most vulnerable to initial corrosive attack.

Alloys which rely on thin films of oxide or other material for their corrosion protection are especially vulnerable because the slip step uncovers a microscopic quantity of bare metal which is highly anodic compared to the surrounding surfaces (Fig. 10.4).

If the metal is able to repassivate quickly, then little danger ensues, but if the passivation time is long enough to allow corrosion of the exposed area to occur and a pit to form, then the criteria for commencement of SCC have been fulfilled.

Even in metals which are not passivated, the formation of slip steps on the surface represents a corrosion problem, for the discontinuity of crystal structure

221

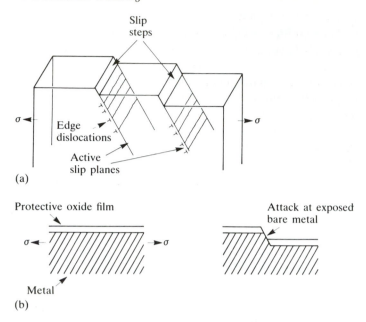

Fig. 10.4 The role of slip steps in environment-sensitive cracking. (a) The formation of slip steps on the surface of a metal by movement of dislocations along active slip planes under the action of tensile stress. (b) A slip step on the surface of a passivated metal creates an active site for the initiation of a stress-concentrating pit.

causes local anodes. Brown [11] places great importance on pit formation in which a porous cap of corrosion product creates a tiny concentration cell. [12] The aggressive species are concentrated within the confines of the cell to form an even more aggressive local environment. The diffusion process operates up a concentration gradient because of the favourable energy change which results from the dissolution process at the exposed bare metal. Much has already been said about the mechanism of pit formation in Chapter 7, all of which applies to this situation.

Considerable evidence has been accumulated that the local environment inside pits and cracks becomes very acidic, sometimes as low as pH 2, a factor which not only accelerates the dissolution processes, but adds considerably to the weight of evidence for the involvement of hydrogen in most, if not all, SCC mechanisms. As pH falls, the thermodynamic tendency for hydrogen evolution becomes greater at constant potential. But the potential is not constant. The creation of an oxygen concentration differential causes an anodic polarisation, which further increases the hydrogen evolution tendency.

The formation of a pit, then, is the usual precursor to stress-corrosion cracking, and probably represents the initiation phase. It is the most unpredictable aspect of this form of corrosion and may occur after several days or many years. Even then, a pit may not have the size and shape necessary to promote a crack in the pit bottom. However, once a pit has been produced with the right conditions, crack growth begins and the propagation phase takes over. This phase of the process is much more quantifiable using concepts developed from fracture mechanics, an aspect which will be discussed later.

Many workers place little importance upon initiation processes, and take the view that in normal engineering materials the fabrication process almost always ensures enough material defects and surface irregularities are present to provide ample initiation sites. In steels, the typical size of defect in a well-made weld is 100–500 μm, yet the calculated critical crack sizes for SCC of mild steel in caustic solution is only 1 μm; for stainless steel SCC in chloride solution, the critical crack size has been quoted as 180 μm [13, 14]. In such cases a treatment of environment-sensitive cracking may deal only with propagation and assume initiation time is effectively zero.

There are many instances of SCC not being initiated at pits, for example, mild steels in nitrates, hydroxides and bicarbonates. The SCC initiators in these cases are localised cracks at the grain boundaries. In all of these environments, the grain boundaries are attacked because of chemical heterogeneity, *even in the absence of stress.*

Acidification at the SCC initiation site does not always occur. Stress-corrosion cracks in mild steels exposed to hydroxides or bicarbonates contain environments essentially the same as those outside cracks, because those solutions are effectively buffered and the solubility of iron is very low. Acidification will only occur when hydrolysis can take place, as with chlorides and steels.

10.3 ENVIRONMENT-SENSITIVE CRACKING: PROPAGATION

Many mechanisms have been proposed for the propagation of cracks in environment-sensitive cracking, but three are generally applicable and will be described here. They are

pre-existing active paths

strain-generated active paths

adsorption-related mechanisms

No single mechanism can account for all the experimental observations of stress-corrosion cracking. As is usually the case where more than one mechanism can be proved, it is likely that each is applicable under different service conditions.

Pre-existing active paths mechanism

In this mechanism, propagation is believed to occur preferentially along active grain boundary regions. The mechanism is essentially the same as for intergranular corrosion, already described in Section 6.2. Grain boundaries may be anodically polarised for a variety of metallurgical reasons, such as the segregation or denudation of alloying elements. It is quite possible that dislocation pile-up can produce the same effect, although this may be less likely where SCC occurs at low stress levels, in which case the role of tensile stress may simply be to keep the crack open, and to allow electrolyte access to the crack tip region. The mechanism can be viewed as predominating in cases where the SCC is governed by electrochemical or metallurgical considerations, rather than the influence of stress. Systems which seem to fit well into this category are shown at the top of Table 10.1.

Evidence in favour of such a mechanism is extensive. Most alloy systems in which grain boundary precipitates are present fail by intergranular cracking. The existence of an active path in unstressed mild steel has been shown by its destruction in boiling nitrate solution when anodic currents are applied. Similar confirmatory evidence correlating metallurgical structure in the grain boundaries with cracking propensity has been produced for aluminium/copper and aluminium/magnesium alloys with appropriate heat treatments. [15]

Strain-generated active paths mechanism

In contrast to the cases of cracking which are dominated by the influence of corrosion, there are many examples, some listed at the bottom of Table 10.1, in which strain is the controlling influence. Such instances have led to the development of the strain-generated active path mechanism. One feature of SCC described in Section 10.1 is that in the absence of stress alone, the alloy is usually unreactive to the environment responsible for the cracking, normally because of the existence of a protective surface film. If crack propagation by dissolution occurs then the growth rate must be greatest at the crack tip, where anodic dissolution occurs, rather than, say, at the sides of the crack, which have been passivated because they have been exposed to the environment for a longer time. The mechanism is thus tightly linked to active/passive behaviour, which in turn has strong electrochemical connections.

The strain-generated active path mechanism is based upon the idea of a strain-induced rupture of the film, followed by metal dissolution at the rupture. The rate of propagation is governed by three criteria: the rate of film rupture, the rates of solution renewal and removal at the crack tip, and the rate of passivation.

Rate of film rupture is determined by the applied strain rate, or in the case of static loading, by the creep rate. **Solution renewal and removal rates at the crack tip** are diffusion-controlled and also governed by the accessibility of the crack tip to the aggressive species. **Passivation rate** is a vital consideration, for if repassivation is very slow, excessive metal dissolution can occur at both the crack tip and sides. The crack widens considerably and blunts, and the usual result is that the crack growth is arrested.

In poorly passivating alloys, general corrosion rather than cracking is to be expected. Conversely, very rapid repassivation leads to slow propagation rates;

it is at moderate repassivation rates that the greatest damage is caused. This concept suggests that the conditions which are most likely to result in cracking are those close to active/passive conditions, found from a potentiodynamic polarisation curve. Figure 10.5 shows two such curves, one for a weakly passivating and one for a more strongly passivating material. At potentials more negative than the free corrosion potential, both types of metals are, in the

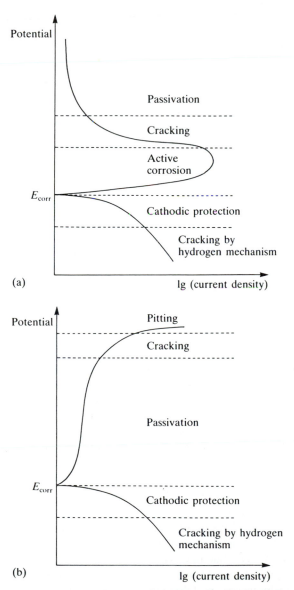

Fig. 10.5 The relationship of cracking susceptibility to the potentiodynamic polarisation curves for: (a) a weakly passivating metal; and (b) a strongly passivating metal.

first instance, not susceptible to cracking. This corresponds with the well-known method of control of SCC by cathodic protection. However, at strongly negative potentials cracking which is usually attributed to hydrogen evolution may be observed. This is particularly dangerous in high strength materials, which are frequently used in situations of high stress, and warns against overprotection by cathodic methods described in Sections 16.4 and 16.5.

Adsorption-related mechanisms

When a weakly passivating metal is at potentials more positive than E_{corr}, general corrosion is observed as far as the passivation potential. It is the region immediately more positive than the passivation potential which represents the intermediate passivating rates most likely to result in cracking. A strongly passivating metal, however, shows rather different behaviour. Passivation occurs at potentials immediately more positive than E_{corr}, and it is not until potential is sufficiently anodic for the film to be destabilised (the transpassive region usually associated with pitting corrosion) that the cracking regime occurs. There is good experimental agreement with this theory and it offers good predictive ability.

In the past it was thought that hydrogen embrittlement could be distinguished from 'pure' SCC because the hydrogen effect could only occur if the specimen was cathodically polarised. Recently, evidence has been found that in the confines of a crack, the solution composition may bear little resemblance to that existing in the bulk. Even though a bulk material may seem to be outside the potential range for the evolution of hydrogen, the combination of pH and potential existing at the crack tip may allow such a cathode reaction. The role of hydrogen in the mechanism of SCC is now considered much more important and care should be exercised in the interpretation of results.

Adsorption-related mechanisms propose that active species in the electrolyte degrade the mechanical integrity of the crack tip region, thus facilitating fracture of bonds at energies much lower than would be expected. In one mechanism, the aggressive ion which is specific for the case in question is thought to reduce the bond strength between metal atoms at the crack tip by an adsorption process which results in the formation of metal–species bonds. The energy used in binding the aggressor to the metal atoms reduces the metal–metal bond energy allowing mechanical separation to occur more easily. It is possible that the specific ion (which is normally unreactive towards the metal) is more reactive because of the increase of thermodynamic energy which occurs in the metal–metal bond as a result of the tensile stress. Figure 10.6(a) is a schematic illustration of a possible mechanism.

A second adsorption-related mechanism is based upon the formation of hydrogen atoms by the reduction of hydrogen ions within the crack. The hydrogen atoms are adsorbed by the metal and are thought to cause weakening, or embrittlement, of the metal–metal bonds just beneath the surface at the crack tip. A number of possibilities exist by which this can occur; three are shown in Fig. 10.6(b). One presupposes the formation of metal hydrides, discrete

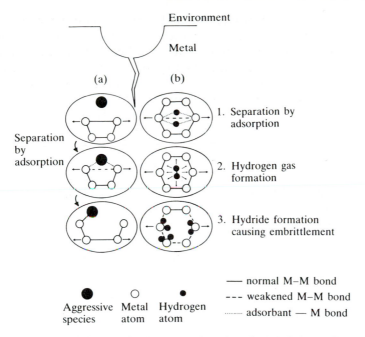

Fig. 10.6 Mechanisms for the propagation of cracks. (a) The aggressive species is adsorbed at the crack tip and causes a reduction in the metal–metal bond strength. (b) Separation (1) by adsorption of hydrogen in locally dilated areas immediately ahead of the crack tip. Formation of hydrogen gas (2) in locally dilated areas or along slip planes. The gas pressure assists in rupture of metal–metal bonds. Formation of metal hydrides (3) causes reduction in metal–metal bond strength and embrittlement of the region ahead of the crack tip.

chemical species which are well known for their brittle nature. In the past, hydride formation has probably been the most favoured mechanism for the general phenomenon of hydrogen embrittlement, though other mechanisms have been proposed. For example, it is also possible that decrease of the bond strength occurs by an adsorption process, similar to that described for the specific ion.

A third possibility is that hydrogen gas is formed in tiny amounts; the thermodynamic tendency for this to occur is very great. It has been shown that atomic hydrogen permeates steel, but combines to form hydrogen gas in cavities. The hydrogen molecule is unable to diffuse through the metal lattice and therefore the pressure in the cavities rises. The extreme pressures which can build up, given sufficient time, can rupture most materials.

Case 10.5: Shortly after start-up of a gas line owned by the El Paso Natural Gas Company in the summer of 1951, a series of breaks was caused by hydrogen penetration of the metal, blistering and failure. The problem was completely cured with the injection of inhibitor into the pipeline in April 1952. [16]

Steels are able to resist between 3000 and 20,000 atmospheres (0.3–2 GPa). The pressure of hydrogen in a defect can exceed this, but crack growth would be expected to occur before such a pressure was reached. Any increase in pressure caused by hydrogen gas in a locally dilated area would augment the existing tensile stress and assist in crack propagation. The mechanism of separation by adsorption is more likely than brittle failure caused by the hydrostatic pressures of hydrogen gas build-up. The exact role of hydrogen in the embrittlement of metals at crack tips is still the subject of speculation.

10.4 ENVIRONMENT-SENSITIVE CRACKING: PRACTICAL ASPECTS

Stress-corrosion cracking is a complex phenomenon. Testing materials for susceptibility to stress-corrosion cracking requires a considerable understanding of all the factors involved. Numerous methods are currently available and a full discussion appears elsewhere. [17] Only two important methods will be discussed because they have a bearing upon the rest of this chapter.

Mechanical tests for stress-corrosion cracking have evolved both for plain and precracked specimens. Methods developed by Parkins have involved the slow straining of smooth specimens at a constant predetermined rate. Tensile test specimens are subjected to different constant strain rates while immersed in a test electrolyte and, after failure, the percentage reduction in area of cross-section is determined. Since a specimen which fails with a small reduction in area is considered to be less ductile than one failing with a large reduction in area, the parameter is a fair measure of cracking susceptibility. A typical set of specimens might produce a graph similar to Fig. 10.7.

At low strain rates, repassivation is fast enough and straining slow enough to have little effect upon the specimens. At high strain rates, the mechanical effect predominates over the corrosion processes; the environment does not have enough contact time with the specimen and the failure occurs by mechanical means only. At intermediate strain rates, usually about 10^{-4} to 10^{-6} s^{-1}, the environmental effect is greatest.

The slow strain rate test method is useful because it always produces a fracture surface which can be examined by traditional fractography; other tests on smooth specimens which do not result in fracture are often less effective. Although the time involved for the test is usually quite short, its major shortcoming is that it is largely qualitative.

Modern quantitative test methods use fracture mechanics to quantify crack growth rates of notched specimens. In Chapter 9 it was shown how a graph of stress intensity against crack growth rate could be used to determine the material property known as fracture toughness (Fig. 9.8). If a series of crack growth measurements are carried out both in the presence and absence of an environment which gives rise to SCC then data will be obtained similar to Fig. 10.8. The data obtained from the experiments in the presence of the corrosive

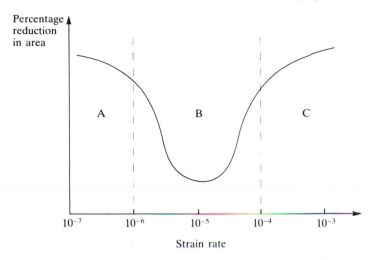

Fig. 10.7 Evaluation of susceptibility to stress-corrosion cracking (SCC) at different strain rates. Measurements of the ratio of reduction in cross-sectional area for tensile specimens statically loaded in air and an aggressive environment show a marked change in region B. This corresponds to a reduction in ductility caused by SCC at the appropriate strain rates.

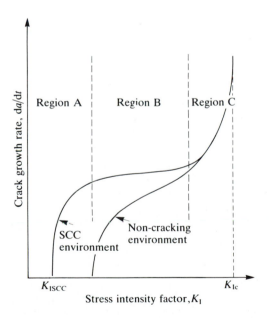

Fig. 10.8 Testing of susceptibility to SCC by crack growth rate measurements. Crack growth rates are plotted against stress intensity factor in air and corrosive environments. Region A: at low values of K_I, the rate is strongly dependent upon K_I. Extrapolation indicates a threshold level, K_{ISCC}. Region B: crack growth is almost independent of stress intensity factor and controlled by corrosion processes. Region C: crack growth occurs by a mechanical mechanism and approaches the fracture toughness, K_{Ic}.

229

environment are usually divided into three regions:

Region A: crack growth is strongly related to stress intensity but drops very rapidly, nominally to zero. Extrapolation indicates the existence of a threshold stress intensity, below which crack growth does not occur.

Region B: in this region there is little, if any, dependence upon stress intensity; crack growth occurs at almost a constant rate, faster than in the control environment.

Region C: here the mechanical duress is so great that there is little effect of the environment. Fracture is strongly dependent upon stress intensity and follows closely the behaviour of specimens tested in the control environment.

The difference in behaviour which results from the presence of the specific environment is clear from Fig. 10.8. For a susceptible material, crack growth rates show a general shift upwards and to the left of the data obtained in an inert environment. However, as the value of stress intensity factor increases, the deviation of the two curves becomes less and approaches a behaviour consistent with purely mechanical fracture.

The discovery of a threshold stress value is an important advance since it modifies the idea that stress must be completely eliminated before control of SCC can be achieved. It is sufficient to reduce the stress levels below the threshold. This significant material property (which relates only to the environment specified) is denoted by the symbol K_{1SCC}, and the threshold stress which relates to it is given the symbol, σ_{TH}.

A major advantage in the determination of a threshold stress is that it allows the calculation of the maximum flaw size which can be tolerated in the material/ environment pair. Fracture mechanics allows the calculation of a tensile strength for any given flaw size, a. Brown [11] has shown that

$$a = 0.2\left(\frac{K_{1SCC}}{\sigma_y}\right)^2 \tag{10.1}$$

Figure 10.9 shows the relationship of tensile strength and critical crack size in schematic form. On the basis of the threshold stress, σ_{TH}, a maximum design stress, σ_{max}, will be evaluated, from which a value of critical crack size is found. If the smallest defect which can be measured by non-destructive examination is a_{nde}, it is vital that $a_{max} > a_{nde}$ to ensure that routine examination will always find cracks that can lead to failure and allow remedial action to be taken before failure occurs.

Experiments on steels in sea-water have shown that K_{1SCC} decreases as yield stress increases, from which it can be deduced that SCC is much more of a problem in high strength steels.

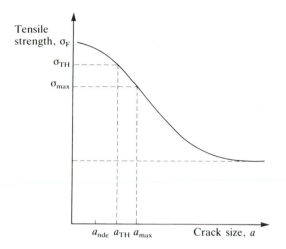

Fig. 10.9 Effect of critical crack size upon tensile strength.

10.5 CORROSION FATIGUE

There are many similarities between corrosion fatigue and SCC, but the most significant difference is that corrosion fatigue is extremely non-specific. Mechanical fatigue affects all metals, causing them to fail at stress levels well below those at which static stress leads to failure. In aqueous environments, it is frequently found that fatigue resistance is reduced. Thus corrosion fatigue is a common, and dangerous, form of corrosion. It is also a very complex discipline, so we can only present the basics.

The stages in the development of fatigue cracks are thought to be as follows:

The formation of slip bands leading to intrusions or extrusions of material

The nucleation of an embryo crack approximately 10 μm long

Extension of the embryo crack along favourable directions

Macroscopic (100 μm to 1 mm) crack propagation in a direction perpendicular to the maximum principal stress, leading to failure

Furthermore, examples of corrosion fatigue may be considered as taking place in one of three different categories:

Active: freely corroding, such as a carbon steel in sea-water

Immune: in which the metal is protected either cathodically (Chapter 16) or with a coating (Chapter 14)

Passive: in which the metal is protected by a corrosion-generated surface film, usually an oxide

The discussion which follows will refer to the conditions of corrosion and crack growth defined above.

Figure 10.8 showed that under SCC conditions, crack growth rates are enhanced at lower stress levels by a greater amount than at values approaching K_{1c}. Under corrosion fatigue conditions the tolerable stress levels leading to similar crack growth rates are lower still. Figure 10.10 shows the fatigue and corrosion fatigue characteristics of a typical low alloy steel in both inert and aqueous sodium chloride environments. In the inert environment the behaviour is as discussed in Section 9.5, but when the aqueous environment is introduced, it is apparent that the effect is greater at lower stress levels; at high stress levels the behaviour is more akin to a mechanism of crack growth by purely mechanical means. The corrosion fatigue curve can be conveniently divided into three regions — initiation, propagation and failure — as was done for the SCC crack growth curve in Fig. 10.8.

An indication of the threshold value for SCC is shown in Fig. 10.10, from which it can be judged that corrosion fatigue can occur at stress levels much lower than those for SCC. SCC crack growth rates in the propagation regime are usually independent of stress intensity factor (parallel to the x-axis). This is not so for true corrosion fatigue, where behaviour is usually in accordance with the Paris Law, eqn (9.12), except when SCC behaviour may be superimposed upon that of corrosion fatigue. In this case the graph would appear similar to the dashed line in Fig. 10.10. The onset of SCC occurs at stress levels

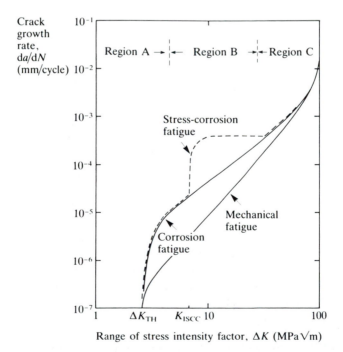

Fig. 10.10   General characteristics of corrosion fatigue curves.

corresponding to the threshold level and causes a further increase in the rate of crack growth. Observation of a plateau on the curve is typical of such a dual mechanism, sometimes called stress-corrosion fatigue, even if the effect may be attributable to a hydrogen embrittlement mechanism.

An example is found in the corrosion fatigue behaviour of structural steels used for offshore platforms. [14] The use of cathodic protection on such structures has been reported as giving favourable corrosion fatigue properties. The metal is in the immune condition because of the applied potential, but is also protected by the formation of calcareous deposits (Chapter 16) which tend to plug incipient cracks. No evidence was found for the lowering of the crack growth threshold, but superimposed upon the crack growth by corrosion fatigue was a plateau which corresponded to the onset of increased propagation rates by hydrogen embrittlement. Experience has shown that offshore platforms have good corrosion fatigue resistance, and concerns that complex stressing leads to breakup of the calcareous film, or overprotection and embrittlement due to increased hydrogen evolution, have been unfounded.

The observation of threshold levels for corrosion fatigue cannot be relied upon. In some systems, it is possible to define such a threshold for use as a design parameter. In many structural metals, it can be as low as $2 \text{ MPa m}^{1/2}$, even in air. The factor which most affects the threshold level is whether the environment causes crack initiation or not. In welded steels it is thought that sufficient defects are always present for initiation processes to have already occurred and for crack propagation to dominate the corrosion fatigue behaviour.

With smooth specimens, the behaviour is quite different because it is necessary for the environment to act conjointly with stress to carry out initiation, propagation and failure. Good evidence of this is provided by the differences observed in the behaviour of smooth and notched specimens in corrosion fatigue endurance (*S–N* curve) tests. When smooth specimens of a low alloy steel were tested in air and sodium chloride solutions there was a severe reduction in the fatigue resistance in the aqueous environment compared to the air (Fig. 10.11(a)). When notched specimens were used, the relative effect of the two environments was much reduced, with the notched specimens behaving more like the smooth specimens in salt-water (Fig. 10.11(b)). From this it was concluded that the main effect of the environment in plain specimens is to introduce stress concentrations with virtually the same effect as a machined notch. Such results highlight the fact that preservation of a good surface finish by effective corrosion protection measures is an effective way of reducing the susceptibility of materials to corrosion fatigue.

The complexity introduced into corrosion fatigue by stress cycling makes the practical investigation of corrosion fatigue behaviour even more difficult than analysis of SCC resistance. In general, it is possible to examine two contributions to the corrosion, one made by the frequency of the cycling, and the other made by the mean stress.

The effect of frequency on the crack growth rate in a structural steel in sea-water is shown in Fig. 10.12. At 10 Hz there is no significant environmental

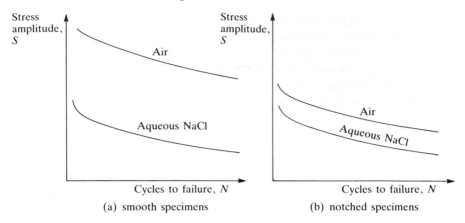

Fig. 10.11   Corrosion fatigue endurance curves for a typical low alloy steel in air and aqueous sodium chloride. (a) smooth specimens and (b) notched specimens.

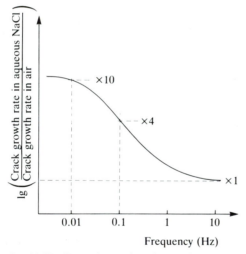

Fig. 10.12   Dependence of crack growth rate upon frequency.

effect, and the crack growth is by mechanical fatigue, rather than corrosion fatigue. For the same stress intensity, the crack growth maximises at about 0.01 Hz, at a rate 10 times greater than in air. This is easily explained because the corrosive environment is allowed a greater interaction time at low frequencies. The importance of carrying out tests at the correct frequency becomes obvious. The frequency with which a marine structure is likely to be stressed by wave action will be in the range 0.1–0.01 Hz and not 10 Hz.

For an environment to have a significant effect upon a fatigue process, sufficient time must be given for corrosion processes to occur. But there are dangers, which become apparent from the following example. A simple fatigue experiment at 25 Hz which runs to a million cycles (a normal number for such experiments) can be completed in 11 hours; at 0.01 Hz the same experiment takes over three years. The use of accelerated testing and extrapolation of laboratory data obtained over short time periods to the tens of years for most design lifetimes requires the utmost caution.

The mean stress is a most important variable because it can take different values while ΔK remains constant (see Fig. 9.9). As has been shown, tensile mean stresses are detrimental to corrosion fatigue resistance if the frequency is in the range of maximum effect. As the mean stress is increased, for the same ΔK (i.e. R increases to more positive values), the crack growth rate also increases. Compressive mean stresses have been shown to be beneficial in carbon steels. Corrosion fatigue resistance was substantially increased in both air and hydrochloric acid by using a compressive mean stress at low frequency. This lends support to the previously well-known beneficial effect of shot peening, a process applied to the surface of metal components which has been shown to result in compressive stresses in a region of the material 50–75 μm below the surface. Other methods of introducing such static compressive stresses into structures or components will probably be an important means of protecting against corrosion fatigue. But in an environment which may cause pitting, this will not help.

It should not be assumed that the traditional *S-N* type testing has been made obsolete by the introduction of more modern fracture mechanics techniques. Corrosion fatigue endurance testing continues to play an important role in lifetime determinations. This is because there are still many occasions when methods of fracture mechanics are not sufficiently accurate to describe the behaviour and a combination of crack growth rate data and endurance testing data provides valuable information. Assuming crack growth rates in accordance with the Paris Law,

$$\frac{da}{dN} = C\Delta K^m \tag{9.11}$$

where m and C are found experimentally, fatigue endurance limits can be calculated for different initial and final crack lengths, a_i and a_f respectively, based upon a typical low value threshold stress intensity for crack propagation of 2 MPa m$^{1/2}$. Using

$$N = \frac{1}{C}\int_{a_i}^{a_f}\frac{da}{\Delta K^m} \tag{10.2}$$

where

$$\Delta K = \Delta\sigma(\pi a)^{1/2}f(a/w) \tag{10.3}$$

endurance values are calculated and plotted on the practical *S-N* curve to give a lower bound for corrosion fatigue behaviour (Fig. 10.13). By such analysis it is possible to estimate the errors which may be present in the fracture mechanical approach because of the adoption of an inaccurate compliance function, $f(a/w)$

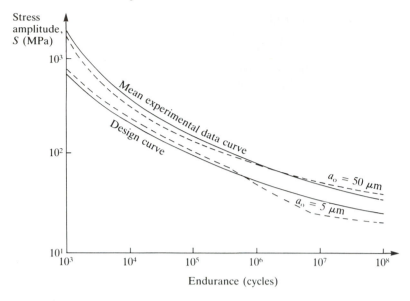

Fig. 10.13   Comparison of integrated crack growth curves with corrosion fatigue endurance curves.

in eqn (10.3), or a failure to describe the growth of short cracks. A design curve which incorporates a safety factor can be defined for a particular case. [18] For example, for a nuclear pressure vessel the safety factors used are 2 for stress and 20 for cyclic life, below the mean of the experimental data.

Table 10.2 lists ranges of typical corrosion fatigue endurance limits reported by Speidel [19] for a number of groups of alloys in salt or sea-water at 10^8 cycles and $R = -1$. By comparison with the discussions in Chapter 15 of alloy performance in saline environments it should be noted that the increased fatigue strength of an alloy can only be exploited if it is first resistant to corrosion by the environment, but the alloy should not be of such high strength that it is susceptible to hydrogen embrittlement. Thus, for steam turbine blades of 12% chromium ferritic stainless steel, the application works well in good quality steam environments, but when the condensate is more aggressive, the corrosion resistance of the steel is poor, and it must be substituted with an alloy of higher performance. In fact, Ti–6Al–4V is used under such conditions.

Despite the advances made in the understanding of corrosion fatigue, it remains one of the most complex and least understood of all forms of corrosion. Some people believe it is becoming critically important to undertake more research into corrosion fatigue in the oil, gas, nuclear, geothermal and wind power industries, otherwise vital resources may be wasted.

Figure 10.14 is a useful schematic diagram which summarises the effect of stress upon crack size. While it should be remembered that there may be considerable overlap of the corrosion fatigue and SCC regimes, the diagram is a good representation of the effects of environment-sensitive cracking.

Table 10.2 Corrosion fatigue endurance limits for alloys in aerated salt solutions of seawater, 10^8 cycles, $R = -1$ (after Speidel [19])

Alloy	Typical corrosion fatigue strength/MN m^{-2}
Cobalt alloys Titanium alloys	475–510
Nickel-base superalloys	360–400
Duplex stainless steels Ferritic stainless steels with >25% Cr	250–320
Austenitic nickel base alloys (600, 800)	170–200
Austenitic stainless steels Martensitic stainless steels Copper–nickel alloys	70–130
Carbon steels Low alloy steels	30–50
Tin, zinc, aluminium and lead alloys	15–30
Magnesium alloys	<15

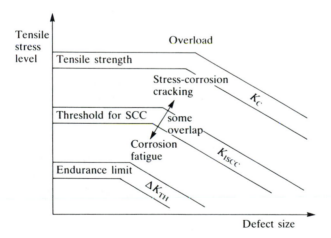

Fig. 10.14 Schematic diagram relating mechanical fracture, stress-corrosion cracking and corrosion fatigue with increasing defect size.

10.6 REFERENCES

1. Moore H, Beckinsale S and Mallinson C E 1921 The season cracking of brass and other copper alloys, *Journal of the Institute of Metals* **25**(1): 35–152.

2. Cowley J 1986 Engineering and safety — a marine engineer's viewpoint, *Transactions of the Institute of Marine Engineers*, vol. 99 paper 6.
3. Parr S W and Straub F J 1927 The cause and prevention of the embrittlement of boiler plate, *The Engineer* **117**: 496.
4. Spahn H and Wagner G H 1976 Corrosion fatigue of steels. In *Bruchuntersuchungen und Schadenklarung, Probleme bei Eisenwerkstoffen*, Allianz-Versicherungs, Munich, Germany, pp. 59–74 (in English).
5. Staehle R W 1993 The corrosion research symposium — 25 years and beyond. Informal presentation at *Corrosion 93, New Orleans LA*, NACE International, Houston TX.
6. MacDonald D D, Begley J A, Bockris J O, Kruger J, Mansfeld F B, Rhodes P R and Staehle R M 1981 Aqueous corrosion problems in energy systems, *Materials Science and Engineering* **50**: 19–42.
7. Collins J A and Monack M L 1973 *Materials Protection* **May**: 22.
8. McReynolds R F and Vennett R M 1975 *Materials Performance* **14**(1): 23.
9. Ford F P 1984 Current understanding of the mechanisms of stress corrosion cracking and corrosion fatigue. In *Environment-sensitive fracture*, edited by S W Dean, E N Pugh and G M Ugiansky, ASTM, Philadelphia PA, pp. 32–51.
10. Parkins R N 1972 *British Corrosion Journal* **7**: 15.
11. Brown B F 1972 A preface to the problem of stress corrosion cracking. In *Stress corrosion cracking of metals — a state of the art*, edited by H L Craig, ASTM, Philadelphia PA, pp. 3–15.
12. Burstein G T, Pistorius P C and Mattin S P 1993 The nucleation and growth of corrosion pits on stainless steel, *Corrosion Science* **35**: 57–62.
13. West J M 1980 *Basic corrosion and oxidation*, Ellis Horwood, Chichester, UK, p. 158.
14. Scott P M 1983 Design and inspection related applications of corrosion fatigue data, *Mémoires et Etudes Scientifiques Revue de Metallurgie*, Novembre 1983: 651–60.
15. Friedrich H, Killian H, Knornschild G and Kaesche H 1994 Mechanism of stress corrosion cracking and corrosion fatigue of precipitation hardening aluminium alloys. In *Modelling aqueous corrosion*, edited by K R Trethewey and P R Roberge, Kluwer Academic, Dordrecht, The Netherlands, pp. 239–60.
16. Fisher L E 1993 Corrosion inhibitors and neutralizers, past, present and future. Paper 537 in *Corrosion 93 plenary and keynote lectures*, edited by R D Gundry, NACE International, Houston TX.
17. Parkins R N 1994 Stress-corrosion test methods. In *Corrosion*, edited by L L Shreir, R A Jarman and G T Burstein, Butterworth-Heinemann, Oxford, pp. 1:215–42.
18. Scott P M, Tomkins B and Foreman A J E 1983 Development of engineering codes of practice for corrosion fatigue, *Journal of Pressure Vessel Technology* **105**: 155–162.
19. Speidel M O 1981 Influence of environment on fracture. In *Proceedings of 5th international conference on fracture, Cannes, France*, Pergamon Press, Oxford, ICF5 vol. 6, 2685–704.

10.7 BIBLIOGRAPHY

Brown B F 1977 *Stress corrosion cracking control measures*, NACE International, Houston TX.

Dean S W, Pugh E N and Ugiansky G M 1984 *Environment-sensitive fracture*, ASTM, Philadelphia PA.

Devereux O F, McEvily A J and Staehle R W 1972 *Corrosion fatigue*, NÁCE International, Houston TX.

Gangloff R P and Ives M B 1990 *Environment-induced cracking of metals*, NACE International, Houston TX.

Gibala R and Hehemann R F 1984 *Hydrogen embrittlement and stress corrosion cracking*, ASM International, Ohio.

Lyle F 1988 *Environmentally induced cracking: the interaction between mechanisms and design*, NACE International, Houston TX.

McEvily A J 1990 *Atlas of stress corrosion and corrosion fatigue curves*, ASM International, Ohio.

Ouchi H and Kobayashi J 1993 *Stress corrosion cracking characteristics of titanium alloys and a high strength low alloy titanium in seawater.* Paper 286 in *Corrosion 93, New Orleans LA*, NACE International, Houston, TX.

Scott P M 1994 'Corrosion Fatigue'. In *Corrosion*, edited by L L Shreir, R A Jarman and G T Burstein, Butterworth-Heinemann, Oxford, pp. 8:143–83.

Sedriks A J 1990 *Stress corrosion cracking test methods*, NACE International, Houston TX.

Staehle R W, Hochmann J, McCright R D and Slater J E 1977 *Stress corrosion cracking and hydrogen embrittlement of iron base alloys*, NACE International, Houston TX.

11 CORROSION MANAGEMENT

Treppenwitz — loose translation: those things or statements which we wished we had done or said beforehand, especially if something later went wrong. (Oliver W Seibert, *Materials Performance*, April 1978)

It is easy to be wise after the event. Engineers seem prepared to spend long hours on stress calculations and styling at the expense of materials selection or design features to reduce corrosion. Later they are amazed, and even offended, by the corrosion problems generated amidst the bad design features and arrays of incompatible materials they have specified.

The Thames Barrier is a large engineering project, opened in 1984, to prevent large areas of London suffering flooding during a tidal surge. The width of the barrier from bank to bank is 520 m and it consists of nine concrete piers with gates between them. Each of the four main gates measures 61 m across and weighs 3,700 tonnes. The structure uses large amounts of stainless steel. The late Lionel Shrier, one of our great corrosion scientists, once lamented that he had been asked for advice on corrosion control features of the flood barrier only *after* its design had been completed.

In Chapter 1 we reported that the cost of corrosion to nations is very great. One conclusion of the Hoar Report [1] was that most corrosion failures were avoidable and that improved education was a good way of tackling corrosion avoidance. More than 20 years later, the number of types of corrosion has not increased, the understanding of corrosion mechanisms is far greater than ever, and there has been an enormous increase in the amount of available corrosion information, but it is generally agreed among corrosion engineers that, in society at large, the numbers and severity of corrosion problems are not diminishing. Are engineers to blame?

Case 11.1: Perhaps we may gain a hint from a story about a corrosion engineer in Canada who was employed to solve thousands of annual faults in underground gas pipelines. He diligently went about his duties, year by year reducing the number of failures until there were virtually none, at which point his employers decided that there was no longer any need for his services.

This is a true story but it appears in many forms as a corrosion conference joke. If we are too good at our jobs we will be unemployed. Management often fails to understand the complexity of corrosion control, unaware it involves far more than merely correcting basic problems in specific areas of plant.

Corrosion control refers to the general process of remediation and has often been used interchangeably with corrosion management. We believe that corrosion control is essentially focused on methods in materials and environments. But effective corrosion control actually requires a complete management strategy involving people as much as equipment. It is the former approach, with its limited scope, which is largely the reason for the continued high cost of corrosion failures.

In this book we promote the term **corrosion management** to include people in the widest sense. We propose that corrosion management should be the term for a systems engineering strategy to improve the performance of engineering systems by specifically including people. Thus, while the adoption of higher performance materials and better understanding of the fundamental problems is beneficial, improved corrosion performance must be based on better management and design methodologies which account for the complex interactions between materials, environments and people.

Figure 11.1 is a Venn diagram which presents a precise definition and distinction between corrosion control and management and, above all, provides a stronger focus for improved performance of engineering systems by reducing premature failure and waste of resources. There are growing signs that if the present levels of economic growth are unsustainable into the twenty-first century, new laws will require ever more economic and efficient use of materials and energy. An important future strategy for improving the situation will replace error-prone human decisions with those from more reliable computers. In this chapter, after describing the accepted principles of corrosion control, we shall broaden the discussion to include people in corrosion management and we shall adopt a positive forward-looking stance which will involve some new ideas.

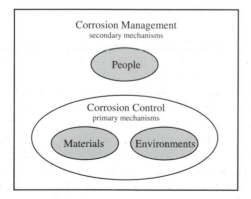

Fig. 11.1   Venn diagram illustrating the relationship between corrosion management, corrosion control, materials, environments and people.

11.1 CORROSION CONTROL

Corrosion control is a process in which humans are very much in control of materials and environments and, by correct application of sound principles, can regulate the rate of corrosion, keeping it within acceptable, or at least predictable, limits for the life of the structure. Corrosion control may be exercised in a number of ways which can be generalised under the following headings:

Modification to the design
Modification to the environment
Application of barrier coats
Selection of materials
Electrical methods

Each of the chapters which follows concentrates on one of these topics, but design is always of overriding importance. The control method should always be considered as part of the overall design concept for the system, taking its place alongside other design parameters such as stress calculations, creep and fatigue, and fabrication techniques. But corrosion control does not stop with design. Its effectiveness must be monitored throughout the life of the system as it performs its design functions, including periods spent in storage or transit, as well as in service.

The possible types of corrosion attack and the consequent rate of metal loss are strongly dependent upon the environment. Small changes to pH, temperature, dissolved oxygen levels, flow rates, pollutants and other factors can radically alter the nature and severity of corrosion. In selecting a control method, particular attention must be paid to possible changes in the environment during the life of the structure, or brief periods of exposure to particularly hostile environments which may occur during fabrication, erection, operation or maintenance.

Damage may only become apparent months or years after corrosion is initiated. In Chapter 7 we saw the example of pitting corrosion in the walls of copper tubes in a domestic hot water system. Prior to dispatch from the factory the tubes had been annealed. Lubricating fluid residues in the bores turned to carbon, which caused pitting at breaks in the cathodic film. In a world increasingly dependent on lean performance, it is not good enough for a designer or maintainer to settle for anything less than maximum system lifetime.

Having introduced a system of corrosion control, rigorous checks must be applied during manufacture, storage and in-service use to ensure that subsequent modifications do not upset the design philosophy. Sometimes an apparently desirable modification to cure persistent corrosion failure can drive the corrosion to a more inaccessible part of the system, difficult to maintain.

Case 11.2: A fitter, tired of replacing the mild steel shaft in a sea-water pump every two years, installed a shaft of stainless steel. Within six months, the shaft failed due to crevice corrosion beneath the impeller. The remainder of the pump was badly pitted having lost the sacrificial protection afforded by the mild steel.

A corrosion protection system may be superfluous, depending upon the nature of the task to be undertaken by the equipment. There is little point in providing an externally applied corrosion control system on the blades of earth-moving machines where abrasion damage will be much faster than corrosion losses. But earth-movers still require protection during storage, rest periods and transit. Such protection is often neglected, particularly in remote geographical regions or under adverse weather conditions.

The traditional justification for the type of corrosion control system balances simple cost against the required life of the structure, the capital and maintenance costs and the scrap value of the materials. But the true cost of failure is much greater than the replacement cost of the damaged parts. Public loss of confidence in a motor manufacturer can be catastrophic once it gains a reputation for poor corrosion resistance. The incidental cost of failure can far outweigh the direct cost.

Case 11.3: When corrosion caused the malfunction of a signal unit at Clapham Junction on the railway system serving London, a repair was effected at a very small direct cost, but the disruption to the railway commuter service was enormous and entailed thousands of lost working hours.

When choosing a corrosion control system, engineers will still aim for maximum cost-effectiveness, but will be increasingly influenced by other factors. It will be necessary to avoid contamination of the environment by corroding metals, and there will be the overriding need for conservation of resources and reliability of performance in safety equipment or weapons systems. Corrosion control will increasingly be considered as part of the wider discipline of corrosion management in which a greater appreciation of the need to reduce human errors and the introduction of more effective machine intelligence will provide the improved efficiency demanded by future generations. These new topics will now be discussed.

11.2 PRIMARY AND SECONDARY CORROSION MECHANISMS

In the first half of this book we have identified and provided **primary mechanisms** for the major forms of corrosion. In engineering situations, all these forms of corrosion result in premature failure of the component or system. In

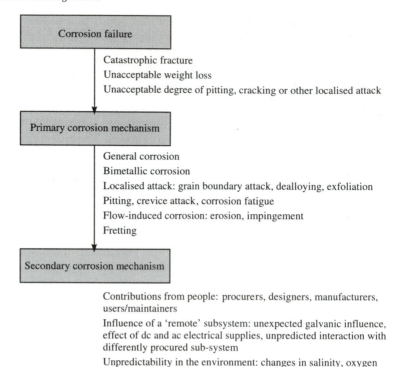

Fig. 11.2 Primary and secondary corrosion mechanisms.

principle, corrosion control focuses only upon the remedies to these clearly identified problems. In reality, the corrosion failure is most often caused by another agency, a **secondary mechanism** (Fig. 11.2). These secondary mechanisms are often overlooked or poorly understood, and are very largely responsible for society's low success at combating corrosion. Tackling the secondary mechanisms involves corrosion management.

Case 11.4: A group of marine biologists carrying out experiments on mussels suspended a series of mussel cages from lightships by means of steel ropes. After placing a dozen cages in their marine locations, the scientists left them for many weeks, only to return and find the ropes dangling freely. The cages had been lost to the bottom of the sea. Part of the fixing to the cage involved making a loop in the rope by passing it around an eye-former and crimping the end of the rope to itself. Unfortunately, the selected crimping material was aluminium.

As we learned in Chapter 5, aluminium forms an active galvanic couple with steel and quickly corrodes in such circumstances. Once sufficient material had dissolved, the crimp was useless for supporting the weight, the loop of rope around the eye came undone and the cage was lost. The scientists' programme was ruined because of a stupid choice of material.

In this very simple example the *primary* corrosion mechanism was bimetallic corrosion with a well-understood mechanism. At the beginning of Chapter 1 we reported that these instances of corrosion have been understood for two centuries, yet they continue to occur with remarkable frequency, contributing to the continued cost of corrosion to society. But in Case 11.4, there was an influence from a human source, i.e. a designer made an incorrect material selection. Thus **three** elements were present in the failure: the **materials**, the **environment** in which they operated and the **people** who interacted with the system. The corrosion *control* principle is not to use the galvanically unsuitable aluminium in this application. The corrosion *management* principle is to ensure that the designer does not make a mistake. This distinction is arrived at by taking a **systems engineering** approach. The definition of the system has been broadened and explicitly *includes* people.

People are inextricably linked to engineering, as Henry Petroski has so beautifully described in his book, *To Engineer is Human*. For our purposes, we shall think of four kinds of people who affect the lifetime of an engineering system: the procurer, the designer, the manufacturer and the user or maintainer. Figure 11.3 summarises the main contributions which each kind of person makes to the success or premature failure of a system. The roles of people are very difficult to define precisely so we have to simplify them somewhat. Sometimes a supplier can be identified as the procurer, obtaining an engineering system because it seems adequate for a certain application, but sometimes coming unstuck.

Case 11.5: A stainless steel (type 316) cage constructed to house some undersea equipment was found to have suffered severe attack after only months of immersion. The cage was towed at the end of a steel rope behind a ship, but the equipment in the cage was self-contained, i.e. not operated from the ship. No obvious galvanically incompatible materials were present.

This case is an example of failure diagnosis viewed with our broadened systems engineering approach and brings into the discussion the role of experts. Figure 11.4 summarises the case and illustrates the differences between an expert and a non-expert approach. The non-expert approach is essentially material-focused and involves an extensive period of experiments and data-gathering, followed by review and assessment. It is thus extremely intensive of time and money, relying upon the traditional scientific investigative approach adopted in the absence of experience or other knowledge.

The expert approach, on the other hand, is system-focused. The expert defines the system as cage + contents + steel rope and the environment as sea-water. The expert is tempted to conclude, as did the suppliers, that the material can, in isolation, suffer severe attack in a very short time, but knows that there are very few materials which could have performed so badly. At this stage, two

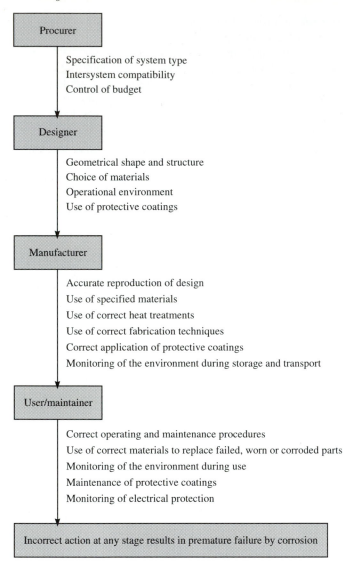

Fig. 11.3   People factors: how people can influence the corrosion performance of systems and contribute to the success of the system or its premature failure.

possibilities exist: the corrosion is caused by a serious galvanic effect with other (usually large) cathodes, or the corrosion is electrolytic action caused by stray currents. Once the expert discovers the material is 316 stainless steel, experience tells that no stainless steel can suffer so badly in such a short time by galvanic action alone. The expert concludes that corrosion was by the second mechanism, although the source of the stray current has not been identified.

Now there are two new possibilities: the stray current is from the power supply of the cage contents, or the stray current is from shipboard power

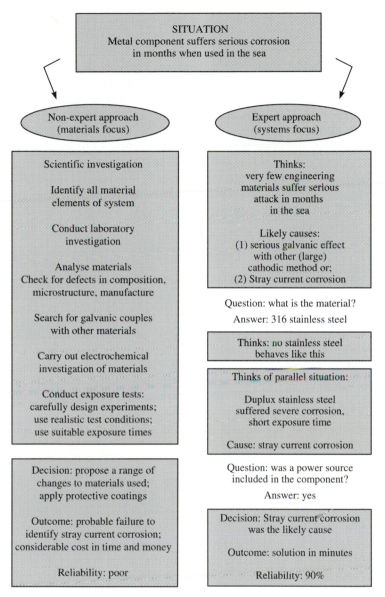

Fig. 11.4 Corrosion failure diagnosis, a comparison of the expert and non-expert approaches (Case 11.5).

supplies leaking down the wire rope. If the first case were correct, the designer should have been clearly aware of the damage which can be caused when as little as 1 A passes through metals immersed in sea-water and should have ensured that the cage was totally isolated from the contents. If the second case were correct, although the designer would have performed his job correctly, a human would have failed to consider that ships frequently leak current from their onboard electrical systems. The supplier should have installed the system so as to be insulated from stray current from the ship. In either case, the situation is clear: a secondary mechanism is involved, the system has been incorrectly specified and the possibility of current leakage has been overlooked by a human. In summary, the expert has provided a very reliable assessment of the cause, in a matter of minutes, whereas the non-expert, focusing on the materials and the environment, may well fail to identify a stray current mechanism and there is a high chance that the wrong decision will be made. Stray currents are discussed in more depth in Section 12.6, see also Case 12.23.

In this example, there was a real chance that the *primary* mechanism would not be diagnosed, not to mention the secondary mechanism. Experts sometimes make mistakes and are not always available. Thus, it is possible that future corrosion failures will be diagnosed by computers. This topic will be addressed again in Section 11.4.

11.3   THE PEOPLE FACTOR

Engineers sometimes use the methodology of **fault-tree analysis** (FTA) to quantify elements of failure in systems. The so-called **top event** is a corrosion failure and all events which occur during the life of the system are carefully itemised, described and given a numerical factor which enables a computer to calculate the expected system lifetime. How do we begin to do this for corrosion?

Mathematically we can express the probability of system failure as

$$p_{sf} = p_m \times p_e \times \text{factor}_p \tag{11.1}$$

where p_{sf} is the probability of system failure, p_m is the probability factor due to materials, p_e is the probability factor due to the environment and factor$_p$ is the contribution to failure from people. Probabilities take values between 0 and 1. An event that is certain to happen has a probability of 1, whereas an event that will never happen has a probability of 0. When factor$_p$ > 1, it is said to be an **aggravating factor**, enhancing the risk of failure; when 0 < factor$_p$ < 1, it is said to be an **inhibiting factor**, reducing the risk of failure.

The influence of people in a corrosion (or any other) failure is extremely difficult to predict, as it is subject to the vagaries of human decision-making. Nevertheless, the impact of engineering upon our society has been remarkable. Most well-designed engineering systems perform according to specification, largely because the interaction of people has been tightly specified and is properly managed throughout the whole life of the system. A particularly notable example is the extremely high safety record of the aerospace industry, which has adopted the most stringent engineering protocols for management, design, manufacture and maintenance. It has reached extremely high levels of safety and reliability, and those tragic accidents which rarely occur these days

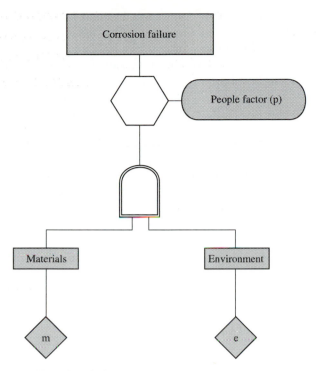

Fig. 11.5 Generic fault tree representation of a corrosion failure.

are usually a result of freak combinations of circumstances, rarely caused by engineering or corrosion failure.

Mechanical engineers work to minimise mechanical failures in which neither people nor the environment are assumed to play a part. Their task is comparatively straightforward; mostly they optimise the performance of the materials alone. Corrosion engineers have long realised that their jobs are rather more complicated because of the complex ways in which environments can interact with the materials and result in corrosion failure. The fatigue which causes concern to the mechanical engineer becomes corrosion fatigue once the environment is implicated (see Chapters 9 and 10). But, although humans have been clearly involved in many case histories, rarely have the roles of humans been implicitly incorporated into the failure process. This is formalised in Fig. 11.5, which illustrates a generic corrosion failure using fault tree analysis. The meaning of the symbols is defined in accordance with Boolean logic, so this kind of structure can be programmed into a computer and used to analyse risk of system failure. Very large plants have been successfully managed in this way.

Case 11.6: The issue of pipeline reliability and safety of an ageing pipeline system is receiving increased public and regulatory attention. Record gas volumes on pipeline systems have resulted in operations close to design capacity all year round, increasing the management emphasis on reliability and risk issues. In Canada, the Nova Oil Corporation has used such a methodology very successfully to analyse pipeline failures (outages) caused by stress corrosion cracking across thousands of kilometres of pipeline. [2]

A cursory glance at the literature reveals extensive work on 'people factors', but it seems that most of the work deals with the effect of engineering systems on people rather than the influence of people on engineering, as expressed by factor$_p$. This is undoubtedly due to the difficulty of quantifying the effect, combined with the perception of people as observers and outside systems, rather than integral parts of them.

The latest forms of engineering, which *do* try to account for people-based decision-making, are based upon such disciplines as fuzzy logic, [3, 4], Bayesian logic, [5] influence diagrams [6, 7] and approximate reasoning principles [8] as part of research into artificial intelligence. [9] There is much research to be carried out in this area, as well as in the development of new machine-intelligent management systems.

11.4 KNOWLEDGE-BASED SYSTEMS

Many of the case histories quoted in this book suggest the possibility that, to overcome the large human contribution to corrosion failure, machine intelligence might be used to provide more reliable information or decisions. The rapid development of accessible computing power in the 1980s has led to the use of computers in every sphere of engineering. Knowledge-based systems (KBSs), also known as expert systems (ESs), have been developed in a number of areas involving corrosion control, but have experienced difficulties which have retarded their development. [10] The extraordinary power of hardware has not been matched by the power of software, particularly due to incompatibility between rule- or logic-based software tools and human intelligence. Transfer of knowledge from the expert brain to the computer, a process known as **knowledge elicitation**, has rarely proved straightforward; experts able to perform complex tasks seem unable to understand or communicate how they do them. Many KBS projects have failed through unforeseen problems of this kind. Not surprisingly, decisions made by computer have frequently been met with scepticism from humans, yet the track record of humans is far from good. Analysis of incidents involving the transport of dangerous substances in Alberta [11] showed that over 40% were directly attributable to human error, rather than mechanical or corrosion failure. Similarly, Meister [12] collected a large body of data relating to failures in a wide range of military systems. He also reported that, on average, 40% of reported failures were the result of human error. This recognition that people are ultimately responsible for systems failure again highlights the need to include secondary rather than primary mechanisms in our corrosion management strategies.

There is growing evidence that the implementation of well-designed KBSs can reduce the often erratic influence of humans, [13] although up to now they have been extremely system-specific and contribute in comparatively constrained design situations. The machine simulation of human knowledge is beyond the scope of this text, but, in short, is assigned one of three levels:

Skill-based knowledge and action
This reacts to simple signals according to preprogrammed instructions. Algorithmic in nature, this lends itself particularly well to automation of processes; human operators can be replaced by microprocessor-controlled devices and other human thought processes can be replaced by present computer software.

Rule-based knowledge and action
Perceived information activates human response based upon rules; performance is goal-driven. At this level, a human might be replaced by an ES or KBS. However, unfamiliar situations cannot be handled because the rules apply only to predefined situations.

Formal reason-based knowledge and action
This is triggered by sensed information that indicates an unfamiliar situation. Human experts often invoke heuristic and other high-level intelligent responses in such situations. At present, there are few machine-based intelligences that can cope. Research is heavily focused in this area and artificial neural networks have been successful when input and output can be represented numerically.

So far, the successful application of machine intelligence has largely been confined to the lower levels of human reasoning. KBSs do have a significant role to play at all four levels of Fig. 11.3, and new systems that apply the above techniques have a much better chance of success.

KBSs for Procurement

Most engineering subsystems result from the efforts of a team of design engineers. The team works together on the overall subsystem but each member is responsible for one aspect. Within that hierarchical layer of function, there may be a number of subsystems, each of which fits into a larger system under the control of another team of design engineers. Sometimes the various elements of the mother system are not created by engineers working for the same company. Significant management problems are created when the various subsystems are integrated.

In the UK Ministry of Defence the process of acquiring new ships and submarines is managed by the Procurement Executive, a management organisation comprised of engineers and naval architects who oversee all aspects of the task. The efficient design and construction of a large system such as a ship or submarine is particularly unusual in engineering because the primary function of the system is efficient operation in a strongly conducting medium — sea-water. This introduces electrical coupling of many geographically remote components, with all the associated corrosion risks. Thus, the responsibilities of the procurer of the ship or submarine weigh far more heavily in the avoidance of premature failure than they might in other branches of engineering. Indeed, it could be argued that in the light of the continued series of problems experienced by naval systems, the procurer is often more to blame than the designers of the subsystems. [14] In the United Kingdom, the downsizing of the Ministry of Defence procurement infrastructure, coupled with the trend towards independent design of subsystem components by contractors, is leading to a significantly increased risk of corrosion failure because a shortage of sufficiently expert personnel will degrade the successful integration of subsystems into the whole.

It is the constraints imposed by the procurer in terms of system specification and budget control, which, when coupled with a lack of appreciation of intersystem compatibility, combine to create premature failure or reduced operating efficiency, as much as any error in the subsystem design or inadequate operating procedure. The value attributed to $factor_p$ is likely to be just as high when applied to a procurer as to an operator. The introduction of reliable KBSs is a vital prerequisite to the more effective management of materials performance in these systems, and knowledge elicitation for procurement purposes is a crucial component.

KBSs for Design

There are many ways in which knowledge structuring for design KBSs can be formulated and the disciplines involved are extremely broad. Examples include

Chemistry: selection and preparation of coatings and inhibitors

Electrochemistry: principles of cathodic and anodic processes and protection

Physics: surface treatments and hot corrosion

Statistics: lifetime predictions

Mechanical engineering: structural integrity, stress and thermal analysis, hydrodynamics and fluid flow

Metallurgy: materials selection, heat treatment and welding

Management: long-term cost and risk evaluations

One particular example is in the design of multimetallic systems common in marine systems. As we saw in Section 5.2, materials selection in condensers must take into account many factors, such as the need for sufficient mechanical strength, thermal efficiency and corrosion resistance. The relevant mechanisms that may be present include the effects associated with flow, ranging from localised attack under stagnation deposits of sand, silt and fouling to erosion corrosion at high flow rates. Ideally, careful design and manufacture should minimise the risks of crevice corrosion, but due to both design and economic constraints, galvanic effects between the tube bundle, tube-plate, header and associated pipework often seem unavoidable.

In order to avoid such problems, the designer may use design [15] and material protocols. Specific military protocols are often adopted in naval systems to cope with perceived greater performance objectives in military as opposed to civilian engineering systems. Physical scale-modelling techniques may be used to model current in multimetallic systems. Dimension And Conductivity Scaling (DACS) [16] is used to design cathodic protection systems for the US Navy. Three-dimensional problems of complex geometry usually require analysis by finite difference (FD), finite element (FE) or boundary element (BE) methods. Until recently the large computational requirement has limited their inclusion in manageable KBSs, but this may well be alleviated by developments in computer hardware. A somewhat simpler method [17] predicts potential distributions by microcomputer, and is ideal for incorporation into present KBSs. It can be used for systems of up to about 800 mm diameter. It deals with various configurations of pipes and condensers and allows easy alteration of variables such as the number, length and radius of tubes within the bundle, the header radii, the header length, and the solution conductivity. With so many contributions to the design process needed from so many disciplines, it is inevitable that the human factor will be aggravating. At present, a KBS that could deal satisfactorily with just one of these design decisions would be a most welcome development.

KBSs for manufacturers and operators

KBSs may be used at all stages of the manufacturing process such as interfacing with complex materials property databases or robots for computer-assisted manufacture; indeed it is at the lower levels that computers *are* already being used to good effect. As we have seen, KBSs may be used on a day-to-day basis for many different aspects of operation and maintenance scheduling. But the most effective KBSs will have been rendered invisible, subsumed within automated systems for such tasks as providing much more effective shipboard impressed current cathodic protection under a variety of operating conditions. This invisiblity helps to overcome people's distrust of machine intelligence and is a goal for all knowledge-based systems. For until they gain people's trust their workplace acceptance will be limited.

11.5 CONCLUDING REMARKS

A thesis has been presented that many engineering systems continue to suffer lengthy downtime because of life-limiting corrosion failures, despite increased education and great improvements in materials and the understanding of failure mechanisms. The aggravating influence of humans was identified as the major factor, representing a complex management problem, continued risk of premature failure and a serious drain of valuable resources in the coming decades of tight budget control.

Principles of corrosion management have been defined in which a broad-based systems engineering approach involving all aspects of human factors is necessary to reduce the risk of corrosion failure. An integral part of that is the development of new reliable knowledge-based systems, which have much to offer in reducing the erratic and aggravating influence of humans. New KBSs must be constructed as tools for all those involved with the creation of efficient engineering systems; managers in particular have much to gain. Other KBSs can reduce the aggravating human influence by replacing the unreliable and error-prone human once they have attained levels of reliability necessary for invisible incorporation into the systems themselves.

The five chapters which follow describe five different elements of corrosion control principles: design, control of the environment, barrier coatings, material properties for correct selection and electrical techniques. However, Fig. 11.1 should be borne in mind at all times, for ultimately it is people and their corrosion management principles that will determine the success or failure of engineered systems.

11.6 REFERENCES

1. Hoar T P 1971 *Report of the committee on corrosion and protection*, HMSO, London.
2. Roberge P 1994 Eliciting corrosion knowledge through the fault-tree eyeglass. In *Modelling aqueous corrosion*, edited by K R Trethewey and P R Roberge, Kluwer Academic, Dordrecht, The Netherlands, pp. 399–416.
3. Kosko B 1992 *Neural networks and fuzzy systems*, Prentice Hall, Englewood Cliffs NJ.
4. Durkin J 1994 *Fuzzy logic*, MacMillan, New York, pp. 363–403.
5. Durkin J 1994 *Bayesian approach to inexact reasoning*, MacMillan, New York, pp. 305–31.
6. Howard R A and Matheson J E 1989 Influence diagrams. In *The principles and applications of decision analysis*, edited by R A Howard and J E Matheson, Strategic Decisions Group, Menlo Park CA, vol. II, pp. 720–62.
7. Shachter R D 1988 Probabilistic influence and influence diagrams, *Operations Research* **36**: 589–604.

8. Bhatnagar R K and Kanal L N 1986 Handling uncertain information: a review of numeric and non-numeric information. In *Uncertainty in artificial intelligence,* edited by L N Kanal and J F Lemmer, North Holland, Amsterdam, vol. 4, pp. 3–26.
9. Rich E and Knight K 1991 *Artificial intelligence,* 2nd edn. McGraw-Hill, New York.
10. Roberge P 1994 Expert systems for corrosion prevention and control. In *Modelling aqueous corrosion,* edited by K R Trethewey and P R Roberge, Kluwer Academic, Dordrecht, The Netherlands, pp. 129–40.
11. Hammond S P 1994 Dangerous goods in Alberta — movements, incidents and trends. In *Proceedings of 1st DND symposium/workshop on risk evaluation and assessment, Kingston, Ontario, Canada,* Royal Military College, Kingston, Ontario, pp. 87–99.
12. Meister D 1977 Human error in man–machine systems. In *Human aspects of man-made systems,* edited by S C Brown and J N T Martin, The Open University Press, Milton Keynes UK, pp. 299–324.
13. Trethewey K R and Roberge P R 1994 Development of a knowledge elicitation shell for improved materials performance of marine systems. In *Computers in corrosion control — knowledge based system,* edited by P R Roberge, P Mayer and W F Bogaerts, NACE International, Houston TX, pp. 63–76.
14. Trethewey K R 1992 An overview of current naval marine corrosion problems and their solutions. In *Proceedings of INEC92,* Plymouth UK, Institute of Marine Engineers, London, pp. 20-1 to 20-9.
15. Dillon C P 1987 Performance of tubular alloy heat exchangers in seawater service in the chemical process industries, Materials Technology Institute, St. Louis, MO.
16. Tighe-Ford D J 1994 A systematic approach to the design of warship impressed current cathodic protection systems. In *Modelling aqueous corrosion,* edited by K R Trethewey and P R Roberge, Kluwer Academic, Dordrecht, The Netherlands, pp. 381–98.
17. Astley D J 1988 Use of the microcomputer for calculation of the distribution of galvanic corrosion and cathodic protection in seawater systems, In *Galvanic corrosion,* edited by H Hack, ASTM, Philadelphia PA, pp. 53–78.

11.7 BIBLIOGRAPHY

Petroski H 1992 *To engineer is human: the role of failure in successful design,* Vintage Books, New York.
Sage A P 1992 *Systems engineering,* John Wiley, New York.

12 CORROSION CONTROL BY DESIGN

> Now in the building of chaises, I tell you what, There is always, somewhere, a weakest spot.
> (Oliver Wendell Holmes, *The Deacon's Masterpiece*)

In far too many engineering structures the 'weakest spot' is the lack of design consideration given to corrosion control. It may seem to an onlooker as though many structures have corrosion designed into them.

The principles of good design to minimise corrosion problems have been known for many years. The introduction to Chapter 1 shows how the most basic errors lead to bimetallic corrosion, errors first remedied in the 1760s but still committed more than 200 years later. Some of the reasons were discussed in the previous chapter. It is not just a question of engineers making simple errors, but part of a much wider situation in which complex interactions between materials, environments and people create apparently unpredictable outcomes.

Designers of engineering systems have to acquire many skills; knowledge of corrosion control forms but a small part (see, for example, Fig. 15.3). However, many would say that corrosion, the most significant life-limiting factor in the vast majority of cases, is too often given scant regard.

The reliability of the corrosion control method selected can influence other design parameters by

Ensuring that a structure has adequate safety margins to fulfil its function for the design life of the structure;

Allowing overdesign to be minimised, resulting in thinner sections, lower weight and reduced capital cost.

Case 12.1: Corrosion occurred in a buried oil pipeline 360 km long, which had a 200 mm external diameter. A wall thickness of 6.3 mm was adequate to cope with the stress levels imposed on the pipe, but to allow for corrosion it was increased to 8.3 mm. This corrosion allowance used an extra 3000 tonnes of steel and caused a loss of 4% carrying capacity in the line.

Careful consideration of the whole structure at the design stage will predict many of the areas of the system which are likely to corrode. The designer can then make provision for easy inspection, maintenance or replacement of corroded sections at these points. Indeed, the designer, acknowledging that corrosion is inevitable, may deliberately cause certain sites to become the anodes in the total system, thus giving sacrificial protection to those areas which are less accessible (see the discussion on pumps, Section 15.8).

In this chapter we will focus on good and bad design practice. The principles are not listed in any order of merit, and their relative importance will vary from project to project.

12.1 HOW LONG DO I EXPECT IT TO LAST?

While it might seem very easy to determine the lifetime of a component or structure, it is actually very difficult. This is because there are so many things that can go wrong along the way.

Case 12.2: A farmer was puzzled that his cows were refusing to cross the farmyard to the milking parlour. Soon afterwards, the milking machine broke down. Aluminium is used to carry current in the UK electricity grid, but copper cables are for internal domestic and industrial use. Therefore, a joint must be made between two dissimilar metals (Fig. 12.1 (top)). The electricians who came to repair the milking machine found that, contrary to recommended practice, the aluminium/copper

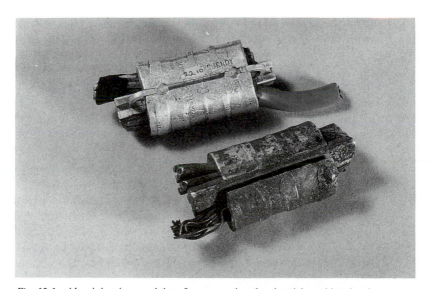

Fig. 12.1 Aluminium/copper joints for connecting the electricity grid to local power supplies. The top joint is in good condition and functions perfectly; the bottom joint was removed from a farm (Case 12.2). The joint was badly corroded and had a very high resistance which made it malfunction.

257

joint had been made on the outside, not the inside, of the building. Corrosion of the couple had caused a build-up of aluminium oxide (Fig. 12.1 (bottom)). The joint developed such a high resistance that the return path for the power ran out of the parlour, across the yard to the nearest pole and up the earthing strap to the overhead line. The cows had obviously been experiencing a potential difference between front and rear hooves; the farmer wore rubber boots and felt nothing!

Having established the life to be expected from a component or structure, this must be compared with the life of the corrosion control system. If the control system has a shorter life than the structure then the method of renewal should be considered at the design stage, and provision should be made for inspection, maintenance and replacement. Should the control system of part of the structure fail for any unforeseen reason before the life required for the whole structure, the question must be asked, Will the difficulty or cost of replacing the part become the life-limiting factor for the whole?

Application of a coating may be routine in a factory where the temperature and relative humidity can be easily controlled, and where the standard of surface preparation and the quality of the coating can be closely monitored. Renewal of the coating in an exposed situation under adverse weather conditions could be a very different proposition.

Case 12.3: The 500 mm square box section of a 3 m long steel strengthening beam was successfully coated in a factory by dipping. It was a very different matter to renew the internal coating after the beam had been installed in a lattice girder bridge spanning an estuary.

If it is not possible to renew the protection system then the original application must be capable of withstanding all the hazards of fabrication and installation which occur after it has been applied, as well as lasting the lifetime of the structure.

Case 12.4: Railings were installed at a car park located about 500 m from the sea by welding together a number of sections of prefabricated galvanised steel tube. These should have given a maintenance-free life of 12–15 years, but rusting occurred around each of the welds only weeks after installation. No post-welding protection had been applied to compensate for the galvanising burnt off during fabrication.

When the time for replacement arrives, the performance of the existing structure and materials should be considered before changing to more expensive or more corrosion-resistant materials.

Case 12.5: A lead-lined wooden tank used to mix sulphuric acid and zinc chloride gave satisfactory service for over a hundred years. When it was replaced, a tank made from the same materials was unacceptable to the owners despite the long service of the original tank. [1]

If a plant is to be scrapped at the end of its life, it is uneconomic for the life of some elements to be much longer than the life of the plant as a whole. Fitting a stainless steel exhaust system to a motor car may not be cost-effective if the exhaust lasts much longer than the car. Conversely, the fitting of mild steel exhausts with a life of 2–3 years is clearly not satisfactory.

12.2 THE EVER-CHANGING ENVIRONMENT

Components will be exposed to a wide variety of environments during the different stages of manufacture, transit and storage, as well as the daily and seasonal variations experienced in service. If the structure is mobile the changes in exposure conditions will be even greater.

The rate of corrosion or deterioration of protective coatings applied to the metal will be affected by changes in factors such as

relative humidity
temperature
pH
oxygen concentration
solid or dissolved pollutants
concentration
electrolyte velocity

As far as possible, these variations in the environment must be identified at the design stage. Clearly, steps will be taken to control the in-service corrosion, but special protection may be required during a particularly hazardous, albeit temporary, phase in production, transit or storage.

Coal tar enamels, bitumen and many tape-wound coatings are often used to control the corrosion of steel pipes buried in the earth. But if, prior to burial, the pipes are stored or transported in bright sunlight, the coating can degrade, leading to a shorter in-service life than predicted. To guard against this deterioration, the coating may be overpainted with high density emulsion, superfluous once the pipe is buried.

Corrosion damage during storage is a common problem. It can be caused by the microclimate generated inside packing cases and storerooms, in systems such as the drained water-cooling circuits of stored petrol and diesel engines, or in boilers and condensers at power stations during shutdown. It may also occur in components stacked in warehouses when the free circulation of air is impeded. General climatic changes also generate damage in unheated warehouses which do not have humidity controllers.

The volatile constituents of wood pose particular hazards for metals stored in wooden containers. While this is dealt with more fully in Section 13.2, it should be noted that all closed wooden crates must have a sealed lining to prevent deposition of acid films on stored components.

Case 12.6: An example of the generation of local climatic conditions, the significance of which was not appreciated until considerable damage had been caused, involved cold-drawn brass tube. The tube was stored in a shed adjacent to a workshop. Birds roosted on the rafters inside the shed and the ammoniacal content of their droppings caused severe stress-corrosion cracking of the tubes.

A factor which can affect in-service performance is the interaction between adjacent installations, or between apparently unrelated parts of a complete system. Units which would give an entirely satisfactory corrosion performance when considered as separate entities have failed catastrophically owing to environmental effects caused by their relative positions on a site or in a plant. This is an example of the secondary corrosion mechanism described in Section 11.2. When two or more design offices are responsible for the individual parts of a system, the people factor becomes a serious aggravating influence on the risk of corrosion failure. Good corrosion management requires care to ensure there is sufficient liaison and understanding between them, so that the parts are mutually compatible with each other and with installations which already exist on the site.

Case 12.7: A copper lightning conductor earthing a chimney was separated from an aluminium roof by a vertical gap of 6 m. When it rained, or mist precipitated on to the conductor, a small amount of copper dissolved in the water as it ran down over the metal. When the water dripped on to the roof panels, ion exchange plated out the copper on to the aluminium, causing severe corrosion of the roof. The problem was eliminated after consultation between the electrical department and the plant design office; the conductor was moved to the other side of the chimney.

Case 12.8: An aluminium aerial support (Fig. 12.2), was quite adequate for the designed task until it was placed just behind the funnel of a ship. Acid produced from the burnt fuel oil removed the paint from the support, while carbon deposits from the smoke coated it with a very effective cathode. Corrosion soon rendered the assembly unsafe.

In our society, rules and regulations proliferate. Considered in isolation, one set of safety regulations may be consistent and beneficial, but severe hazards can be produced when they interact with those of another organisation.

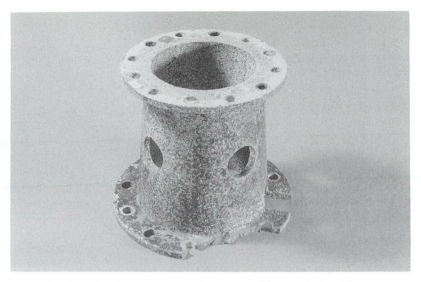

Fig. 12.2 Corrosion damage to a cast aluminium aerial support, placed in an atmospheric environment containing soot, acid exhaust gases and moisture (Case 12.8).

Case 12.9: Health regulations require milking machines to be washed regularly, while electrical safety regulations require them to be earthed. On a Devon farm, thorough washing caused corrosion between a machine's copper earthing strap and its stainless steel body. The electrical resistance at their joint became so high that the earth failed to function. During milking a fault developed in the machine, and a cow was electrocuted.

12.3 MINIMISE BIMETALLIC CORROSION CELLS

It is not true to say that all dissimilar metal couples must prove disastrous. There are countless examples of bimetallic (and multimetallic) couples which have given many years of useful service, although the metals involved are widely separated in the galvanic series. Corrosion of the coupling will occur only if a galvanic cell is formed. Recall that a basic wet corrosion cell consists of four parts; if one part is eliminated, corrosion ceases. Although two dissimilar metals provide the possible anode and cathode in a bimetallic coupling, corrosion control is achieved by preventing access of an electrolyte to the joint. Alternatively, the joint may be insulated by breaking the coupling to stop the flow of electrons between the two metals.

Case 12.10: The aluminium valve body shown in Fig. 12.3 has bronze inserts in the top face to take the spindle, and securing bolts for the valve cover.

Fig. 12.3 Aluminium valve body with bronze inserts. There is no corrosion around the bronze inserts in the top face where the gasket prevented the ingress of electrolyte. On the flange there is heavy corrosion where there was free access of electrolyte to the brass securing bolts (Case 12.10).

Although this would appear to be most unsatisfactory, no corrosion has occurred around the bimetallic couples where only three parts of the cell are present – the gasket and valve cover effectively prevent any electrolyte from reaching the metals. At the flange, however, where brass bolts were used to secure the two halves of the valve, the free access of water completed the cell and caused severe galvanic corrosion of the aluminium.

Wherever two different metals are in contact, bimetallic corrosion will always remain a possibility should the remaining parts of the cell develop in service.

Case 12.11: Cold liquid flowing in a steel pipe which passed through a heated room caused condensation on the outside of the pipe. This gave rise to general corrosion of the exterior of the pipe as the paint film deteriorated. The pipe was adjacent to and in contact with a brass valve and also suffered severe bimetallic corrosion on the external surface. Lagging the pipe to prevent condensation of the electrolyte cured the problem.

Sometimes the cell develops through unexpected routes, either over long distances, where the conducting path is hidden and escapes attention, or through alternative routes which bypass the careful arrangements of the designer. Tunnel vision is dangerous because it concentrates the designer's attention on a single area.

Case 12.12: A dry dock was constructed from steel-reinforced concrete with the steel buried well inside the concrete. In the bottom of the dock a number of iron supports were mounted to assist in the location of ships entering the dock. Around the top of the inside of the dock a 90/10 cupro-nickel pipe system was installed. The pipe and the supports were in contact with the steel reinforcing bars inside the concrete. When the dock was flooded, the sea-water covered the pipe and completed the circuit to the iron, which corroded vigorously.

Case 12.13: In a plumbing system a mild steel pipe had to be coupled to a copper one. A ceramic insulator was used to insulate the pipes from each other. However, the carefully engineered assembly was attached by metal brackets to a wall panel made of sheet metal. The panel bypassed the insulated joint and allowed electrons to flow from the steel to the copper. The steel pipe leaked in a very short time. [2]

In practical engineering it is impossible to eliminate every bimetallic cell. Where it is necessary to accept a cell in a design, several steps can be taken to minimise the corrosion damage.

Anodes should always be kept as large as is practical in the particular component or location to reduce the current density.

If the electrolyte flows through the system, the anodes should be kept upstream from the cathodes to prevent ion exchange plating out local cathodes on to the anode. Local cathodes cause pitting of the anode (see Chapter 7 and Case 12.8). Indeed, in a flowing system, it is possible to introduce short pieces of pipe which are anodic to all the metals in the rest of the system (Fig. 12.4). They act as sacrificial wasters by corroding to protect the other metals. It is common practice to put a valve on each side of the waster and fit it with flanged joints,

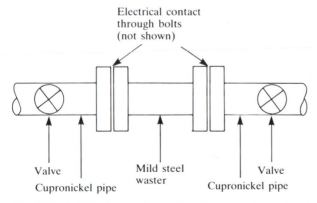

Fig. 12.4 Sacrificial waster inserted in pipe system. The waster is made anodic to the remainder of the system.

263

then it can be removed without draining the whole system. Wasters should always be located in positions where they are accessible and there is room to replace them.

The electrolyte can be modified to make it less aggressive. This will be discussed in Section 13.3.

In all cases where dissimilar metal joints are inevitable the first aim should be to isolate both sides of the joint from the electrolyte by applying a barrier coating. Alternatively, the metals should be insulated from each other to prevent the flow of electrons across the joint.

12.4   AVOID DIFFERENTIAL-AERATION CELLS

While bimetallic corrosion is destructive, cells caused by oxygen differentials in the electrolyte are more widespread and cause greater damage. In Section 7.3 we discussed the development of the differential-aeration cell. The zone of the electrolyte that is low in oxygen produces an anode on the metal surface, the zone that is richer in oxygen produces a cathode, even though the materials are identical. This effect was demonstrated in Expt 3.11. An oxygen differential can develop in any situation where water is in contact with a surface. Therefore, every effort should be made to prevent the ingress of moisture or free-standing water to areas where it will be held for long periods on the metal surface. Potential moisture traps should be sealed, fitted with drainage holes, or ventilated to dry off the water. Surfaces in free contact with water should be protected with barrier coats or cathodic protection systems.

Examples of differential-aeration cells are legion. Some of those which cause the most damage are described below.

Crevices

Any point at which two metal surfaces are separated by a narrow gap is a possible cell. Moisture enters the gap, often drawn in by capillary action. Where the liquid is in contact with air, the oxygen is replenished but the centre of the water film becomes impoverished in oxygen and corrosion occurs at that point. Crevices are formed behind spot-welded overlays or butt joints, under the rim of sheet metal which has been folded to give a smooth outer edge, at bolted or riveted joints and at shingled or overlayed plates. Crevice cells have also been formed beneath loose inert fastenings, string or wire wrapped round metal supports (Expt 3.15) and behind tallies and labels improperly secured to metal surfaces. Rubber grommets have produced severe crevice corrosion on stainless steel shafts and couplings. Some aspects of good and bad design are shown in Fig. 12.5.

Crevice attack used to be found in many areas of the modern car, although better designs have been adopted in recent years. However, corrosion is still the most significant limit (apart from mechanical damage) on vehicle life. It occurs in seams, box sections, the bottom of doors, behind overlay trims and fasteners, and under the folded edges of boot (trunk) lids and doors. Additional cells are created by owners who embellish the vehicle with fittings such as spotlights, advertising stickers behind which water accumulates, retrofitted spoilers and other individual trims. The site of the corrosion is usually inside the structure of the vehicle and penetrates through the metal to the external surface. As a result it is often called **inside-out corrosion**.

Debris traps

Debris which will absorb or hold water, such as mud, loose corrosion product, leaves, old rags and paper, set up differential-aeration cells. The corrosion occurs out of sight beneath the debris, which forms a poultice on the metal surface. If inside-out corrosion is not to occur, box sections should be designed so that debris cannot accumulate in the section. The risk of damage is minimised by rounding internal corners and edges, by easy access for sweeping and cleaning, in particular drainage covers and filters. An untidy dirty system is more likely to suffer corrosion than a clean well-maintained system (Fig. 12.5(f)).

Case 12.14: A mud poultice was trapped in the rear of an open channel section which supported the shock absorber on the front wheel of a sports car. Corrosion penetrated the bottom of the channel which distorted when the vehicle gently hit a grass verge. The distortion caused loss of steering control and a much more serious crash ensued. The vehicle was 10 years old and the corrosion should have been found during routine inspection before the accident. For several years before the accident, the vehicle manufacturers had galvanised the underframe of this model, including the shock absorber supports.

Case 12.15: The painted open-ended mild steel pillar sections of a bus body were failing within 3–7 years service because of inside-out corrosion. In some cases complete penetration had severely weakened the structure. Rectification involved the replacement of the pillars and entailed the removal of many body panels. The differential-aeration cell started beneath a 15 cm mud poultice which held water and road de-icing salts on the internal surface. The manufacturer solved the problem by swaging the bottom end of the pillar until it was almost closed, leaving an opening to provide drainage. The closure was then orientated so that spray and mud were deflected from the remaining opening. The modified pillars were constructed from zinc-coated steel and stoved after painting. The additional cost to the manufacturer was only 1% of the cost to the user in repairing the damage; the safety improvement was incalculable. [3]

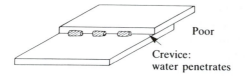

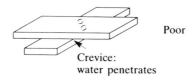

Poor

Crevice:
water penetrates

Poor

Crevice:
water penetrates

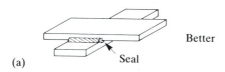

Better

Seal

(a)

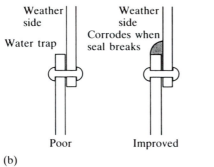

Weather
side

Water trap

Weather
side
Corrodes when
seal breaks

Weather
side

Water
drains
off

Poor

Improved

Best

(b)

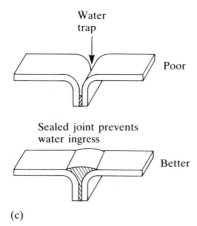

Water
trap

Poor

Sealed joint prevents
water ingress

Better

(c)

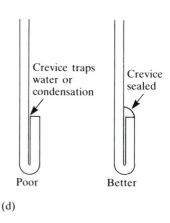

Crevice traps
water or
condensation

Crevice
sealed

Poor

Better

(d)

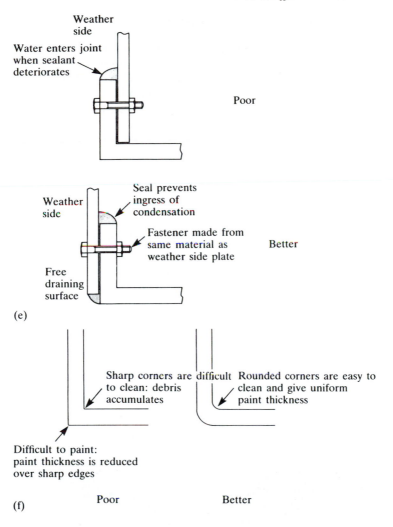

Weather side

Water enters joint when sealant deteriorates

Poor

Seal prevents ingress of condensation

Weather side

Fastener made from same material as weather side plate

Better

Free draining surface

(e)

Sharp corners are difficult to clean: debris accumulates

Rounded corners are easy to clean and give uniform paint thickness

Difficult to paint: paint thickness is reduced over sharp edges

Poor   Better

(f)

Fig. 12.5 Typical sites susceptible to crevice corrosion, and methods for reducing it. (a) Tack welds or other methods for localized fixing of metal plates leave crevices which water can penetrate. Continuous welds exclude water by eliminating crevices. (b) Riveting of vertical plates can create crevices unless the geometrical arrangement is correct so that rainwater runs off the joint, not into it. A sealant should also be used to prevent ingress of water by capillary action. (c) Potential water traps occur at round-edged joints. Careful, continuous sealing is necessary. (d) Folded sheet steel, commonly found on cars, provides ideal sites for crevice corrosion. Once again, these should be carefully sealed. (e) Correct geometrical arrangement is essential for the elimination of crevices in structures exposed to the weather. Note the possibility of dissimilar metal corrosion between plate and fastener. (f) Rounded edges and corners are always preferable to sharp ones, especially when a paint coating is to be used for corrosion protection.

Inadequate drainage and ventilation

If light rain or spray falls on a bare steel surface, rings of rust will be found after the water has evaporated. Each droplet acts as a differential-aeration cell and the rust ring develops where the iron(II) ions from the anode meet the hydroxyl ions generated on the cathode (see Expt 3.11 and Fig. 7.5). If the surface is free-draining, or there is adequate ventilation to dry the water droplets rapidly, the corrosion damage will be limited. Even on painted surfaces, there will be damage if the droplets persist for long periods. The risk of paint failure beneath the droplet will rise, followed by pitting corrosion at the centre of the droplet.

Increased damage is often found on the bottom surfaces of a structure, where the ventilation is likely to be less efficient, because the area is protected from prevailing air currents. Water droplets will remain on the surface until they grow sufficiently large to fall under gravity. Severe corrosion has been found on the bottom surfaces of the outside of tanks, on the undersides of overhanging units on steel bridges and towers, as well as on the lower surfaces of girders, beams and crane jibs. In such cases, adjacent vertical and top surfaces which are exposed to prevailing air currents have remained in reasonably sound condition (Fig. 12.6).

All channels and box sections should be free-draining so that water will not sit in the section. The internal surfaces and the edges of any drainage holes should be covered with long-lasting barrier coats. Ventilation, either along or through the section, will also help by keeping the surfaces dry. If a channel is closed by welding after the coating has been applied, any damage to the coating, internal as well as external, must be made good (Fig. 12.7).

Fig. 12.6 Increased corrosion on the bottom of oil storage tanks owing to poor ventilation, which left water droplets suspended from the surface for prolonged periods. The free flow of air on the sides of the tank kept the surfaces drier and free from corrosion.

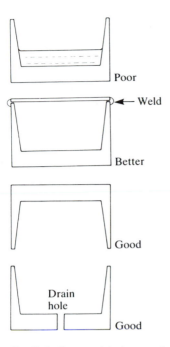

Fig. 12.7 Improved drainage and ventilation reduce the risk of corrosion in box sections and channels.

Water-absorbent soundproofing and insulation

Porous materials are often used to absorb sound and to reduce noise emission levels. They may also act as sponges and absorb water. Oxygen differentials in the water held within the sponge then generate differential-aeration cells on metal surfaces. Pads of foam rubber have been placed in the ends of internal box sections in cars to absorb the drumming sound created in the section when the car is moving. Condensation and spray from wet roads soak the rubber and start inside-out corrosion of the section wall.

Case 12.16: The doors of an expensive saloon car were made from thin sheet steel. Naturally, when the door was closed it produced a cheap-sounding clang. In order to produce a more satisfyingly luxurious clunk, much more in keeping with the car, the doors were lined with sound-absorbent foam. It was a common occurrence to find rows of blisters in the external paintwork of the door after only two or three years' service. The blisters were produced by corrosion from differential-aeration cells, formed beneath the water-absorbent foam on the inside of the door skin, which penetrated the metal.

Lagging on pipes can produce a similar problem, as Case 12.17 shows.

Case 12.17: Severe corrosion was found on the external surface of a lagged steel pipe which fed a factory heating system. The corrosion occurred inside a building where it was impossible for water to have dripped on to the outside of the lagging, nor was there any evidence of leakage from the pipe. It was found that water was dripping on to the lagging at a point several metres away. The lagging acted as a wick and conducted the water along the pipe. However, the heat from the pipe prevented the water from penetrating to the metal surface at the point of ingress, but as the pipe cooled along its length the water penetrated further through the lagging and wetted the metal surface at the point where the corrosion was found.

Water-absorbent materials should either be eliminated from all positions where the relative humidity will exceed 60% (see Section 13.2) or the outer surface should be fully sealed to prevent the ingress of water. Sealing should include all joints and repairs to the lagging.

12.5 TANKS AND PIPE SYSTEMS

There are several factors to be considered when designing tank and pipe systems for the storage or transport of electrolytes. Galvanic and differential-aeration cells, discussed in Sections 12.3 and 12.4, may develop because of the various constructional materials used, or at crevices in joints or behind gaskets.

Additional problems arise when electrolyte is trapped in the system by badly sited drain taps or poorly designed bends and junctions in the pipes. Some examples of these are shown in Figs 12.8 and 12.9. Drain taps must always be positioned so the system empties completely, and all bends should be rounded with smooth transitions through T-joints and into tank bottoms. In most cases, trapped electrolyte will evaporate, exacerbating the corrosion problems as the concentration of dissolved aggressive ions and the electrical conductivity of the solution increases. However, some electrolytes, notably sulphuric acid, may absorb water from the atmosphere. The rise in the level of the liquid increases the area of potential corrosion. If the more dilute electrolyte is not capable of

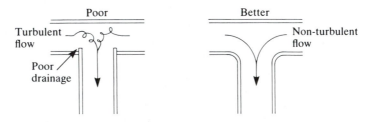

Fig. 12.8 Poorly designed joints introduce water traps and turbulence in pipe systems.

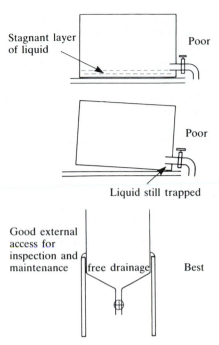

Stagnant layer
of liquid

Poor

Poor

Liquid still trapped

Good external
access for
inspection and
maintenance

free drainage

Best

Fig. 12.9   Improved drainage enables tanks to be emptied completely and reduces the
risk of internal corrosion. The external arrangements are also improved.

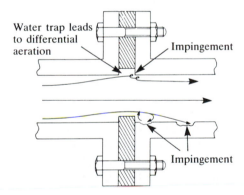

Water trap leads
to differential
aeration

Impingement

Impingement

Fig. 12.10   Badly fitted gaskets produce water traps, turbulence and cavitation damage
in pipe systems.

sustaining any passive film which may be protecting the metal surface, pitting
corrosion will occur, followed by general wastage of the metal wall.

Any obstruction in a pipe system may cause turbulence which will produce
erosion corrosion by cavitation or impingement (see Chapter 8). Turbulence can
be caused by poorly fitted gaskets, burrs, weld metal or solder protruding into
the pipe bore at joints, bends which have too tight a radius, or any other
disturbance to flow. Figure 12.10 shows the effect upon flow of a badly fitting
gasket (see also Case 8.1 and Fig. 8.1).

271

Fig. 12.11   A cast zinc alloy temperature sensor used in the cooling system of a motor car (Case 12.18).

Case 12.18: Figure 12.11 shows a cast zinc alloy temperature sensor used in the cooling system of a motor car. Attached to each end were rubber hoses which fitted over the metal. Flow was from left to right. The constriction on the flow as the water passed from the hose into the sensor caused turbulence which severely eroded the material downstream. It is also likely that copper ions from the copper radiator replated out of the water and on to the inside of the sensor, causing accelerated corrosion through galvanic action.

Cavitation damage has been found on the bores of pipes in which the liquid appears to be stationary. Vibrations in the tank or pipe wall produced by adjacent machinery, pumps or even the flow of liquid in other sections of the system set up transverse pressure oscillation in the layer of liquid adjacent to the wall. The pressure changes produce cavitation and pitting attack on the wall. Pipe runs, valves and pumps must be well clamped and damped to minimise vibration along pipes.

Case 12.19: Severe pitting corrosion was found on the outside surface of a diesel engine wet cylinder liner. The flow of coolant over the liner had been carefully smoothed and regulated to avoid cavitation damage, but the vibration in the liner wall caused by the force of the gas ignition and the passage of the piston up and down the liner created very large pressure changes in the liquid boundary layer. These pressure changes led to pitting damage by cavitation, which penetrated to one-eighth of the wall thickness (see Fig. 8.5).

272

Many materials used to manufacture valves, pumps and pipes have a limiting velocity for flow of electrolyte over their metal surfaces. If this velocity is exceeded, protective passive films may be removed and the metal corrodes and erodes rapidly. This will be covered more fully in Chapter 15. Liquid velocities in pumps, through valves and around bends can be much higher than in the straight runs of pipe. Although most metals have *maximum* limiting velocities, some have *minimum* values which must be maintained in the system.

An example of minimum velocity is found in the stainless steel/sea-water combination. Sea-water is very aggressive to the chromium oxide film on stainless steel, and pitting corrosion usually results in static systems. However, the material performs better in *flowing* sea-water. If the metal is specified for condenser tubes, a minimum velocity of 1.5 m s^{-1} of well-aerated sea-water is required to maintain the film. It is normal practice to select stainless steels containing up to 25% chromium and 6% molybdenum for use in such situations: the molybdenum addition reinforces the surface film.

Where aggressive electrolyte must flow through a system, make sure it can be thoroughly flushed through after draining down; this is to remove or neutralise aggressive ions. If the system is to be left unused for a prolonged period it may be advisable to treat it with a film former, which coats the surface with an inert barrier layer, or an inhibitor. Always follow the manufacturer's instructions and limitations on use.

There have been many instances of stainless steel giving excellent service over prolonged periods while oxygenated fast-flowing sea water passed through the system. However, massive crevice corrosion failure has occurred immediately after shutdown, even though the system has been drained. When water is trapped in crevices, inside pumps, below impellers, around shaft bearings or valve stems, and under gaskets, dissolved chloride rapidly destroys the protective chromium oxide film on the metal surface, leading to failure.

Case 12.20: Stainless steel has been used for ship propellers with success because of their generally good performance in well-aerated turbulent sea-water. However, when stainless steel was used for bow thrusters, serious localized attack was observed. In contrast to the propellers, the bow thrusters were located in shielded positions underneath the hull and at right angles to the propellers. Since they were only used for short periods while the ship was manoeuvring in harbour, long periods of comparatively low oxygen, stagnant conditions were possible in the bow thruster.

12.6 STRAY CURRENT CORROSION

Metal structures often carry unsuspected electric currents. These currents may be induced from neighbouring electrical conductors, e.g. in cast iron water mains by adjacent power cables when the two are buried close together to supply services in urban areas. They may arise whenever current finds a path of resistance less than the path envisaged by the designer, e.g. from current leakage when buried metal structures are used as earth conductors to ground electrical distribution systems, or during welding in industrial fabrication and repair work. In very long systems, such as pipelines running over many tens of kilometres, they have even been attributed to electromagnetic induction (leading to telluric currents) produced by the earth's rotation. Collectively, these are known as **stray currents**. [4, 5] Sometimes, stray current corrosion is called **electrolytic action**. Stray currents can frequently be held responsible in cases where metal has mysteriously corroded within weeks or months, although the source of the stray current can be difficult to identify.

Severe corrosion damage occurs at the point where *conventional* electric current leaves the structure. In Chapter 4 we saw that anodes are sites on a metal surface where electrons are released to flow through the metal, while ions conduct the current through the electrolyte. By definition, the *conventional current passes in the opposite direction to the electron flow*. Therefore, where the conventional current leaves the metal, the electron flow is *into* the metal and that site becomes the anode, as shown in Fig. 12.12(a). In normal corrosion an internal energy difference between the anode and cathode generates a potential and gives rise to the current. But in stray current corrosion the current is produced by an external source and the corrosion rate is dependent on this imposed current, not on differences between metal surfaces. This gives rise to two important points:

The corrosion rate is not dependent upon the rate of oxygen supply to the cathode, often rate-determining in normal corrosion processes.

Although coatings reduce the area which corrodes, the current density and the corrosion rate increase at any breaks in the coating.

Because of the large physical separation which is usually found between the anode and cathode in stray current corrosion systems, the cathode products do not form films on the anode to stifle the corrosion. However, chlorides, nitrates and other anions in the soil often migrate to the anode site to give corrosion products not normally found in natural corrosion cells. Where the conventional current enters the metal from an electrolyte, cathodic reactions occur and the resulting alkali may damage some metal surfaces or barrier coats.

The electrolyte may be free water, such as a river or the sea, or water contained in the soil. For buried structures, the magnitude and site of the corrosion may vary with the climatic conditions. Corrosion may cease in dry weather, only to recommence after the next fall of rain. Changes in soil conductivity because of dissolved ions will also alter the severity of the attack. For these reasons, pipelines are often buried in a carefully selected medium known as a **backfill** in an attempt to better control the environment.

In the days when trams were a common sight in city streets, very large

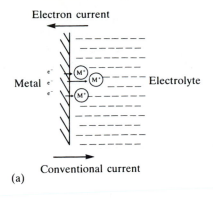

Electron current

Metal

Electrolyte

Conventional current

(a)

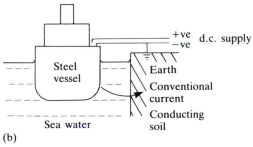

+ve
−ve d.c. supply

Steel vessel

Earth

Conventional current

Conducting soil

Sea water

(b)

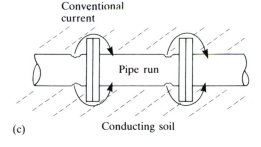

Conventional current

Pipe run

(c) Conducting soil

Fig. 12.12   Stray current corrosion: (a) development of an anode by an imposed current; (b) corrosion caused by incorrect wiring of a welding unit; and (c) conventional current flow around joints with high electrical resistance in a pipe system.

electrical currents were often measured in metal systems carrying water and other services. In Brooklyn in the early 1900s, 70 A was measured in a cast iron water main. [6] Today, electric-powered mass transit systems have returned to our cities and reintroduced the threat of stray current corrosion of adjacent buried services. Considerable efforts have been expended to insulate the systems. [5]

Case 12.21: A 65 km multiple pipeline route crossed under four sets of railway lines which were electrified on the 750 V dc third rail system. The pipelines were protected by impressed current cathodic protection at four substations. After the final pipeline was installed, the cathodic protection failed to control the corrosion resulting from the stray currents, which flowed from the pipeline to the rail system's negative conductor via the soil in which the pipe was buried. To prevent corrosion damage the pipelines and the negative rail were electrically bonded so the current did not have to leave the metal at any point. This in turn produced two other interesting effects. (1) At times, the conventional current flowed from the railway to the pipe down the bonded joint. This caused corrosion at other sites. A diode rated at 150 A was included in the rail bond to ensure that current only flowed from the pipe to the rail. Note the rating of the diode! (2) By placing suitable resistors in the individual bonds to each separate run of pipe, the potential of the pipe was limited to −0.85 to −2.5 V CSE, thus providing its own cathodic protection system for the whole 65 km of pipe. The cathodic protection substations were disconnected. [4]

Case 12.22: Because of joint leaks, a flanged and bolted, underground, stainless steel pipeline was being replaced with an all-welded system. It rained during the night, and a portion of the pipeline was submerged; it was the portion already laid in the trench but not backfilled. The next day, stray currents from the welding machine 'drilled' thousands of holes through the wall of the submerged pipe as the current found an easier return through the water than by following the welder's clamped-on earth connection. [1]

Similar problems have affected ships undergoing weld repairs when dc electrical supplies for welding units have been taken from a shore-based generator. The current sometimes finds an easier return from the hull through the water in which it floats and the earth to the grounded terminal of the generator. The result is thinning of the hull (Fig. 12.12(b)). In serious cases, hulls have perforated and ships have sunk at the quayside.

Case 12.23: A 750 mm diameter Ferralium sealing ring failed after about a year's service in a high pressure marine system. Ferralium would normally be expected to perform well in marine systems, even systems containing dissimilar metals. The sealing ring was observed to have a number of serious surface pits, and sectioning revealed extremely long subsurface tunnels, occasionally meeting the surface. The tunnels followed the direction of the circumferential grain structure imposed by the forging process (Fig. 12.13). The severity of this type of attack could only have been caused by a large stray current. Even allowing for larger areas of possible cathodes, no galvanic effect could result in so much dissolution in such a short time.

Increased resistance across flanged and bolted joints in pipelines has caused stray currents to bypass the joints, flowing either through the soil or into the

Fig. 12.13 An example of the severity with which the electrolytic action from stray current can destroy a component in months (Case 12.23). This section of a 750 mm diameter Ferralium 255 sealing ring in a marine system showed a number of serious surface pits, but when sectioned, the material was found to contain long subsurface holes running parallel to the orientation of the grains.

liquid in the pipe if it is a conductor. The current often returns to the pipe on the other side of the joint. Engineers have been surprised to find a repeated pattern of corrosion at each joint along a pipe run. In each case the same side is badly corroded and the other side is uncorroded. This effect is shown in Fig. 12.12(c).

Methods of controlling stray current corrosion include

Electrical bonding of the structure so the current does not leave the metal at any point.

Using impressed current cathodic protection.

Attaching sacrificial metal pieces to the buried structure.

Sacrificial metal pieces are attached at the points where the conventional current leaves the system. These should be orientated in the direction of the current flow in the surrounding medium, carrying the current well away from the structure. The sacrificial pieces should be inspected regularly and replaced as required.

12.7 REFERENCES

1. Seibert O W 1978 Classic blunders in corrosion protection, *Materials Performance* **Apr**: 33–7.

2. Seibert O W 1983 Classic blunders in corrosion protection revisited, *Materials Performance* **Oct**: 9–12.
3. Department of Industry 1979 *Controlling corrosion: case studies*, Pamphlet 5, HMSO, London.
4. Allen M D and Ames D W 1982 Interaction and stray current effects on buried pipelines — six case histories. In *Proceedings of cathodic protection theory and practice — the present status*, Institute of Corrosion, Coventry UK.
5. Gundry R (ed) 1994 *Stray current corrosion in mass transit rail systems*, NACE International, Houston TX.
6. Evans U R 1948 *Metallic corrosion, passivity and protection*, Edward Arnold, London, p. 37.

12.8 BIBLIOGRAPHY

Landrum R J 1989 *Fundamentals of designing for corrosion control: a corrosion aid for the designer,* NACE International, Houston TX.

Pludek V R 1977 *Design and corrosion control,* Macmillan, London.

Szeliga M J 1994 Stray current corrosion: the past, present, and future of rail transit systems, Paper 300 in *Corrosion 94, Baltimore MD*, NACE International, Houston TX.

13 CORROSION CONTROL BY ENVIRONMENTAL CHANGE

What fool hath added water to the sea ...

(William Shakespeare, *Titus Andronicus*)

O! When mine eyes did see Olivia first, methought she purged the air of pestilence.

(William Shakespeare, *Twelfth Night*)

Corrosion is the reaction between a metal and its environment, therefore any modification to the environment which makes it less aggressive will be beneficial in limiting the attack upon the metal. But what constitutes a reduction in aggressiveness? Very often the effect on corrosion of a given environmental change, whether deliberate or natural, is not obvious. Depending on the circumstances, an increase in oxygen concentration may increase the corrosion rate or decrease the corrosion rate.

Case 13.1: A series of stainless steel heat exchangers, cooled with recirculating treated water, had given over 10 years' problem-free service. However, in an emergency, untreated river water was used to cool the units for 48 hours. Several weeks later, five units failed because of stress-corrosion cracking.

Case 13.2: Copper-based alloys, which are known to give excellent service in sea-water systems on ships, have often failed during the first few months of service. The culprit is a thin film of sulphide which develops on the metal surface during exposure to polluted estuarine water in the final stages of fitting out the vessel. The sulphide is cathodic to the copper alloy and produces severe pitting corrosion at breaks in the film. Once the normal passive film has developed after exposure to fresh sea-water over a period of time, the alloys become immune to this form of attack. It can be controlled during the fitting-out stage by adding 0.01% of an inhibitor, sodium dimethyldithiocarbamate, to the polluted sea-water.

Environments are fundamentally either gaseous, liquid or solid phase so we will consider three situations.

The bulk of the environment is gaseous. Normally, this will be air in the temperature range $-10°C$ to $+30°C$. Some of the methods used to reduce corrosion rates in the atmosphere are

(i) Lower the relative humidity
(ii) Eliminate volatile components given off by surrounding materials
(iii) Change the temperature (but see Section 13.1)
(iv) Remove contaminants (including solid abrasive particles), deposits which will form cathodes (e.g. soot) and aggressive ions.

The material is immersed in free water which contains sufficient ions to be an electrolyte. Modifications to electrolytes include

(i) Lowering the ionic conductivity
(ii) Altering the pH
(iii) Uniformly reducing the oxygen content
(iv) Changing the temperature.

The metal is buried in soil, dissolved minerals forming the electrolyte. Control is normally by cathodic protection or surface coating, but the environment can be made less aggressive by using backfills to improve drainage, control pH and change the conductivity.

Corrosion processes in all these environments can be controlled by adding *inhibitors* to reduce the rate of attack. Much of the development work on corrosion inhibitors was accomplished from the late 1940s to the late 1960s, and is still used used today. [1]

Shakespeare talks of diluting the sea and removing pollutants from the air. Environmental regulation is now moving towards the requirement for as little change to the natural environment as possible. Levels of emission into the atmosphere must be generally reduced; oxides of carbon, sulphur and nitrogen, chlorofluorocarbons (CFCs), and organolead compounds are just some of the substances considered harmful to the planet. Factory discharges into rivers and seas frequently destroy the natural balance, while the addition of large quantities of chemicals to the soil in modern agriculture gives increased yields at the expense of additional run-off into streams and rivers. In this context, the extensive use of inhibitors for corrosion control in the natural environment is limited. But where the environment within a system can be considered separately from the natural environment, change is not only practical but often desirable. The limited space may be the atmosphere in a warehouse, storeroom or a packing case. It may be a box section in a structure. And it may be the electrolyte in a tank, or in a closed loop system such as a cooling circuit for an engine or a processing plant. All these situations are amenable to this form of corrosion control.

13.1 ATMOSPHERIC CORROSION

Before examining the control of atmospheric corrosion, let us consider some of its causes.

Most important is the presence of water from rain, mist or condensation owing to a high relative humidity; their effects are different. Heavy rain can be beneficial by washing away the pollutants which have collected on a metal surface. Care should always be exercised in the design of a structure to ensure that rainwater drains away freely and there is sufficient ventilation to dry all surfaces (see Chapter 12).

Mist and condensation constitute the corrosion hazard from the atmosphere, wetting all surfaces, external and internal. The thin films of water they produce may not drain away; instead they remain on the surface until dried by air currents or increases in temperature. A thin invisible film of water is all that is necessary for attack to begin. Many metals — iron, steel, nickel, copper and zinc — corrode if the relative humidity exceeds 60%. If it exceeds 80%, the rust on iron and steel becomes hygroscopic (water absorbent) and the rate of attack is further increased.

The thin film of moisture which forms from mist or high relative humidity is easily saturated with oxygen from the atmosphere, therefore the cathodic reaction, either oxygen reduction or hydrogen evolution, is not the rate-determining step in the corrosion process. The rate and severity of the attack is usually determined by the conductivity of the electrolyte. And the conductivity depends upon the level of dissolved contaminants, which vary from area to area. Carbon dioxide produces slightly acidic solutions in rural areas; sulphur dioxide, sulphur trioxide, nitrous compounds, hydrogen sulphide and ammonium ions contaminate industrial areas; chloride ions occur at marine locations (see Table 15.1).

Temperature affects atmospheric corrosion in two ways. First, there is the increase in reaction rate which normally follows an increase in temperature. An approximate rule is that the rate of a reaction approximately doubles for every $10°C$ rise in temperature. However, at higher temperatures, the solubility of oxygen is reduced and it has been suggested that this reduction lowers cathodic reaction rates, limiting the corrosion. In thin films with a good oxygen supply from the atmosphere, the limiting effect will be small.

Second, changes in temperature affect the relative humidity and can cause dew-point condensation. If the temperature falls below the dew-point, the air becomes saturated with water vapour and free water droplets condense on any exposed surfaces. The condensation forms on all surfaces which are cool enough, internal as well as external. The droplets can collect in water traps to produce pools of free electrolyte in sheltered areas, causing problems inside a structure where corrosion might not normally be expected.

Dew-point condensation is responsible for much of the damage to exhausts on cars and flues carrying burnt fuel gases. If the gas stream temperature falls below its dew-point before it is released to the atmosphere, condensation forms. The fuel gases usually contain sulphur oxides and nitrogen compounds. These dissolve in the condensed water and form aggressive electrolytes which, together with the galvanic effect of any carbon deposits with the base metal, lead to the rapid corrosion failure which originates on internal surfaces.

Sulphur trioxide produces sulphuric acid, a particularly aggressive electrolyte.

Increased sulphur trioxide also raises the dew-point of the gas, causing the condensate to form at higher temperatures and thus exposing greater lengths of the exhaust or flue to the electrolyte. Vanadium, another common fuel impurity, acts as a catalyst for the conversion of sulphur dioxide to sulphur trioxide in a combustion chamber; again this increases the possibility of early failure in the exhaust system. Flue gas containing no sulphur trioxide has a dew-point in the range 38–46°C; 5 ppm of sulphur trioxide raises this to 100°C, and 40 ppm will give a dew-point of 168°C.

Solid particles carried in air currents can be abrasive to paint finishes and protective films on metal surfaces. The damaged areas are often first to corrode when an electrolyte forms on the surface.

Aircraft are prone to corrosion damage from abrasive particles. Near the ground, damage to paint films is caused by sand and dust swept up in the turbulence created by the aircraft. Very fine ice particles, and to a lesser extent rain, cause damage at altitude. Corrosion ensues over the damaged areas when the aircraft enters moister climates. Epoxy-based paints for aircraft are more vulnerable than those based on polyurethane, particularly to fine ice particles, which puncture tiny holes through paint.

Abrasive particles ingested into gas turbines can remove the high temperature coatings applied to blades, resulting in hot or high temperature corrosion (Chapter 17).

When the relative humidity is high, miniature differential-aeration cells can be generated beneath dust or grit particles which adhere to metal surfaces. The result is a surface pock-marked with tiny rust spots over the corrosion pits. Stainless steel, used for decorative façades on buildings in busy towns and cities, has often suffered extensively in this way and required overcoating with transparent oils, lacquers or varnishes to stop the formation of the cells (see Case 7.3.). Some particles, such as soot, may act as active cathodes, generating local cells to pit the metal surface.

13.2 CONTROL OF ATMOSPHERIC CORROSION

The most effective way to minimise atmospheric corrosion is to remove the atmosphere under vacuum and seal the components in impervious envelopes. The protection is only effective while the integrity of the covering is maintained. Should it be punctured with even a small hole, corrosion will commence. It is prudent to include drying agents inside evacuated sealed covers as an additional protection.

Corrosion begins on many common metals at a relative humidity of >60%, so the air in warehouses and storerooms can be heated to keep the relative humidity below 60%. This does not remove water vapour from the air, so condensation will still form on any surface that cools the atmosphere below the dew-point. If you wear spectacles you will know how lenses mist over when walking into a warm room from a colder one. The metal temperature must be considered in heated storerooms. Light corrosion often affects metal objects, freshly introduced into temperature-controlled warehouses, before the metal attains the air temperature of the room. Serious problems arise if water collects on poorly ventilated internal surfaces where drying is slow. The same effect will happen if goods are packed and hermetically sealed in warm processing areas then stored or transported in colder environments. Condensation forms and is trapped inside the sealing.

Water vapour can be removed from the air before circulation in a warehouse. Chilling the air by passing it over surfaces cooled well below the working temperature of the warehouse precipitates sufficient water to keep the relative humidity below the critical level at the operating temperature of the store area. **Freeze-drying** can be used to produce very low residual water contents if required. To maintain an average relative humidity below 60%, it will be necessary to keep the general level much lower, perhaps at 30–40%, to allow for the exchange of air through open doors and by natural ventilation. Comfortable air conditions for people require relative humidity 40–65%. The slightly lower levels for warehouses are not thought to be harmful to health, although staff may experience some dryness in the throat over prolonged periods.

In packing cases or stores, where the expense of freeze-drying is not justified, desiccants are used to dry the air. Again, for maximum effect, the atmosphere must be contained by sealing both the components and the desiccant in impervious envelopes, or by minimising the rate of air exchange in the room. The desiccants should be non-corrosive to the metals involved, and preferably they should be cheap and easy to handle. In practice, the commonest ones are silica gel and activated alumina; molecular sieves are used to obtain very low humidity. Colour changes indicate the deterioration of the drying capacity of the desiccant, which can be reactivated by heating in an oven.

Case 13.3: The suspension bridge across the River Severn in the United Kingdom (Case 1.7) which carries the main road link between London and South Wales consists of 2,800 m of box-girder road-deck, divided into 14 sealed sections. Bags of silica gel, each containing 4.5 kg, are distributed over the floor inside each section at a rate of 0.2 $kg\,m^{-3}$. Silica can absorb up to 33% of its own weight in water. Monitoring is enabled by the inclusion of a sealed window with a weighing balance placed against it. A bag of gel placed on each balance can be seen from the outside and replaced when its weight reaches 5.9 kg. Each treatment lasts about three years.

The packaging for a desiccant requires careful consideration. Air should be free to circulate over the drying agent, while spillage is prevented. In storerooms and other large stable spaces, open trays of desiccant are often used, but in packing cases and other containers that are liable to be inverted, closed perforated metal boxes or porous bags are used.

Contaminants can be removed by scrubbing the air with fine water sprays before it is introduced into the system. The large surface area of the fine spray enables gaseous impurities to dissolve quickly, while solid particles are swept from the air stream. Clearly, the scrubbing process should take place before the air is dried.

An alternative method of protecting steel components during transport and storage is to use **vapour phase inhibitors (VPIs)**. As their name suggests, VPIs are volatile components that spread to occupy the free space in a container. Then they dissolve in moisture films, where they inhibit the corrosion. Although VPIs are beneficial for ferrous metals, they can increase the rate of attack on other materials; if non-ferrous metals, paints or plastics form part of the structure, VPIs should be used with great caution.

VPI compounds consist of a volatile organic cation, usually an amine, and an anion which acts as an inhibitor. On the metal surface, the cation produces a thin adsorbed film which has two important functions: (1) it is hydrophobic and (2) it controls the pH of any moisture layer which forms on its surface. Carbonate and nitrite are the anions in two typical VPIs. They are carried with the volatile cation to deposit on the metal surface. Should free water form on the surface, or the component become immersed so that the cation film breaks down, the anion acts as a normal inhibitor (Section 13.3) to control the rate of corrosion by polarising the electrode reactions. Immersed in water, the anion usually gives the same level of protection as its sodium salt.

For steels and aluminium, two common VPIs are dicyclohexylamine nitrite (DCHN) and cyclohexylamine carbonate (CHC). DCHN has the lower vapour pressure and takes longer to produce an effective surface film, but it maintains protection for a much longer period. It has a vapour pressure of 0.027 Pa at 25°C, giving a pH of 6.8 in water. One gram saturates 550 m^3 of air and renders it relatively non-corrosive to steel. In secure packaging at room temperature it will inhibit corrosion for several years. However, it can cause increased attack on some non-ferrous metals, plastics, paints and dyes. Zinc, magnesium and cadmium are particularly affected. CHC has a high vapour pressure, 21.3 Pa at 25°C, and produces a pH of 10.2 in water. The higher vapour pressure yields a protective coating much more quickly than does DCHN, but for a given weight of inhibitor the system has a much shorter life. CHC is most useful in containers or stores which are opened periodically; the VPI can be renewed regularly and the protection is rapidly restored once the unit is closed. It has no inhibiting effect on the corrosion of cadmium, but increases the attack rate on copper, brass and magnesium. Plastics, paints and dyes may also deteriorate in a CHC atmosphere.

DCHN-impregnated paper is widely used in precision engineering to wrap tools, measuring equipment and other close-tolerance components. The slightly greasy paper gives protection for many years, providing the packaging is not disturbed. CHC is normally inserted in open trays or porous containers.

Mixtures of the two VPIs are available. CHC rapidly builds a protective film, whereas DCHN contributes the longer-term control. Other VPIs are borates for zinc and chromates for copper or copper alloys, but they do not appear to be widely used.

Vacuum seals or desiccants are used to protect electronic and optical equipment, but the greasy film produced by VPIs prohibits their use.

Many organic materials — wood, plastic and paints — give off aggressive vapours which can initiate and contribute to corrosion damage on adjacent metal surfaces. Maleic acid, glycol and styrene have been desorbed from glass-reinforced plastics; wood can emit acetic acid vapours, a problem when wooden cases are used to store or transport metal objects.

Cellulose is the principal constituent of wood. It consists of long chains of sugar molecules which contain basic hydroxyl groups, some of which are combined with acetic acid in the form of esters. These groups can react with water to release free acetic acid:

$$CH_3COOR + HOH \rightleftharpoons ROH + CH_3COOH \tag{13.1}$$

$$ester + water \rightleftharpoons alcohol + acid \tag{13.2}$$

The equilibrium for the reaction always means that the moisture in wood is slightly acidic. However, the main corrosion hazard arises from the volatile nature of the acetic acid which escapes from the wood to fill the surrounding atmosphere with acid vapour. An excellent electrolyte is formed when this vapour condenses on to a metal surface.

Open wooden crates which allow free circulation of the air can reduce the degree of damage. Desiccants or VPIs are used in confined spaces where the microclimate can become heavily charged with the acid vapours, or when it is necessary for objects to be entirely enclosed within the crate. All wooden crates should be lined to keep the vapour away from the metal. Polyethylene, tar paper, bitumen-coated kraft paper and even zinc sheets have been used as liners. The desiccant or VPI is placed inside the lining material, which should be carefully sealed along all joints. The most severe attack occurs on mild steel, zinc, cadmium and magnesium alloys; titanium, tin and austenitic stainless steels suffer least damage.

Corrosion from acidic vapours is always a possibility when wood and metals are in close proximity and there is insufficient ventilation to remove the volatilised acid.

Case 13.4: Steel wire cables wrapped on wooden drums had been so badly corroded after a period in store that they were condemned as unsafe for use as securing strops.

Preservatives and fire retardants applied to wood can also increase the risk of corrosion damage. Diammonium phosphate, ammonium sulphate and borax mixtures form the basis of many proprietary fire retardants; copper chrome arsenic is a common fungicide. These compounds can release ammonia, soluble sulphate and copper, all of them corrosion hazards. Ammonia is particularly harmful to copper and some copper alloys because it leads to both general wastage and the likelihood of stress-corrosion cracking. Sulphate ions produce very aggressive electrolytes and copper can form effective cathodes after plating out by ion exchange on base metals.

Case 13.5: An unusual case of vapour corrosion happened on the inside of a copper belfry roof in Denmark. The roof was supported on wooden beams which had been treated with an ammonium-based fire retardant. Condensation on the inside of the copper roof leached the ammonium salts from the wood then dripped on to a concrete floor. The alkali in the concrete released ammonia vapour from the condensate, which rose into the roof space and attacked the copper roof. The initial cuprammino corrosion product decomposed to sulphate/carbonate and released ammonia to continue the attack. A thickness of 0.8 mm was penetrated in three years.

13.3 MODIFICATION OF THE ELECTROLYTE

Recall from Section 4.2 that there are four basic processes involved in the corrosion of a metal:

The anode reaction

The cathode reaction

Ionic conduction through the electrolyte

Electron conduction through the metal.

In this chapter we are concerned with the efficiency of the first three processes; the electron conduction through the metal is not normally affected by the electrolyte.

The rate at which the metal corrodes is controlled by the slowest of the cell processes; the metal cannot corrode and produce ions faster than the cathode can consume the resulting electrons, or faster than the electrolyte can transport the current by ion conduction. See the discussion at the beginning of Section 4.8 concerning rate-determining steps.

The properties of an electrolyte, which can be varied to limit its aggressiveness to a metal surface, were given in the introduction to this chapter. The presence of any dissolved ions will affect the corrosion rate by some or all of the following:

Changing the conductivity of the electrolyte

Attacking or strengthening passive films on the metal surface

Changing the pH.

Case 13.6: Conditions which arise outside of the normal operational regime can often inadvertently cause serious problems. Figure 13.1 shows one half of the power turbine disc of an Olympus TM3 marine gas turbine made of a 10.5% chromium steel known as FV448. After shutdown, condensation combined with small residues of ingested sea-salt to cause

Fig. 13.1 Top half of the power turbine disc (880 mm diameter) of an Olympus marine gas turbine that was written off. Pitting corrosion (not visible) at the blade roots was caused after shutdown by condensation combined with ingested sea-salt.

submillimetre pitting attack at the blade roots sufficient for the disc to be written off. This necessitated the introduction of freshwater rinsing after shutdown to wash out salts and greatly reduce the conductivity of any electrolyte which might be present.

A thin film of protective oxide forms on the surface of most metals exposed to air at room temperature. When the metal is subsequently placed in an electrolyte, the anion concentration plays a significant role in the behaviour of the electrolyte towards this protective film. On mild steel, the anions are aggressive at low concentrations and attack the film. At higher concentrations they may either become inhibitive or act in such a way as to plug any pores in passive films. An inhibitive anion is adsorbed on to the metal at weak points in the film. It then suppresses anodic dissolution of the metal and allows oxide formation to take place, strengthening the film. The inhibitive role of some anions is very weak and they can be considered aggressive towards oxide films. The most powerful of the aggressive ions towards mild steel surfaces are sulphates, thiosulphates, sulphites and thiocyanates. Chlorides are also very aggressive, but nitrates are only mildly aggressive.

On the other hand, dissolved cations which are more noble than the metal exposed to the electrolyte can plate on to the metal surface to set up local galvanic cells which often result in pitting attack.

Case 13.7: In Cornish tin mines, now disused, water often contained large quantities of dissolved copper ions. The ions instantly formed metallic copper layers whenever the waters touched ironwork in pumps and pipes. Production was a constant battle against bimetallic corrosion resulting in pump breakdowns and pipe failures. In contrast, the waters acted as a preservative to the timbers, often over 70 years old.

Reducing the concentration of ions in an electrolyte will reduce the efficiency of the ionic current transport and change the anodic processes on the metal surface. A change from sea-water to distilled water, for example, not only reduces the rate of attack upon the anode, but also the conductivity of the electrolyte and reduces the maximum distance which can separate the anode and cathode in an active cell.

In general, inhibitors affect either the anodic or cathodic processes. **Anodic inhibitors** increase the polarisation of the anode by reaction with the ions of the corroding metals to produce either thin passive films, or salt layers of limited solubility which coat the anode (Fig. 13.2). For iron and steel, two types are important: inhibitors which require dissolved oxygen to be effective, such as molybdates, silicates, phosphates and borates, and inhibitors which are themselves oxidising agents such as chromates and nitrites.

Cathodic inhibitors affect the oxygen and hydrogen cathodic reactions. For oxygen:

$$2H_2O + O_2 + 4e^- \rightarrow 4OH^- \tag{4.22}$$

the inhibitor reacts with the hydroxyl ion to precipitate insoluble compounds on the cathode site, thus blanketing the cathode from the electrolyte and preventing access of oxygen to the site. The most widely used inhibitors of this type are salts of zinc and magnesium, which form insoluble hydroxides; calcium, which produces insoluble carbonate if the water chemistry is adjusted (see below); and polyphosphates.

For hydrogen:

$$2H^+ + 2e^- \rightarrow 2H \rightarrow H_2 \tag{4.20/21}$$

the evolution of hydrogen is controlled by increasing the overvoltage (polarisation) of the system, as shown schematically in Fig. 13.3. The salts of

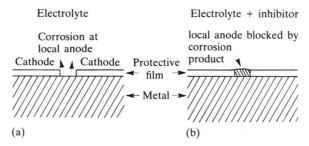

Electrolyte   Electrolyte + inhibitor

Corrosion at   local anode blocked by
local anode   corrosion
Cathode   Cathode   Protective product
                     ← film →
                     ← Metal →

(a)   (b)

Fig. 13.2 The effect of concentration of anodic inhibitor on corrosion rate. (a) Corrosion at a local break in the oxide film on the metal surface. The surrounding film is the cathode, the bare metal is the anode. (b) The anion in the anodic inhibitor reacts with the metal ions in solution to seal off the anodic site.

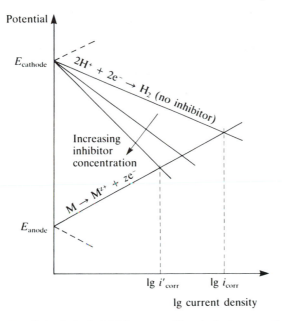

Fig. 13.3 Cathodic inhibitor controlling the hydrogen evolution reaction. As the concentration of inhibitor increases, the cathode reaction is polarised. The mixed corrosion potential, E_{corr}, moves to more negative values, closer to the rest potential of the metal redox process, E_{anode}, and the overall corrosion rate, i_{corr}, falls to i'_{corr}.

metals such as arsenic, bismuth and antimony are added for this purpose to form a layer of adsorbed hydrogen on the surface of the cathode. Owing to the toxicity of these metals, organic compounds which achieve the same objective have now been developed. Inhibitors which control the evolution of hydrogen gas from the surface may allow hydrogen atoms to diffuse into steel and cause hydrogen embrittlement (Chapter 10).

Cathodic inhibitors are said to be **safe**. Even if too little inhibitor to eliminate all the cathodic sites is added, the active cathode area, and hence the corrosion rate, is reduced, as shown schematically in Fig. 13.4(a). On the other hand, anodic inhibitors may be **dangerous** if the addition of too little inhibitor fails to eliminate all the anode sites. Using our rule that increasing cathode/anode area ratio leads to increased corrosion of the anode, we predict an increase in the corrosion rate on those anodes which are left. The most vigorous attack occurs just before complete inhibition is achieved (Fig. 13.4(b)). An anodically protected system may change from a condition of no attack to deep localised pitting attack, as a result of a slight dilution of the inhibitor concentration. In summary, if a partially passive state can persist for a significant time, local increases in anodic attack will occur when

Insufficient inhibitor is added to the electrolyte.

The electrolyte is diluted after the inhibitor is added.

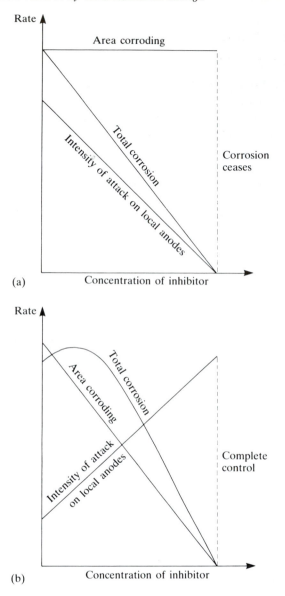

Fig. 13.4 Safe and dangerous inhibitors. (a) Cathodic inhibitors are 'safe'. An increase in the concentration of inhibitor, even below the critical amount, reduces the cathodic area and the current demand from the anodes. Although the area corroding remains unaltered, additions of the inhibitor below the critical concentration will reduce the corrosion rate. (b) Anodic inhibitors are 'dangerous'. An increase in the concentration of inhibitor, even below the critical amount, reduces the anodic area. The cathodic area is unaffected, therefore the current demand from the anodes that are still active increases to reach a maximum value just below the critical concentration. The corrosion rate rises on the active anodes and produces severe pitting attack.

Large concentrations of depolariser ions, such as sulphate or chloride, are introduced to reduce the effectiveness of the inhibitor present in the solution.

The inhibitor fails to penetrate 'dead legs' in the system.

Fortunately, with many modern anodic inhibitors this is not the case; anodic inhibitors are often more efficient than cathodic ones. Modern systems reduce the risk from the dangerous anodic inhibitors while profiting from their greater efficiency by using a combination of both anodic and cathodic inhibitors. The cathodic inhibitor retards the overall corrosion rate and allows the anodic inhibitor to seal off the anode sites at lower concentrations than would be needed if it acted alone. Such **synergistic** inhibitors are typified by the chromate/polyphosphate/zinc system.

The concentration of all inhibitors should be monitored regularly and additions made to maintain effective control.

Many inhibitors are organic molecules with groups of atoms which are adsorbed and desorbed from the metal surface. The bulky molecules limit the diffusion of oxygen to the surface or they trap the metal ions on the surface, reducing the rate of dissolution. It is claimed that as little as 0.2% agar-agar in distilled water will reduce the rate of corrosion to only 2.7% of the rate without agar-agar. Thus, many organic inhibitor molecules contain polar groups such as sulphur, nitrogen, hydroxyl or phosphorus; they have sizeable molecular weight and often contain aromatic rings. A list of some inhibitors and their uses is given in Table 13.1.

The effects of temperature, pH and dissolved oxygen content are interrelated. Figure 13.5 shows the effect of pH and oxygen content on the corrosion rate of steel at 25°C. In Section 13.1 it was stated that the rate of reaction

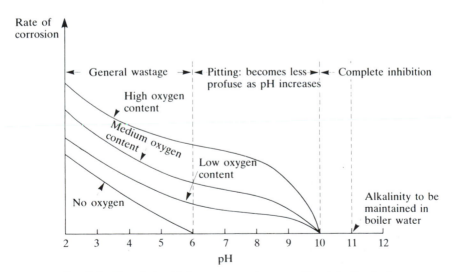

Fig. 13.5 The relationship between corrosion rate, pH and the dissolved oxygen content of water at 25°C and 1 atmosphere pressure.

Table 13.1 Some corrosion inhibitors and their uses

Inhibitor	Use	Metals protected	Typical concentration
Sodium nitrate	Cooling water	Steel	0.05%
	50% sea-water	Steel	10%
	Engine cooling	Steel, cast iron	up to 1%
Sodium benzoate + sodium nitrite	Heating systems	Lead/tin joints and cast iron in same	benzoate/nitrite ratio >7:1
Sodium hydrogen phosphate Polyphosphates Phosphonates	Cooling water Boilers Sea-water (with sodium nitrite) Road de-icing	Steel Steel, copper, zinc Steel	1% 10 ppm 10 ppm
Borates	Engine cooling Glycol cooling systems	Steel Steel	1% 1%
Sodium silicates	Potable water Oilfield brines Sea-water Cooling waters	Steel, copper, zinc Steel Zinc	10–20 ppm 0.1% 10 ppm 20–40 ppm
Chromate Dichromate + phosphate	Desalination plant Sea-water	Steel Steel	5 ppm 5 ppm + 50 ppm
Organic amines, e.g. octadecylamine diamines, alkyl pyridines, imidazoline	Vapour condensate in boilers Oil production and refining	Steel	Variable
Organic sulphur compounds, e.g. phenylthiourea, mercaptans, sulphides	Sulphuric acid	Steel	Typically 0.003–0.1%
Hydrazine	Oxygen scavenger (high temp)	Steel	As required
Sodium sulphite	Oxygen scavenger (low temp)	Steel	As required
Tannins		Most metals	Variable
Sodium benzotriazole	Neutral solutions Chloride solutions		
Mercaptobenzothiazole		Copper	Variable

approximately doubled for each 10°C rise in temperature. The same is true for totally immersed conditions. However, as the temperature is increased, the solubility of oxygen decreases. In contact with air, the solubility of oxygen at 25°C is approximately 8.5 ppm in tap-water and 6.5 ppm in sea-water. At 60°C, the solubility in tap-water falls to 5.6 ppm, while at 100°C it is all removed.

Case 13.8: In the mid 1960s it was found that sea-water containing oxygen levels less than 20 ppb (0.02 ppm) would corrode carbon steel at a rate of about 0.025 mm y^{-1} as long as the pH remained about 8 and velocities were less than 3 m s^{-1}. De-aeration to this concentration was subsequently adopted in sea-water floods all over the world as the primary means of achieving acceptable corrosion rates with bare carbon steel equipment and has been successful, providing flow rates are not excessive. [2]

Case 13.9: The first Royal Navy submarine, *Holland I*, was launched in 1902 and sank in 1913 while being towed for scrap. It lay forgotten on the seabed until 1982 when it was lifted and transported to a museum in Portsmouth. All those involved with its retrieval remarked on the lack of corrosion on the inside, a phenomenon explained by the very low levels of oxygen. The hull was made of wrought iron and had very few perforations. On the outside, it had corroded from a thickness of 11 mm to a mean thickness of about 6 mm, a wastage rate of about 0.06 mm y^{-1}. After a very successful restoration, the submarine is now on display. [3]

In large volumes of electrolyte, the diffusion path from the free air/liquid interface to the cathode will be much longer than in a moisture film formed on a metal surface. Hence the level of available oxygen at the cathode will fall, and limit the corrosion rate.

If the pH $\geq$ 7, the oxygen reduction cathodic reaction is written as eqn (4.22), but oxygen reduction when pH $<$ 7 is better written as

$$O_2 + 4H^+ + 4e^- \rightarrow 2H_2O \tag{13.3}$$

At 50°C in tap-water, the corrosion penetration rate for low carbon steel falls from 7.5 mm y^{-1} at 6 ppm oxygen, to 1 mm y^{-1} at 1 ppm oxygen, although as little as 0.1 ppm oxygen can cause corrosion rates to increase in a flowing system. Problems arise when the level of dissolved oxygen varies in different parts of the system; differential-aeration corrosion may then occur.

Reducing the level of dissolved oxygen uniformly to very low levels can reduce the corrosion rate to negligible proportions if the pH is neutral or alkaline. This is seen in domestic central heating systems which use copper pipes and steel radiators, and in car engine cooling circuits where cast iron blocks are linked to copper radiators. The bimetallic corrosion in both systems is high just after they have been filled or refilled with water. However, the cathode reaction quickly consumes the dissolved oxygen, a thin layer of black magnetite forms on the steel surface, eqn (13.4), and the corrosion rate falls.

$$3Fe(OH)_2 \rightarrow Fe_3O_4 + H_2 + H_2O \tag{13.4}$$

Should the system leak, the replenishment of the oxygen supply in the make-up water renews the bimetallic attack. Hydrogen periodically has to be bled from central heating systems.

The importance of maintaining low oxygen levels can be seen in the following case histories.

Case 13.10: Some sectors of the American motor industry suffered severe corrosion problems in engine blocks when the plastic overflow bottles in the cooling circuits cracked and allowed fresh oxygen to dissolve continuously in the cooling water.

Case 13.11: Hastelloy B, a nickel–molybdenum alloy designed for use in reducing environments, was specified for the manufacture of a pump in a process line transporting 35% sulphuric acid. The pump failed in two weeks. The sulphuric acid was passed through a scrubbing jet which aerated the system; had this aeration not occurred, the material might have proved satisfactory. Hastelloy C, a nickel–chromium–molybdenum oxidation-resistant alloy, worked well (see Section 15.7).

13.4 CONTROL OF AQUEOUS ENVIRONMENTS

The control of electrolyte quality is particularly important for water used in industrial heating and cooling systems, steam-generating plant and condensers. The corrosion rate in these systems increases as the pH falls and as the oxygen content rises. Water may be used in once-through processes, in which minimum control is exercised owing to cost, or in recirculating systems for which continuous monitoring is required with careful control of make-up water.

In low temperature once-through systems using fresh water, a low cost inhibitor may be used, or the composition of the water may be adjusted to produce a thin protective scale on the surface of the metal. The scale is mostly composed of calcium or magnesium carbonates. It should be of negligible thickness and self-healing. It must not grow so that it impedes or blocks the flow of water, or significantly alters the heat transfer characteristics of the system. If the scale is damaged and does not self-heal, corrosion can occur on the bare metal surface. The production and maintenance of such a scale requires careful balance of water chemistry. The solubility of calcium carbonate in water is low, and the film is precipitated from the bicarbonate ions produced from dissolved carbon dioxide:

$$CO_2 + H_2O \rightleftharpoons H_2CO_3 \rightleftharpoons H^+ + HCO_3^- \tag{13.5}$$

If calcium salts are present:

$$CaCO_3 + H_2O + CO_2 \rightleftharpoons Ca(HCO_3)_2 \qquad (13.6)$$

On heating or if there is a slight deficiency of carbon dioxide in solution, causing the pH to rise, the reaction is driven to the left and calcium carbonate precipitates.

The ability of the water to form a scale is given by the **Langelier index**, which depends upon the pH, temperature, concentration of calcium salts and total dissolved solids. The calculation and application of the index is beyond the scope of this book, but a description of its use will be found in most water treatment books. If there is just too little dissolved carbon dioxide to keep the bicarbonate in solution, calcium carbonate deposits on the cathodes, owing to the slight rise in pH which accompanies the cathode reaction. If the film is complete and adherent, then the metal is effectively separated from the water by a large barrier to oxygen diffusion. No scale is formed if there is too little carbon dioxide. Excess carbon dioxide causes the scale already formed to redissolve in the acid solution, leaving the metal unprotected.

In recirculating steam-generating plant principally made of steel, the water is treated to maintain the pH above 11 and remove scale-forming salts. The main feed water consists of distilled condensate from steam which has already passed through the system. The impurities in the condensate will be carbon dioxide, oxygen and dissolved salts, mainly sodium-based, which have been carried over by the steam. The additional water for 'making-up' the system will contain calcium and magnesium salts (producing the hardness value) and further quantities of dissolved oxygen and carbon dioxide. In marine installations and those using estuarine water as an initial source, large quantities of sodium chloride may also be present. Let us first consider the effect of these impurities on the corrosion of a steam generator.

Carbon dioxide is very soluble in cold water, producing carbonic acid, eqn (13.5), pH of 5.5–6.0. When the water is boiled, the gas is ejected and passes through the system, redissolving in any condensate. Condensate water in any part of the system will always have a pH well below that required to maintain a passive film on steel surfaces (Fig. 13.5).

Oxygen will increase the efficiency of the cathode reaction in the alkaline conditions which always exist in steel boilers. It can also cause oxygen or air-bubble pitting if it is ejected from the water as the temperature rises and is allowed to collect in the system. When the boiler is working hard, the turbulence carries the gas out of the unit, but in quiescent periods or at shutdown, it rises to collect in air traps or on the top surfaces of the boiler and it is then that oxygen scavengers may be advantageous. The development of an air bubble pit is shown in Fig. 13.6. The oxygen-rich area at the base of the bubble becomes a cathode in the alkaline water, while the surrounding metal corrodes. The corrosion product settles on the bubble wall. Once the bubble is completely encased in corrosion product (iron hydroxide), ingress of further oxygen to the bubble is prevented, although iron (II) ions can still diffuse. Once the oxygen in the bubble is consumed, the area at its base can no longer support the oxygen

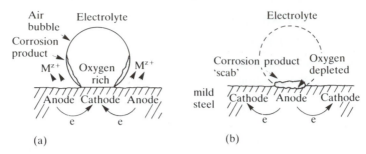

Fig. 13.6 Air-bubble pitting. (a) A differential-aeration cell is generated around the bubble. The oxygen-rich area at the base forms the cathode and the surrounding metal forms the anode. Iron (III) hydroxide corrosion products form on the bubble wall. When the wall is totally encased in corrosion product, the diffusion of oxygen into the bubble ceases, although metal ions can still diffuse through the wall. (b) Once the oxygen content at the bubble base falls below that of the surrounding water, the differential-aeration cell reverses. The small anode at the bubble base develops into a pit. The wall collapses to form a scab over the pit, below which differential aeration continues to drive the cell.

reduction cathode reaction and becomes an anode. The surrounding metal assumes the role of the cathode — the cell has reversed. Generally, the skin on the bubble wall collapses to seal off the new anode and differential-aeration corrosion continues beneath the scab to produce a deep pit.

Other forms of scabbed pits develop on steel, again assisted by differential aeration below the corrosion product.

The dissolved salts of magnesium and calcium precipitate from the water as it evaporates, producing layers of scale on the metal surfaces. As the scale thickens, heat transfer rates are reduced, leading to a loss of efficiency, increased risk of buckling or distortion and an increased risk of ash deposits forming on the hotter areas on the fire side of the boiler tubes.

The treatment of feed and top-up water for steam generators is a complex procedure which depends upon the level of impurities present and the operating conditions of the unit, especially the temperature and pressure of the steam.

Amines are a group of organic compounds which have a considerable inhibitive effect on corrosion. A particularly large group are the non-volatile, synthesised, high molecular weight, film-forming amine inhibitors based on imidazoline. Amines are added to control the harmful effects of carbon dioxide. Neutralising amines, such as morpholine, are added in very small quantities. They are volatile and weakly alkaline. The alkaline vapour condenses and dissolves in the condensate to neutralise the carbonic acid which forms as the carbon dioxide redissolves. Filming amines, such as octadecylamine, are volatile but insoluble in water. The vapour is driven off by steam but condenses on the cooler metal surfaces to form an oily hydrophobic layer which separates the metal from the condensate.

Case 13.12:   The first corrosion inhibitor was tested in sour West Texas oil wells in 1946 and in a sour gas line in Texas in 1949. The system handled 10 million cubic feet $(300,000 \text{ m}^3)$ of gas per day and it was reported that maintenance due to corrosion was very expensive. The use of Kontol 118, an amine-based inhibitor, reduced the corrosion by 99%.

Oxygen can be reduced to low levels by mechanical means, and de-aeration is a technique frequently applied in oilfields. If dissolved oxygen concentrations in the water exiting the deaerator are higher than desired, the remainder is removed with oxygen scavengers such as sodium sulphite or ammonium bisulphite. [2] Hydrazine can also be used to scavenge oxygen chemically and is preferred in high pressure systems with high steaming rates. None of the reaction products is harmful to the metal system, but hydrazine is toxic, especially in confined spaces:

$$N_2H_4 + O_2 \rightarrow 2H_2O + N_2 \tag{13.7}$$

Hydrazine decomposes if it does not react with oxygen:

$$3N_2H_4 \rightarrow 4NH_3 + N_2 \tag{13.8}$$

The ammonia acts in a similar way to neutralising amines in combating the effect of carbon dioxide, but hydrazine and ammonia can be corrosive to copper and copper alloys in condensers. Therefore, the excess concentration of hydrazine is limited to very low levels, typically 0.1 ppm. The sulphite is used in low temperature or closed-down systems, but excess sulphite should be avoided as it is aggressive to steel surfaces. In modern plant, the maximum level of dissolved oxygen is kept below 0.03 ppm, whereas in high pressure power stations oxygen in the feed-water is kept below 0.007 ppm. Control of pH is achieved by adding either sodium hydroxide, in conjunction with sodium hexametaphosphate, or sodium orthophosphate. In hard waters the ortho-phosphate precipitates sludge, which necessitates periodic 'blowing down' of the boiler to remove the residue. In marine practice, Navy Boiler Compound is used to maintain the quality of the water in boilers. It consists of 38% sodium carbonate, 48% disodium hydrogen phosphate and 13% cornstarch. Sodium carbonate decomposes under the boiler pressure and temperature to give sodium hydroxide, which provides the alkalinity to keep the pH above 11. Disodium hydrogen phosphate, in the presence of the carbonate, precipitates the calcium and magnesium ions. Starch acts as a catalyst and may produce flocculent precipitates of the neutralised salts. In seriously contaminated boilers, sodium carbonate (soda ash) is used alone.

Stainless steel closed loops are used in nuclear power steam generating systems. The water in these loops is passed through ion exchange columns to remove ionic contamination, and hydrazine is used to scavenge oxygen. One of the main causes of failure in the stainless steel loops has been stress-corrosion cracking. As early as 1967 it was well known that the chloride and oxygen content of the water passing around the loop controlled the time to failure of the metal. Sulphate ion contamination resulting from carry-over of the ion exchange resin also induces stress-corrosion cracking of sensitised stainless steel in the presence of oxygen. In these systems, the efficient operation of the ion exchange unit and the oxygen scavenging are of paramount importance.

Today, steam boiler corrosion is less important in the marine industry because of the conversion to gas turbine propulsion, but the subject of boiler corrosion is becoming increasingly important in the application to incinerators for waste disposal. Although the disposal of waste by burning is as old as civilisation, the idea of controlled burning on a large scale was first proposed by Alfred Fryer in Manchester. In 1874 a 'destructor' for household refuse was built and he was granted a patent two years later. Odours resulting from the destructor were tackled by means of a 'cremator', patented in 1885, and soon this combination became known as an 'incinerator'. [4]. Most were refractory-lined furnaces that discharged through a smoke stack to the atmosphere. Some early units made use of energy recovery by having a boiler downstream of the furnace. A steam generator of this type entered service in Hamburg in 1893. Only by the early 1970s was it recognised that chlorine is the primary cause of corrosion in these systems. [4]

For a detailed description of scale and deposit control, and relevant inhibitors, see Wilkes [5] and Laronge, [6] which also contain excellent lists of references for water treatments.

Case 13.13:   From a corrosion standpoint, acidising is an intriguing technique used in the oil industry to solubilise water-insoluble salts which would otherwise restrict the flow of oil and hence limit production. In 1895 Herman Frasch suggested the idea of pumping hydrochloric acid into oil-producing limestone to act upon the rock some distance from the borehole. By dissolving the limestone, oil flow would be stimulated. An experiment was carried out using 65 barrels of hydrochloric acid with an initial pressure of 900 psi (6 MPa). After 40 days of acidising, the pressure required had fallen to 150 psi (1 MPa); oil production had risen by 300% and gas production by 400%. But the early optimism was short-lived, because the hydrochloric acid simply destroyed the steel components of the rig. After 3 months, most wells using the acidising technique were shut down. One might think this enough to bury the idea, but the promise of increased production enabled research to continue. By 1932 the technique was back in favour, supported by the use of inhibitors. The first generation was based upon arsenic compounds. Approximately 0.5% of arsenic calculated as As_2O_3 was added to 15% HCl. Uninhibited acid in a well would cause $3.3 \, g \, m^{-2} \, day^{-1}$ steel weight loss. A galvanising inhibitor reduced it to $1.7 \, g \, m^{-2} \, day^{-1}$. But the arsenic inhibitor limited loss to only $0.016 \, g \, m^{-2} \, day^{-1}$, making the acidising technique economic. Arsenic inhibitors have now been replaced by materials which are largely organic. [7]

13.5   SOIL AS AN ENVIRONMENT

The corrosion rate for metals buried in dry soil is negligible. However, as the moisture content rises, the corrosion rate for buried metal is controlled by the conductivity, pH, oxygen content, aggressive ion concentration and biological activity in the soil. The nature of the soil (sand, silt or clay) and the values of these parameters can vary considerably. Long pipelines or deep oil and gas well casings may pass through several different types of soil which generate anodes and cathodes on the metal surface, often separated by large distances.

The conductivity and pH depend upon the mineral content of the soil and any ions which are washed into it. Fertiliser spread on fields, or de-icing salts washed from the roads will affect the corrosivity of the surrounding soil.

Case 13.14:   A buried stainless steel pipeline carrying steam condensate passed under a road. The pipeline gave excellent service, except for the areas directly beneath the two edges of the road. Here, de-icing salts applied to the road in winter ran into the soil and caused chloride stress-corrosion cracking of the hot pipe within two years.

Differential-aeration cells can be set up between the top and bottom surfaces of pipes, or down buried piles and well casings owing to the generally lower oxygen content at deeper soil depths. The same effect can happen where metal systems pass under roads, which are relatively impervious to oxygen. However, the moisture content can also be reduced and the practical outcome may be difficult to predict; indeed it may vary over a period of time depending upon the amount of rainfall.

Pipelines which have been buried for a long time can become encased in corrosion products which retard the rate of corrosion. Should replacement lengths of pipe be installed in the old line, the new metal is anodic to the corrosion product on the old. Severe pitting corrosion occurs at holidays (small pinholes in a paint coating) in the coating on the new pipe, which often fails in a much shorter period than the original ones. Indeed, in clay soils there were several instances where the old mild steel gas pipes corroded away completely, leaving the gas flowing through holes in the impervious clay. The problem only became serious when natural gas, supplied at a higher pressure, was substituted for low pressure town gas.

Case 13.15:   The first crude oil line was laid in 1865, the first gas line in 1876 and the first product line in 1930. In 1923 the common practice was to cover in bitumen without wrapping or reinforcement. The need for mechanical reinforcement was soon apparent and the use of hair felts and asbestos became common. In 1938 one of these early lines after 13 years of service was changed over to natural gas. It was estimated that only 10% of the gas reached its destination because of corrosion leaks. [8]

Today, buried structures are normally protected, either with barrier coats, paint, bitumen or concrete, or by cathodic protection. These methods are discussed fully in Chapters 14 and 16. Short runs of pipe are often placed in inert ducting, sealed to prevent the ingress of water which cannot drain away. On longer runs in trenches, suitable backfills can ensure good drainage and control of pH. Granite chips are used to improve drainage, limestone to control pH.

Correct backfill is very important around anodes used in cathodic protection systems. The backfill provides a uniform environment for the anode and ensures an even reaction over the whole of the active surface. It also prevents contact with soil which might passivate the anode and reduce its effectiveness. For zinc or magnesium anodes, a mixture of 75% gypsum, 20% bentonite (clay) and 5% sodium sulphate is used. For small anodes the backfill and the anode are placed in perforated bags, whereas a loose filling is used for larger ones. The backfill attracts moisture and increases the conductivity of the area around the anode. It prevents local corrosion of the anode by aggressive elements in the soil, and it prevents passivation of the anode surface. Impressed current anodes are surrounded with carbonaceous material, usually a high carbon coke mixed with 5% slaked lime. This has a low resistivity: less than 50 Ω m. Owing to the efficient current transfer from the anode to the backfill, most of the reaction is on the outer edge of the backfill, increasing the apparent anode size and the anode life.

13.6 REFERENCES

1. Fisher L E 1993 Corrosion inhibitors and neutralizers, past, present and future. Paper 537 in *Corrosion 93 plenary and keynote lectures*, edited by R D Gundry, NACE International, Houston TX.
2. Patton C C 1993 Are we out of the Iron Age yet? Paper 56 in *Corrosion 93 plenary and keynote lectures*, edited by R D Gundry, NACE International, Houston TX.
3. Waite D F and McKendrick R D 1984 *Holland I — an underwater miracle, The Metallurgist and Materials Technologist* 16(2): 70.
4. Krause H H 1993 Historical perspective of fireside corrosion problems in refuse-fired boilers. Paper 200 in *Corrosion 93 plenary and keynote lectures*, edited by R D Gundry, NACE International, Houston TX.
5. Wilkes J F 1993 A historical perspective of scale and deposit control. Paper 458 in *Corrosion 93 plenary and keynote lectures*, edited by R D Gundry, NACE International, Houston TX.
6. Laronge T M 1993 Water in boilers — a retrospective and a prospective. Paper 54 in *Corrosion 93 plenary and keynote lectures*, edited by R D Gundry, NACE International, Houston TX.
7. Cizek A 1993 A review of corrosion inhibitors used in acidizing. Paper 92 in *Corrosion 93, New Orleans LA*, NACE International, Houston TX.
8. Sloan R N 1993 50 years of pipe coatings: we've come a long way. Paper 17 in *Corrosion 93 plenary and keynote lectures*, edited by R D Gundry, NACE International, Houston TX.

13.7 BIBLIOGRAPHY

Anon 1985 *Corrosion in sulfuric acid*, NACE International, Houston TX.

Bogaerts W F and Agema K S 1992 *Active library on corrosion*, CD-ROM for DOS-PC, Elsevier, Amsterdam.

Bregman J J 1963 *Corrosion inhibitors*, MacMillan, New York.

Fitzgerald J H 1993 Evaluating soil corrosivity then and now. Paper 4 in *Corrosion 93, New Orleans LA*, NACE International, Houston TX.

Hausler R H 1988 *Corrosion inhibition — theory and practice*, NACE International, Houston TX.

Lewis B, Lichtenstein J, Enloe C L and Moore J L (eds) 1993 Underground corrosion control. NACE International, Houston TX.

Manning M I 1988 Effects of air pollution on structural materials, *Metals and Materials* **4**: 213–17.

Nathan C C 1973 *Corrosion inhibitors*, NACE International, Houston TX.

Pollock W I and Steely C N 1990 *Corrosion under wet thermal insulation*, NACE International, Houston TX.

Putilova I N, Balezin S A and Barannik V P 1960 *Metallic corrosion inhibitors*, Pergamon, Oxford.

Raman A and Labine P 1993 *Reviews on corrosion inhibitor science and technology*, NACE International, Houston TX.

Romanoff M 1989 *Undergound corrosion*, NACE International, Houston TX.

TPC Publication 1990 *Cooling water treatment manual*, 3rd edn. NACE International, Houston TX.

Wilkes J F 1993 A historical perspective of scale and deposit control. Paper 458 in *Corrosion 93 plenary and keynote lectures*, edited by R D Gundry, NACE International, Houston TX.

14 CORROSION CONTROL BY COATINGS

There is no way to success in our art but to take off your coat, grind paint and work like a digger on the railroad, all day and every day.

(R W Emerson, *The Conduct of Life: Power*)

From the time of the Industrial Revolution, the first iron marine structures were coated with natural materials such as coal tar, asphalt and oil-base coatings. It was not until the late nineteenth century that varnish was made from natural resins and oil, and that paint was available which, when pigmented, was called enamel. In 1928 an oil-soluble phenolic resin paint was developed and additional water and corrosion resistance was obtained. The modern age of using paint for corrosion resistance had begun. Yet the spread of this concept throughout the industrialised world took time. In the 1950s, for example, many World War II tankers were being used to transport oil products around the world mostly in uncoated tanks. Corrosion was measured in tons of scale removed from the tanks during annual repair; it was so severe the bulkheads may have lost 50% of their thickness in 7 years.

Fact: In the United States it was estimated that in 1975 about 40% of the steel production was used to replace parts which had rusted and 92 million gallons of maintenance paint for use on steel was sold.

Today, coatings of many different types are applied to metal surfaces, mostly to separate the environment from the metal, but often to control the micro-environment on the metal surface. Coatings can be organic or inorganic, metallic or non-metallic; the range of possibilities is vast.

The technology of paint and its application is changing fast, spurred on by ever rising costs of energy, raw materials and labour, but most of all by the need for environmentally acceptable products and production methods. Pollution has become a problem in a conservation-conscious society. It has been calculated, for example, that about 360,000 tonnes of volatile organic compounds are released into the atmosphere every year as a result of the application of paint. The paint industry has responded by rapidly replacing organic solvents with 'high solids' water-based formulations for less evaporation of undesirable materials. Dangerous lead additives have mostly been replaced with harmless materials such as titanium dioxide, a white pigment additive.

There has also been a recent tendency to brighten up industrial plant with more colourful paint schemes on the basis that people work more efficiently and safely in a more attractive environment, enhancing the owner's image and generating good public relations.

The protection afforded by coatings is remarkable in view of their usually small thickness. A typical paint coating is between 25–100 μm thick. Heavy duty in severe environments, such as offshore use or water treatment and chemical applications, may require a combination of coatings several millimetres thick, but a lacquer inside a can of food may be as little as 5 μm. Table 14.1 lists some of the many factors to take into account.

We have already seen (Section 12.2) that temporary coatings may be applied to protect the metal, or even the service coat, during a hazardous stage of fabrication, storage or transit. Many cars are now coated with water-soluble temporary paint to protect them from minor damage during delivery to a showroom. When the car is safely in the dealer's hands, the paint is removed in a modified car-wash. A black gloss temporary paint has also been used to show up small dents in car body panels. The imperfections can then be knocked out and the temporary paint removed.

Table 14.1 Considerations for optimum selection of a coating

Key aspects	Considerations
Project requirements	Nature of structure
	Lifetime required
	Time available for application
	Primer applied in factory or on site
	Access for maintenance
	Special requirements, e.g. contact with food
Exposure conditions	Climate
	Atmospheric or immersed
	Cathodic protection
	Contact with chemicals
	Impact and abrasion
	Surface temperature
Surface preparation	Techniques and staff available
	Contamination removal
	Ability to achieve required grade
	Surface profile attainable
	Access available
	Environmental aspects
Application and safety	Techniques and staff available
	Expected climatic conditions
	Access available
	Time and number of coats required
	Environment and safety
Economics	Available budget
	Can economies be made

Many modern applications are found for coil-coated steel. This steel is factory-produced by unrolling a massive steel coil and passing it through a paint sprayer. The steel is immediately baked to dry and cross-link the paint, after which it is recoiled and is ready for transport to a fabrication process. Refrigerator, freezer and washing machine panels can be pressed out of the sheet and are ready for immediate assembly. Alternatively, items such as guttering or railings can be formed on-site by installers.

The discussion in this chapter tries to summarise the main features in an enormous, rapidly changing, worldwide industry. It covers not only paints, but plastic, concrete and metallic coatings.

14.1 PAINT SYSTEMS

A coating is in fact a complex combination of materials. The complete coating is referred to as a **paint** or **coating system**, and it will consist of one or more layers, known as **coats**, each of which is carefully selected for the task. The first coat, applied to bare metal or concrete, is a **primer**; the last coat is a **finish** or **topcoat**. Any other coats are **intermediate** coats.

A paint consists of

A vehicle: the liquid that gives the paint its fluidity and dries or evaporates to form a solid film.

A pigment: suspended in the vehicle, the pigment controls the corrosion reaction, or the rate of diffusion of the reactants through the dry film.

Additives or fillers: these accelerate the drying process or better enable the dry coating to withstand the working environment.

The vehicle may dry by one of three processes. There may be **evaporation** of one of its solvent constituents. There may be **chemical change**, mainly **oxidation**, of its liquid constituent, linseed oil, for example. The paint dries from the outer surface inwards and is applied in a number of thin coats to build a thick layer.

And there may be **polymerisation**, a chemical reaction between the vehicle and a **curing agent** mixed into the paint just before application. The curing agent is kept in a separate container and the paints are referred to as **two-component** or **twin-pack systems**. The paint polymerises to form a cross-linked structure, its properties determined by the extent and nature of cross-linking. The curing process commences immediately the agent is added . After mixing, there is very little time before the paint becomes unsuitable for application. This is called the **pot life**.

When paint has dried, the remaining solid portion of the vehicle forms the **binder**. It holds the pigment in place, keys the film to the surface and provides a barrier to restrict the passage of water, oxygen and aggressive ions to the metal surface.

Although pigments do give colour to the dry paint film, they fulfil two much more important roles. First, in a primer coating, the pigment controls the corrosion process at the metal surface, either by inhibiting the reactions or providing sacrificial protection to the substrate metal. Second, in the finish coats, inert pigments increase the length of the diffusion path for oxygen and moisture penetrating the film. This is to delay the onset of the corrosion process and to slow the reaction rates. The mechanism of corrosion control by pigments is complex and only a brief summary of the processes can be given here.

When clean steel is exposed to dry air, a film of iron (III) oxide forms on the surface. This film is impervious to the diffusion of iron ions and protects the underlying metal. However, water (even the thin layers deposited from the atmosphere when the relative humidity exceeds 60%) breaks down the integrity of the film and rusting occurs. The iron (II) ions formed by the dissolution of the iron react with the hydroxyl ions produced on the cathode to form $Fe(OH)_2$ in the first instance, and finally rust, $FeO(OH)$.

Some basic pigments, notably lead compounds but also those of zinc and cadmium, inhibit the corrosion reaction on steel by stabilising the iron (III) oxide. When they are ground with linseed oil, the lead compounds form soaps which deposit metallic lead on the steel surface. Here oxygen reduction proceeds more easily and generates a sufficiently high current density to maintain iron (III) film production until the oxide coating on the steel surface thickens and it becomes impervious to iron ion diffusion. Lead salts are more efficient inhibitors than zinc and cadmium, but lead and cadmium are particularly toxic and their use is becoming more and more restricted; lead-based paints must never be used where they will come into contact with food or drinking-water, and people who handle the paint, or who clean or remove old lead paint, should be given frequent medical checks.

Another common class of pigments contains chromate salts which yield chromate ions in solution after water has penetrated through the binder. The chromate ions react with the air-formed iron (III) film to produce a complex chromium–iron oxide (a spinel), which forms an impervious barrier over the metal surface, as well as acting as an anodic inhibitor. Chromate salts are thought to be carcinogenic, so paints containing them should be treated with care; obtain advice on their handling from the manufacturers.

Some soluble lead pigments increase the pH of any water which penetrates the binder, and this inhibits the corrosion reactions on the steel. Zinc phosphate, either alone or with red lead, also produces a tightly adherent barrier layer.

Anodic metallic pigments can protect the steel sacrificially. To be successful, there must be electrical contact between the particles of the pigment, and between the pigment and the steel surface. Formulations based on zinc and aluminium are now successful in marine environments and other aggressive locations. The initial penetration of water through the binder causes zinc, for example, to corrode sacrificially, producing zinc hydroxide. Carbon dioxide also diffuses through the binder to react with the hydroxide and form a carbonate. The zinc corrosion product fills the pores in the paint film and produces an impervious, compact and adherent layer. A high zinc content in the dry paint film is essential if the first part of the reaction is to succeed in blocking the pores;

once the film is sealed the loss of electrical contact between the pigment particles is not so important.

Micaceous iron oxide, often termed 'MIO', occurs as thin plate-like particles which can be used to strengthen organic films, giving better durablility and improved resistance to ultraviolet radiation in strong sunlight. The particles also hinder the diffusion of oxygen and water through the film. Glass beads such as quartz fulfil a similar function and can be built up into thick coats.

Small additions of other materials are made to modify the properties of the paint. **Organic metal salts** are added to accelerate the drying of the paint. **Antioxidants** prevent skins forming on the paint surface while it is still in store, but must not interfere with the drying of the applied paint film. **Surface-active agents** help to disperse the pigment uniformly through the paint, and to prevent segregation as the paint dries. **Thixotropic agents** reduce runs and drips in the wet coat.

14.2 PAINT CHARACTERISTICS

A dried paint coating is expected to give a long life during which it will restrict the access of air, moisture and aggressive ions to the metal surface. While many paint films are impermeable to ions such as chloride, sulphate and carbonate, no paint film forms a complete barrier to either oxygen or water. These, in time, always penetrate to the metal surface and paint films cannot, therefore, inhibit the cathode reaction.

Paints which are exposed to the atmosphere rely upon pigments to control corrosion (see Section 14.1). Such paints must be tolerant to wide variations and rapid changes in the ambient conditions. Not only will the temperature of the surface fluctuate throughout the day but thermal expansion of the substrate will stress the paint skin. Changes in the relative humidity will introduce wetting and drying cycles which can cause the paint to swell or crack with water uptake. Ultraviolet radiation will degrade the paint surface. Aggressive ions in polluted atmospheres may directly attack the paint or reduce the pH of rainwater falling on the painted surface, leading to chemical changes in the pigment or binder which will decompose the coating. As the paint ages, the continued oxidation of the binder and loss of gloss will increase the permeability of the paint and result in erosion of the coating and more rapid loss of pigments.

Paint systems which are used under water may, or may not, have a pigmented primer, e.g. an inorganic zinc, but the top coats rely upon very low water absorption and transmission coefficients to restrict the access of the electrolyte to the metal surface. The bonds between the paint coats and from the paint to the metal substrate must be strong and continuous over the whole surface to prevent damage by osmosis (Section 14.3). A high quality surface preparation is required to ensure the paint wets the whole of the metal surface as it is applied.

An ability to resist attack by alkali is a most important property for paints applied to immersed structures which are protected by impressed current cathodic protection (Section 16.5). The cathode reaction on the metal surface releases hydroxyl ions which can soften many paints and lead to stripping and undercutting of the film. This increases the area of metal requiring protection from the impressed current system. The large shields which surround impressed current anodes are to protect the area in the immediate vicinity of the anode from this excessive alkali production.

The thickness of the dried paint film should be uniform over the whole of the surface to be treated, including the edges and corners of plates, bolts, exposed threads, rivets and joints. However, surface tension tends to reduce the thickness over sharp corners, as shown in Fig. 14.1. Rounding of edges helps to maintain the film thickness but special formulations are now available to reduce this effect.

A failure at any point in the coating will concentrate corrosion in that area and lead to further deterioration as differential-aeration corrosion spreads beneath the paint film. If a stone chip in the paintwork of a car is neglected, not only will a rust spot form in the chipped area, but the surrounding, apparently sound, paintwork will easily peel off to reveal rust between the paint and the metal. Thus, flexibility in coatings, to resist impact and other mechanical or abrasive damage, is a vital property in many applications.

The vagaries of the climate and the conditions under which most painting is carried out require that a paint should also be very easy to apply using a variety of methods such as brushing, rolling, spraying and dipping. It should dry very quickly to facilitate the application of additional coats, to prevent rain, dust and debris from spoiling the surface, and on maintenance painting, to reduce downtime and costs. It is usual to grit blast a surface before painting, in order to remove rust and other loose surface deposits, but there are now formulations claimed to be suitable for application to lightly rusted surfaces which do not require preblasting. Other paints can be applied in damp conditions or when rain is expected soon after painting.

Paint should be easy to touch up or repair, be resistant to attack by fungi and bacteria, and should have a good long-lasting appearance. This is a good reason

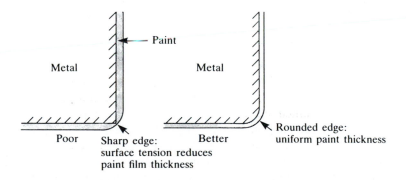

Fig. 14.1 The effect of surface tension on paint film thickness over a sharp corner.

for the application of multiple coats; in general, the greater the thickness, the greater the protection. Small pinholes in a paint coating are known as **holidays**. These can be very dangerous because large cathodes (the painted surface) cause high corrosion rates at small anodes (the holidays). Occasionally, in systems containing a bimetallic couple, only the anode is painted because, it is thought, only the anode will corrode. However, the corrosion rates at holidays are even greater than usual because of the combination of the area effect and the bimetallic effect.

14.3 PAINT TYPES

Paint systems are divided into broad generic groups named according to their use or their chemical nature. Table 14.2 summarises types of coatings, properties and applications for a wide range of modern coating systems.

Primers

The primer is the foundation of a paint system for steel. It has the largest influence on the corrosion protection properties of the coating system and must provide good adhesion to the surface as well as a good base for topcoats.

Modern primers are both lead- and chromate-free. They can be applied to suitably prepared dry steel, damp steel, non-ferrous metals and galvanised steel, as well as old paint systems. Formulations containing zinc phosphate are good for atmospheric exposure, have fast overcoating times and can be used in all seasons because of their low curing temperatures (down to $-10°C$). They can be applied in one coat up to 150 μm and can act as a **tie-coat** between aged alkyds and modern epoxies.

Zinc-pigmented primers are particularly formulated for corrosion resistance and suitable for both atmospheric and immersion conditions. Modern aluminium-pigmented primers are very versatile; in high solids formulations they contain only 10% solvent. With brush application, coats are typically 125 μm, rising to 200 μm with an airless spray-gun. They may even be applied under water with abrasive blasting.

Primers have traditionally been labelled **prefabrication** or **pretreatment** primers. Prefabrication primers are used at the manufacturing stage, whereas pretreatment primers are used as part of a maintenance package to condition the metal surface for good adhesion and performance of the final paint system. They offer only limited corrosion protection to the metal and should be overcoated as soon as the primer is dry. To obtain good results the steel surface should be clean and free from rust; pretreatment primers must not be applied over mill scale, prefabrication primer or old paint. It is recommended that a pretreatment primer should always be used on zinc coatings, particularly when zinc chromate paint will be used as a primer for the final coating system.

Aluminium and its alloys are treated with zinc chromate or zinc-oxide-based pretreatment coats, called **etch primers**.

Oil-based paints

Vegetable drying oils such as linseed or tung oil form the base for these paints. They take a long time to dry by oxidation and it is considered necessary to leave the paint for up to 48 hours between coats and up to 7 days before applying topcoats to primers. Red lead in linseed oil is a typical example of an oil-based primer. Zinc phosphate may be added should a shorter drying time be required. The pigment forms a soap with the oil. These paints are rapidly becoming rare as environmentally acceptable paints with superior performance are made available.

Oleoresinous paints (varnishes)

A drying oil and a natural or synthetic resin are used to form the vehicle in this broad class of paints. The resin improves the drying and binder properties of the coat which are an improvement over the simple oil-based types. Tung oil with 100% phenolic resin paints can be used on structures immersed in water, including ship hulls which are not protected by impressed current systems. Phenolic resin paints resist abrasion but do not tolerate surface dampness during application. Conversely, coal tar resin paints do tolerate some surface dampness but have poor abrasion resistance.

Alkyds

Alkyd paints are widely used and the number of types is unlimited. The basic paint is a polyester, prepared by reacting polyhydric alcohols with monobasic fatty acids and dibasic acids. A typical composition would be ethylene glycol, linseed oil and phthalic anhydride respectively. Long oil paints dry by oxidative polymerisation of the oil and therefore have a high oil content, more than 65%. Those which are dried by stoving (heating) usually do not require such high oil contents and are referred to as short length oils, perhaps less than 50% oil.

Epoxy paints

Epoxy resins (and urethanes) were developed just after World War II, but were not used until the late 1950s and early 1960s. Epoxies are now the primary building blocks of the organic protective coatings industry, used to protect steel and concrete structures in almost every climate and location. Indeed, the combination of epoxy for adhesion and corrosion resistance, and the urethanes for topcoats, is one of today's major offshore and marine coating systems.

Epoxies offer a variety of properties depending upon the molecular structure of the final cross-linked network and its various fillers and pigments. Figure 14.2 shows the basic process of creating an epoxy coating. The basic reaction is very

Table 14.2 Paint types, applications and properties (Courtesy of Sigma Coatings)

Paint type	Percent solids content by volume (approx)	Optimum dry film thickness/μm	Touch dry after	Time before overcoat	Uses and properties
Alkyds					
Fast-dry enamel	32	38	1–2 h	6–8 h	Exterior exposure
Tank-finish aluminium	52	25	45 m	6 h	Exteriors of steel storage tanks
Red oxide primer	42	38	2 h	24 h	Lead-free, general purpose, hard, water-resistant, anti-corrosive
Zinc phosphate primer	54	38	2 h	24 h	Lead- and chromate-free, general purpose for steel in atmospheric exposure, can recoat conventional paint systems
General purpose gloss	55	38	1 h	16 h	Areas intermittently exposed to water immersion and atmospheric conditions
High gloss, high quality finish coat	54	38	1 h	16 h	High gloss enamel for interior and exterior surfaces
Aluminium finishing coat	48	25	45 m	6 h	For dry cargo holds, good impact resistance
All purpose primer	32	25	10 m	30 m	Fast-drying, anticorrosive
Coal Tar Epoxies					
Polyamide cured coal tar	80	400	7 h	24 h	Two components, one coat, high build, outstanding oil and water resistance, cures to 10°C and resistant to 93°C in dry exposure
Polyamine-adduct-cured coal tar	71	125–500	4 h	6 h	Two components, high build, cures down to −5°C when substrate ice-free
Epoxy Coatings					
Solvent-free epoxy (spray or trowel)	100	500–625	2 h	6 h	High resistance against tidal and impact actions, good oil and water resistance, good adhesion to many surfaces
Weather-resistant epoxy acrylic	44	50	2 h	2–3 h	Two components, topcoat, excellent weathering resistance, colour and gloss retention, isocyanate-free

Paint type					Description
Non-skid epoxy	72	500–1000	20 h	24 h	Helidecks, ramps, walkways and concrete floors, resistant to chemical spillage
High build, high solids epoxy	85	75–200	5 h	16 h	Potable water systems, resistant to cathodic disbondment
High solids epoxy primer	73	75–150	3 h	16 h	Tenacious adhesion to steel and concrete as primer for epoxies, urethanes and vinyls, good chemical resistance suitable for food contact
Armour epoxy	100	3000–5000	6 h	24 h	Solvent-free, quartz-reinforced
Zinc-rich epoxy primer	46	25–63	10 m	6 h	Two components, fast-drying, useful for a holding primer for rapid turnaround
High build, anticorrosive epoxy	61	125–150	2 h	3 h	Two components, for land and marine structures, unlimited overcoating time, good impact resistance and flexibility
Multipurpose epoxy primer	52	50–125	30 m	8 h	Two components, holding primer, can be overcoated with many other paints
Epoxy-based adhesion primer/sealer	57	38	1 h	8 h	Two components; for atmospheric and underwater systems, excellent adhesion to and sealing of aged zinc coatings, resists temperatures up to 175°C in dry exposure
Micaceous iron oxide epoxy primer	61	75–150	2 h	3 h	Two components, for land and marine structures
High build epoxy lining	78	250	3 h	8 h	Useful tank coating in two-coat system, resistant to wide range of chemicals
Epoxy high build/finish coat	55	75–150	3 h	10 h	Two components, topcoat on primed steel and concrete for atmospheric conditions, seals inorganic zincs
Solvent-free epoxy	100	300	4 h	24 h	For crude oil/ballast tanks and aliphatic petroleum products
High build abrasion resistant epoxy	88	250–500	3 h	16 h	Amine-adduct-cured, glass-flake-reinforced
Acid-resistant epoxy	55	75–150	3 h	20 h	Two components, polyamine cured
High build, recoatable epoxy	62	150	2 h	12 h	Two components, general purpose

311

Table 14.2 *Continued*

Paint type	Percent solids content by volume (approx)	Optimum dry film thickness/μm	Touch dry after	Time before overcoat	Uses and properties
Solvent-free epoxy	100	>300	5 h	40 h	One coat, long-life protection for steel structures and storage tanks, suitable for ballast, crude oil and aliphatic petroleum products
High build, surface-tolerant epoxy	80	125–200	2.5 h	6 h	For maintenance where blast cleaning is not possible
Phenolic epoxy	60	50–100	45 m	18 h	Two components for low friction on inside of pipes, resistant to crude oil containing H_2S and CO_2
Phenolic epoxy lining	55	75–150		12 h	For sweet and sour crude, brine, processed petroleum products, CO_2- and H_2S-containing media
Phenolic epoxy tank lining primer/intermediate and finish formulations	66	100	2 h	24 h	Two components, good resistance to chemicals and higher temperatures
Heat Resistant Coatings					
Heat-resisting aluminium	28	35	30 m	8 h	Two-coat system or overcoat for inorganic zinc primer, good up to 360°C
Heat-resisting aluminium finish	50	25	60 m	16 h	Good up to 175°C
Moisture-cured inorganic silicate	38	50	60 m	5 h	Finish coat over inorganic zinc primer, good from −90°C to 600°C
Inorganic Zincs					
Water-based inorganic zinc primer	65	100–125	15 m	8 h	Self-curing alkali silicate primer under most paint systems, good from −90°C to 400°C, withstands rain 60 m after application, fair flexibility, good abrasion resistance when cured

Polyurethanes					
High solids aliphatic acrylic polyurethane	60	51–76	1 h	12 h	Two components, unlimited recoatability, excellent resistance to chemicals such as water, oil and solvents, tough, flexible and abrasion resistant, excellent hiding power
Aliphatic acrylic polyurethane	41	50	30 m	4 h	Two components, excellent colour finish, excellent resistance to atmospheric exposure, resistant to mineral oils, paraffin and gasoline
Waterborne Coatings					
Water-borne primer	43	50–75	10 m	2 h	Single-component primer as base for complete water-borne system on steel
Water-borne rust converting primer	43	50–75	30 m	6-8 h	For lightly rusted steel, eliminates the need for blasting, suitable for epoxy topcoats
Water-borne finish	42	50–75	30 m	1 h	For use over waterborne primer; for light to moderate chemical or high humidity exposure; good resistance to mechanical damage
Water-borne aluminium	45	50–75	15 m	2 h	For primer, intermediate or topcoats, excellent aluminium appearance and durability
Miscellaneous					
Polyvinyl butyral etch primer	10	13	5 m	1 h	Adhesion primer for ferrous and non-ferrous metals
Vinyl finish	35	50	60 m	16 h	Durable finish coat over high build chlorinated rubber and vinyl systems applied to interior and exterior surfaces in marine and industrial environments
Water-based decorative wall paint	42	38	15 m	2 h	Flat, decorative finish on walls and ceilings, for interior and exterior use
Low viscosity solvent-free epoxy sealer	100	–	–	16 h	Excellent penetrating sealer and adhesion promoter for concrete and prior to application of coatings, excellent chemical resistance

Epoxide resin component   Hardener

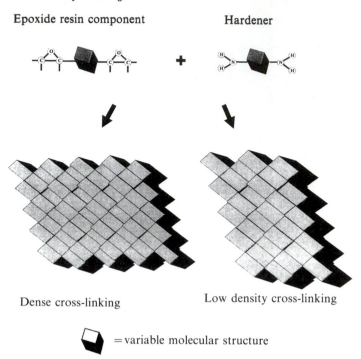

Dense cross-linking   Low density cross-linking

= variable molecular structure

Fig. 14.2 Formation of an epoxy coating. Both resin and hardener molecules can be modified during manufacture to give a range of epoxy coatings with a remarkable variety of properties. Dense cross-linking gives very good chemical resistance, whereas less cross-linking and long molecules gives a flexible product. (Courtesy Sigma Coatings)

versatile, for both resin and hardener can be modified considerably to generate different types of cross-linked structures. Dense cross-linking leads to good chemical resistance, whereas less dense cross-linking produces flexible films.

Six months' corrosion resistance to constructional steel on site can be achieved using fast-drying film in a fast shop application of only 20 μm thickness. Immediate handling is possible and there are no problems with welding or coverage by other kinds of topcoats. Epoxy systems can be used to coat wet-blasted steel even before it is dry, and the latest high solids formulations can be applied by brush and roller, even to surfaces that cannot be blasted. Airless pumps are used for spraying solvent-free epoxies at ambient temperature, an excellent development which conforms to environmental demands.

Excellent chemical resistance is offered by the phenolic variety of epoxy, and self-levelling coatings are applied to concrete floors for smooth chemical-resistant surfaces. For heavy-duty wear-resistant applications, such as jackets of offshore platforms, quartz particles are added for coatings up to 5 mm thick. For pipes, solvent-free powder coatings can be applied to the outside, while epoxy flow coatings on the inside give low friction and reduced pumping resistance.

314

Coal tar epoxy (CTE) paints

Another development of paint technology in the 1960s was the coal tar epoxy (CTE). The combination of inorganic zinc base coat, an epoxy polyamide primer and CTE topcoats forms the most resistant of all marine coatings. It is particularly resistant when combined with cathodic protection because it is resistant to attack by hydroxyl ions produced by the cathode reaction. The CTE paint film is one of the least permeable to water and resists attack by most chemicals. CTEs are used extensively on submerged areas of marine structures, ships, oil platforms and piers, as well as in the extremely aggressive environments found in water treatment facilities.

Case 14.1: Archimedean screw-pumps with a screw diameter as big as 5 m are often used for handling raw sewage at capacities up to $11 \, \text{m}^3 \text{s}^{-1}$. They are typically made from mild steel protected by abrasive blasting, epoxy primer and 300 μm CTE finish. (Courtesy of Sigma Coatings)

CTE paints may eventually be replaced by high solids formulations which do not contain coal tar, a constitutent now considered by many to be environmentally unacceptable.

Case 14.2: In 1986, tests were carried out at the Oxley Creek waste-water treatment plant in Brisbane, where local conditions frequently encounter severe slime growth which requires chlorine treatment and mechanical cleaning. After six years of exposure testing, a high solids epoxy coating in three films totalling 450 μm thickness outperformed similar films of standard CTE paints. (Courtesy of Sigma Coatings)

Polyurethanes

Polyurethanes form a group of paints which set by polymerisation. The reactants are an isocyanate curing agent and an alkyd polyester resin, possibly with pigments added. Air-drying systems come in two packs, whereas stoving paints allow the curing agent to be included with the resin in one pack. The properties of the dry paint depend upon the ratio of alkyd to polyester in the resin. As a group, the paints are expensive and they do not normally tolerate high humidities or dampness during application or curing. A pretreatment primer is usually required to give good adhesion, and this must dry thoroughly before the polyurethane is applied. However, the dry film has a high resistance to water, it is hard, abrasion resistant and retains a high gloss for long periods. Its performance often justifies the extra cost and care needed during application.

Vinyls

A very broad family of paints with many copolymers, vinyls are suitable for a wide range of applications. A common type is polyvinyl chloride/polyvinyl acetate copolymer modified with maleic anhydride. The copolymer is dissolved

315

in a solvent and the paint dries by evaporation of the solvent. However, the dry film is always soluble in the solvent and recoating is easy with new layers bonding to clean old paint. The drying time is short, 2–5 minutes. Vinyls are resistant to oils and greases. They form very effective coatings for steel structures immersed in water. However, the paints are very difficult to apply by brushing, so spraying is the normal method of application. Adhesion to bare steel is poor and a pretreatment primer is recommended. One suitable primer consists of polyvinyl butyral, zinc chromate and phosphoric acid. Surface moisture also impairs adhesion; it is not uncommon for whole areas of coating to peel off if the paint is applied to damp surfaces. Two other problems affecting their storage and use are the inflammability and toxicity of many of the solvents. But if these paints are correctly applied and matched to the environmental conditions, they give excellent service.

Chlorinated rubber

Case 14.3:   Chlorinated rubber paints were first developed in 1930. In 1938 the first high performance and chemically resistant coating system became available, combining chlorinated rubber as primer with a heavy-bodied vinyl intermediate and clear vinyl seal coats. One of its first marine uses was in fish tanks of tuna boats operating out of San Diego. Corrosion in the sea-water tanks was severe and fouled the fish. The vinyl coating not only greatly improved the corrosion resistance but was resistant to the fish oil and made the tanks easier to clean. [1]

Chlorinated rubber paints are produced by dissolving chlorinated rubber in a particular group of solvents (aromatics). After application the evaporation of the solvent deposits the dried film in which there is very little polymerisation. The rubber alone forms very brittle layers and inert plasticisers are added to the paint to yield a tough durable coating which retains a high resistance to acids and alkalis and has good weathering characteristics when exposed to the atmosphere. Adhesion to the substrate and between coats is very good; damaged areas are easily repaired. However, the paints are readily softened by many oils and greases.

Water-based paints

Water has a number of advantages as a base for paint. It is cheap, non-flammable, dries very rapidly and, in general, results in good adhesion to the substrate and between coats. Several paint systems are available, either dissolved in water or as emulsions. Three of them are vinyls, acrylics and epoxides. The vinyls and acrylics are used on masonry for decorative and waterproof coatings. Restriction of the penetration of water into concrete controls the corrosion of steel reinforcing bars (see Section 14.7). They can also be used as topcoats over zinc-rich primers. Epoxide paints are used over inorganic zinc primers as maintenance coats for structural steel, tanks and marine structures, as well as for initial paint treatments. Adequate ventilation must be provided while the paint is drying to maintain a low relative humidity over the surface. If the relative humidity is too high, water is trapped in the paint lattice as it dries, resulting in a poor performance from the coating.

The mass production car industry uses water-suspended paints for the intermediate coating on vehicle bodies. The process is known as electrophoretic coating. The bodyshell is first degreased and dipped in a phosphoric acid priming bath which deposits a tough corrosion-resistant layer of phosphate on the metal surface. It is essential to treat all the surfaces, internal and external, so large access holes are left in sills, box sections and pillars. The phosphate-treated body is immersed in a bath containing a water-dispersed paint formulated to give large, electrically charged paint particles. An array of electrodes is arranged around the body and a potential applied so that the particles of paint are attracted to and deposited on the metal, where they give up their charge. The neutral deposited layer acts as an insulator, and by limiting the potential applied, the thickness of the coat can be controlled. The electrolyte resistance and the distance between the electrode and the metal surface also control the electrodeposited paint thickness. Complicated electrode arrays, and very large holes to increase the throwing power of the electrodes, may be required to ensure an even thickness inside closed sections with limited access. The large holes allow easier ingress of moisture and mud when the car is in service, increasing the risk of inside-out corrosion. The electrodeposited film is baked in a stoving process to form a coherent and adherent paint coating.

After stoving, the external visible surfaces are primed and coated with final gloss coats, while the underbody is treated with an underseal. Additional protection is given to the electrodeposited coats inside box sections and sills by wax injection. The wax forms a film approximately 0.05 mm thick which is drawn into crevices and gaps by capillary action. Poorly applied wax coatings that do not completely fill a crevice by capillary attraction, but which merely seal its mouth, can cause increased corrosion damage. This occurs when an electrolyte enters the crevice by another route and is unable to drain away, owing to the sealing of the normal drainage route.

On small-volume luxury car production the expense of installing an electrophoresis coating line is not justified and many of these cars are hand-sprayed over dipped phosphate and primer treatments.

Inorganic zinc

The development of inorganic zinc coatings has been one of the most important discoveries, enabling as much as 15 years of service for marine structures. The first application of a zinc paint was in 1942, but the process was impractical and it was not until 1952 that a cold spray-gun technique was developed. Instead of heat, an amine phosphate was used as a curing agent. By the late 1950s zinc formulations were recognised as a major coating improvement.

These coatings are essentially a combination of zinc dust and complex silicates, with a binder that is either water-soluble or a solvent-based self-curing system. The dry film is tough, abrasion resistant, adheres firmly to the substrate, and appears to be unaffected by weathering from sunlight, ultraviolet light, rain or condensation. Recoating is easy. Depending upon the formulation and environmental conditions, the paint is capable of giving 10–40 years protection to steel from a one-coat application, protecting the steel sacrificially until the zinc corrosion products have plugged all the micropores and holidays in the coat. It has given excellent protection over extended periods when overcoated with a variety of paint types, such as epoxies, vinyls, chlorinated rubber or polyurethane.

Case 14.4: A coating on a 400 km pipeline, applied in 1942, was still giving good protection in 1984, despite being exposed to brush and grass fires, salt and marine environments, as it traversed bush, salt marshes and marine areas.

Case 14.5: Ten 350,000 tonne tankers were coated inside and out with a post-cured inorganic zinc coating, a combined area of 35 million square feet (3.25 km^2). After five years' service, most were found to be essentially corrosion-free both in the tanks and on the external hull. [1]

Antifouling paints

Antifouling paints are applied to the immersed areas of marine structures, as the final coat in a paint system. They release toxins into the water to prevent living organisms from attaching themselves to the structure. A heavy infestation of barnacles will increase the drag on a ship or the legs of an offshore platform, limiting the speed of the vessel and increasing its fuel consumption, or raising the stress levels in the platform (see Section 16.3).

Copper and tin are two toxins currently in use. The toxin is held as a pigment in the binder network and leached from it. The leach rate from a freshly applied coating may be very high, giving excess toxin. But an older coat is likely to have a surface layer of denuded binder which may impede the diffusion of the toxin. The dried paint should therefore be formulated to give a sufficiently high leach rate throughout the life of the coat, but not so high as to exhaust it before the next scheduled repainting.

In the 1970s a second generation of antifouling paints was introduced, called erodible or self-polishing. The paint matrix is an acrylic with a long molecular chain to which the toxin, tributyl tin, is loosely attached as a side-chain. The paint is impermeable to moisture, but on the surface of the paint water hydrolyses the side-chain to release the toxin. The matrix, weakened by the loss of the side-chain, erodes away to reveal a fresh tin-containing surface. The leach rate remains reasonably constant throughout the life of the paint and the smooth surface reduces drag. Although very effective, these paints have been shown to produce long-lasting local pollution and have now been prohibited in many areas.

The selection of a paint system is a task for a **specialist**. We have discussed only the broad outlines of the characteristics of each group and there are many pitfalls for the unwary. A primer and topcoat, even in the same generic group, may be incompatible; two apparently similar primers may require different pretreatments; a pigment may be incompatible with a substrate, or with some pollutant in the environment. Specialist help should always be sought before embarking on a costly painting exercise in which the actual paint may account for far less than 5% of the charges (see Section 14.5).

Case 14.6: A contract for painting two water tanks specified that the units should be blast-cleaned, pretreatment primed, vinyl primed, given two coats of a vinyl intermediate coat, and an aluminium pigmented topcoat. The fabricator carried out the process as far as the vinyl primer in his factory and erected the tanks on site. The painting contractor was obliged to blast-clean the erection welds and areas damaged in transit. To save money he then applied to the bare steel a vinyl primer which did not require a pretreatment primer, and carried on with the intermediate coats and the topcoat over the whole of the primed surfaces. The tank went into service. He started on the second tank but an inspector stopped him and demanded to know why he was not applying the pretreatment primer to the bare steel surface before the vinyl primer. He was made to comply with the specification. During removal of the scaffold from the second tank it was found that the coating had failed to adhere and was lifting from large areas of the surface before the tank was put into use. Inspection of the first tank revealed good adherence of the paint film and it subsequently gave a satisfactory life. Investigation showed that the contractor's vinyl primer, which he had used on both tanks, was incompatible with the pretreatment primer, hence the failure on the second tank. The fabricator's scheme which used a vinyl primer matched to the pretreatment primer was successful; the contractor's scheme which used a vinyl primer suitable for direct application to steel was also successful, but mixing the two paint schemes resulted in immediate failure owing to the incompatibility of the different coatings.

14.4 PAINT FAILURE

The major causes of failure in a paint system which has been selected to match the environmental conditions are

Poor or inadequate surface preparation.

Application of the paint coating under unsuitable atmospheric conditions or by inappropriate methods.

Some new formulations may be applied over light surface rust, indeed some coatings have a phosphoric- or tannic-acid-based component in the paint, claimed to convert rust to an inert adherent layer, usually oxidising Fe(II) to Fe(III). The paint then keys to the inert surface. However, most paints, particularly those which will be exposed to aggressive environments, benefit from a thorough surface preparation. For outdoor work, wire brushing a surface is not a satisfactory preparation because it leaves contaminants and potentially active corrosion products on the metal.

Satisfactory surfaces for painting are obtained either by chemical treatments, such as pickling by immersion in acid solution, or by grit-blasting to remove grease, dirt, corrosion products and scale. Pickling baths are normally inhibited to stop attack on the metal. Components treated by pickling must be thoroughly washed and dried after treatment to remove all the active chemicals from crevices and holes, and to ensure the paint is able to key to the metal.

Grit blasting was first introduced by the US Navy during World War II with increased demands for greater performance of ship hulls. Until then, most protective coatings had been applied directly over the mill scale on hot-rolled steel. At the same time, a 'wash-prime' (pretreatment primer) was developed which combined a vinyl butyral polymer with phosphoric acid to enhance the adhesion of primers. Grit-blasting can be combined with a water spray to wash away all the debris, dust and water-soluble contaminants which would reduce the adhesion of the paint. The surface profile produced by the grit is very important. If it is too rough, micropeaks will be poorly covered by the paint and premature penetration of the film will occur, as illustrated in Fig. 14.3.

For a satisfactory life, paint must be applied under the correct atmospheric conditions. If the relative humidity is too high, a thin invisible film of water will prevent proper keying to the metal surface and may interfere with the integrity of the dry paint film. Ambient temperature affects the drying or curing time. Solvent evaporation can be slow at very low temperatures and some of the two-component twin-pack paints will not cure below a specified temperature.

Variations in temperature across a component, especially where the paint is heated or stoved in an oven to speed or complete the drying process, can cause solvent which has evaporated from one area to condense on adjacent cooler surfaces which are below the dew-point of the solvent. The paint dissolves in the excess precipitated solvent, which runs off the surface leaving streaks in the finish and thin areas in the protective film (Fig. 14.4).

Many failures of twin-pack paint systems have been attributed to inadequate mixing of the two components before application or to incorrect proportions in

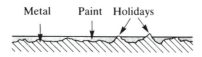

Metal   Paint   Holidays

Fig. 14.3 Paint thickness is affected by roughness of the substrate surface. Grit used for surface blasting should be graded to produce a uniform roughness if holidays are not to occur.

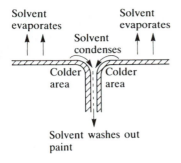

Fig. 14.4 Solvent washout. In a stoving oven the single-thickness skin heats up more quickly than the double-thickness joint. While the joint is below the dew-point for the solvent, condensation occurs in the joint and washes out the paint.

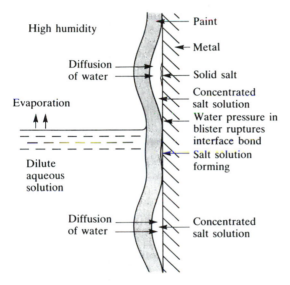

Fig. 14.5 Development of blisters in poorly adherent paint films owing to osmotic pressure.

the mixture. Failures have also occurred when twin-pack paints, which rely upon a cross-polymerisation cure to form the finished coat, have been applied below the minimum temperature for rapid and full curing. This can be as high as 15°C for some epoxy-based systems.

When moisture diffuses through the paint film, salt solutions based on soluble corrosion products or compounds from the paint can form at any point where there is a lack of adhesion between the paint layers or between the paint and the substrate (Fig. 14.5). The paint film acts as a semipermeable membrane; osmotic pressure rises with rising salt concentration at the defect until water is forced through the film to dilute the solution. This increases the water pressure in the defect and blisters the paint while spreading the area of delamination

321

between the layers. Besides ruining the paint finish, the pH of the solution in the blister can become very high and cause caustic burns to anyone who carelessly pierces the blister to let the 'water' out prior to remedying the defective paint scheme.

14.5 PAINTING COSTS

Not surprisingly, when it comes to coating a structure, the paint itself is only a small fraction of the cost. The biggest cost is to erect scaffolding for access to the structure, to maintain the correct temperatures and relative humidities, to clean and dry the painting surfaces, to provide adequate supervision and inspection. Any loss of production, any temporary relocation of staff must also be taken into account.

In the repainting of a large land-based plant, perhaps with large runs of open steel framing, the cost of the paint may be only 5% of the total bill; in marine locations it will be even less.

Case 14.7: For centuries the reclaimed land of the Netherlands has been protected from the sea by large engineering structures. In 1953 work began on the large Delta Works project. One of the first parts to be completed was the Haringvliet Sluices near Rotterdam. The large concrete structure has 17 tidal flow openings, each of which allows a water passage of 20,000 $m^3 s^{-1}$ and has two hollow steel gates, 56 m long, 12 m deep, 400 tonnes weight. Between 1962 and 1974 coal tar epoxy was selected to coat the total area of 200,000 m^2, during which time only planned maintenance was necessary. Since then, a tar-free coating system has been applied, consisting of primer to 75 μm and two coats of high solids, solvent-free epoxy totalling 240μm. Rivets, bolts, welds and sharp edges are coated to 2–3 mm thickness; the flexible coating has excellent coverage and resists the formation of cracks. During preparation, the gates are protected by tents of area 2,500 m^2. These allow the workers to continue in gales up to force 8, while dust from the grit-blasting is sucked from the tents at 8,000 $m^3 h^{-1}$. Meanwhile, dry salt-free air, heated to 15°C above the dew-point, is pumped into the tents at 5,000 $m^3 h^{-1}$ so the reduced pressure inside the tents stops dust escaping. (Courtesy of Sigma Coatings)

Porter found that contractors quoted a price of approximately £6 per square metre for painting girders in 1977, excluding scaffolding, cleaning and transport. [2] Duncan estimated the cost of repainting the tidal zone of a North Sea oil platform at £300 per square metre in 1984. [3] This included such essential items as accommodation on a support ship, safety training on the platform, transport and lost time owing to inclement weather and tide conditions. The paint cost was a trivial amount, but the overall cost for a programme covering three or four platforms could be over £5 million in a single year. Three new developments in particular have contributed to cost reduction:

Development of coated blasting grit which deposits the primer coat as the blasting process continues and hence allows work to continue in damp conditions.

Development of high build, fast-curing topcoats which can be applied at low ambient temperatures.

Development of coatings which build the full thickness in one application.

In these circumstances a higher price can be paid for the paint if it would reliably tolerate wet steel and high relative humidities during application. However, the best quality paint applied to a poorly prepared surface or under unsuitable environmental conditions will give worse performance than a medium quality paint properly applied to a thoroughly prepared surface.

14.6 PLASTIC COATINGS

As coating technology has developed during the twentieth century, it has become increasingly difficult to distinguish between a 'paint' and a 'plastic coating', largely because the end-product of paint technology is essentially itself a plastic coating. The current trend is away from traditional solvent-based paints to high solids, solvent-free coating systems. Although described in paint terminology, contemporary systems involve a genuine polymerisation process and are really just the application of a plastic coating. It is thus becoming more relevant to call them 'coating systems' instead of paints.

Thermoplastic and elastomer coatings are applied to many relatively cheap metal substrates to combine the mechanical properties of the metal with the corrosion resistance of plastics. Like the paint systems described above, plastic coatings have a wide range of capabilities in many different environments, and follow similar procedures for their performance and selection.

In service, under the effect of heat or ultraviolet radiation, some plastics release volatile components which can attack adjacent metal surfaces. For example, PVC releases hydrogen chloride gas, and nylon releases acetic acid.

Plastics are coated on to metals in a number of ways. In **dipping**, the heated component is dipped into a fluidised bed of finely powdered coating material. The powder adheres to the heated surface. Subsequently the component is heated to higher temperatures to fuse the powder into a smooth coat. PVC, nylon and polythene are applied in this way. In some processes a vacuum is used to avoid the entrapment of air in the coating. Masking can be applied to leave desired areas of the surface free from the coating.

Although **spraying** is labour-intensive, sprays can be applied on-site or to large structures which cannot be dipped. It is very easy to apply the coating to selected sections of the structure. Spraying techniques include airless spraying, electrostatic spraying and flame spraying. In electrostatic spraying a high voltage is used to charge the powder, and the component is earthed so the charged powder adheres to its surface. In flame spraying the powder is heated in the spray-gun. Post-spraying heat treatment may be required to produce a sound bond and a smooth coating. Spray coatings are gradually built up to the desired thickness; time is allowed between coats for the material to cure. The pattern of spraying for each coat should be varied to ensure the entire surface is covered and any holidays are sealed by subsequent coats. Spray coatings are easily built up to 10 or 15 mm, although it may be necessary to use an intercoat adhesive layer if there are long periods between applications, say more than 12 hours. **Rolling, trowelling and brushing** are manual techniques used to apply many twin-pack systems.

There is fierce competition to supply plastic coating materials, as intense as in the paint industry. A wide range is available, and most materials are sold under trade names. Some of the basic materials are described below.

Nylon

The low water absorption nylons are preferred. They are easily coloured, do not chip and have good resistance to oils and solvents. Nylon can be used at temperatures up to 120°C. This enables it to be sterilised, which makes it useful in the food processing industry. Coating thickness is normally up to 1 mm and it is applied by dipping or spraying. While a good bond is formed on both steel and aluminium, an adhesive primer may be necessary on copper alloys.

Polyethylene

Polyethylene (polythene) coating is applied either by dipping or spraying, but only in thin films. It is prone to stress-corrosion cracking in some environments, for example, certain detergents, alcohols and silicones. It is used for coating domestic ware, wire shelves and display units, while high density polythene has been applied to ducting, chemical tanks, industrial shelving and pipework.
As we saw earlier, water purification plant operates in a very aggressive environment and high density polyethylene linings have been used with considerable success.

Polyvinyl chloride

Polyvinyl chloride (PVC) is a most versatile coating with properties which can be varied by changing the plasticiser content to meet differing service conditions. An adhesive/primer coating is necessary to give good bonding to the substrate. The material is applied by spraying and fluidised bed techniques, including the liquid plastisol process in which a heated component is dipped into cold liquid PVC. On the hot surface the polymer and plasticiser cross-link to give a gelatinous deposit. This is cured in a higher temperature treatment to yield a tough coating. Coatings up to 12 mm thick are laid down in one treatment. But the powder method is more suitable when a thinner coating is required. For continuous use, the ambient temperature should not exceed 60–70°C.

Polytetrafluoroethylene

Polytetrafluoroethylene (PTFE) is an expensive but highly corrosion-resistant material. It is well known for its high temperature stability (up to 250°C) and its resistance to attack by solvents, acids and alkalis. It does not absorb water. However, corrosion control on a substrate cannot be guaranteed owing to the difficulty of building a sufficiently thick layer to eliminate microporosity in the film.

14.7 CONCRETE COATINGS

The construction industry uses large quantities of steel reinforcing bars and load-carrying girders to strengthen concrete structures. The highly alkaline environment which exists in the concrete inhibits the corrosion of the steel by maintaining a passive film on the metal surface. However, if water, oxygen and carbon dioxide are able to penetrate the concrete, the carbon dioxide reacts with components of the concrete to precipitate carbonate in place of the hydroxides which confer the alkalinity. As the pH falls, the passive film breaks down and the water and oxygen cause the formation of rust, which occupies a larger volume than the metal from which it formed. Thus, the formation of rust generates tensile hoop stresses in the concrete surrounding the steel and the concrete often cracks. (The tensile strength of concrete is much lower than the compressive strength.) Once the concrete has cracked, water accelerates its deterioration by freely entering the cracks or freezing inside. When carbon dioxide reacts with concrete it is called **carbonation**, and the cracking is called **spalling**. Chloride ions from a marine environment or from de-icing salts also contribute to the deterioration. Figure 14.6 shows an example of spalling.

In the 1950s calcium chloride was added to concrete mixes to accelerate the setting of the material. The calcium chloride not only accelerates the setting of concrete, but also the rate of corrosion of the steel, and leads to premature cracking and spalling of the concrete matrix. Such additions are now prohibited.

An impervious dense surface layer of concrete will maintain the alkaline conditions at the metal/matrix interface. Clearly, the exposure conditions will govern the thickness of such a layer, but for structures immersed in sea-water a minimum of 50 mm is recommended. On oil platforms, the recommended minimum thicknesses of concrete above the steel are 75 mm for submerged areas and 100 mm for areas in the splash zone. Often, the permeability and porosity of the concrete is reduced by addition of pulverised fuel ash.

Cathodic protection is often applied to reinforcing bars in concrete to control

325

Fig. 14.6 Concrete spalling to reveal corroded steel reinforcing bars. This effect on a concrete pavement on a bridge deck in the United States was caused by the use of de-icing salts during winter months. (Photo courtesy of Taywood Engineering Ltd)

corrosion, especially corrosion from chloride ions (see Section 16.5), although inhibitors can also be added to concrete to reduce the corrosion of the reinforcing steel.

Sulphur is a serious threat to the breakdown of concrete structures. Organic sulphur comes from the domestic waste and food industries, the chemical industry uses much inorganic sulphur, and hydrogen sulphide is frequently produced in nature through the action of various bacteria (see Section 6.4). Concentrations of H_2S as low as 10–15 ppm can result in sulphuric acid concentrations of 6% in different parts of the system.

14.8 METALLIC COATINGS

Many of the commonplace objects around us are finished with metallic coatings to preserve and give lustre to the basic substrate metal, which provides the strength, rigidity and formability. For example, dustbins are galvanised, cans are coated with tin, and the bright parts of cars are chromium-plated (although many of these have now been replaced with plastics).

The metallic coating interposes a continuous barrier between the metal surface and the surrounding environment. The ideal properties of a metal coating are summarised in five points.

A metal coating should resist attack from the environment to a greater degree than the substrate metal.

A metal coating should not promote corrosion of the substrate at any flaws or deliberately introduced breaks in the coating.

The physical properties of a metallic coating, such as elasticity or hardness, should be adequate to meet the operational requirements of the structure.

The method of application of a metallic coating must be compatible with the fabrication processes used to produce the completed product.

A metallic coating should be uniform in thickness and free from porosity and holidays.

A prerequisite for all metal coating processes is thorough preparation of the substrate surface to remove all surface contamination, grease, oil, dirt and process debris; to remove surface corrosion products; and to control the physical characteristics of the surface.

The simplest method of removing surface contamination is to dip the component in a tank of solvent such as acetone, trichloroethylene, tetrachloromethane or benzene at room temperature. A more efficient process is to use hot solvent or solvent vapour. Many of the solvents are toxic or carcinogenic and most are highly flammable, so the treatment plant must be correctly designed and maintained. Alkaline solutions are also used in immersion baths, followed by thorough washing of the component to remove the solution. The effectiveness of both solvent and alkaline baths can be improved by ultrasonic agitation during the cleaning process.

Acid or alkaline solutions are used to remove corrosion products, according to the metal involved. During cleaning, hydrogen may be generated on steel surfaces and may diffuse into the metal causing embrittlement of the matrix. High strength steels are particularly susceptible to hydrogen embrittlement (see Section 10.3). Grit-blasting is also used to clean steel surfaces before a coating is applied.

Although most of the coating processes can be applied to cast, wrought, polished or grit-blasted surfaces which have been thoroughly cleaned and degreased, spray coatings adhere only to roughened surfaces. A uniform roughness is required and this is obtained by grit-blasting with a grit size appropriate to the substrate metal. On steel surfaces the spray coat is put on immediately after blasting to prevent reoxidation of the metal.

14.9 METHODS OF METALLIC COATING

The following methods are commonly used to apply metallic coatings described in the previous section.

Electroplating

To electroplate a component, immerse it, together with rods or plates of the plating metal, in an electrolyte containing salts of the plating metal. Apply a potential to the cell so that the component becomes the cathode and the rod or plate becomes the anode. Ions of the plating metal will deposit from the solution on to the component surface while more ions dissolve from the anode. Using correctly formulated solutions and anodes, alloys as well as pure metals can be plated. Good practice produces coatings of controlled thickness, a fine grain size and relative freedom from porosity. **Throwing power** describes the ability of a plating solution to produce an even coating thickness as the distance between the anode and the component surface varies when plating complicated shapes. Chromium has poor throwing power and requires complicated anode arrays to give an even plating thickness, especially on curved or convoluted surfaces, or in blind holes. Hydrogen embrittlement can occur during the plating process and post-plating heat treatment may be specified to diffuse the hydrogen and prevent cracking of the substrate.

Hot dipping

In hot dipping, the structure is dipped into a bath of molten coating metal. A good metallurgical bond is formed with the substrate, owing to interfacial alloying. There is less control over the coating thickness in the dipping process; the coat tends to be thicker on lower surfaces and thinner on top. However, all surfaces exposed to the molten metal are coated. The process is limited to low melting point metals such as tin, zinc and aluminium.

Spray coats

To produce spray coatings, wires of the coating metal are fed into a torch where they are melted and blown out under pressure as fine droplets. The droplets, travelling at $100–150 \text{ m s}^{-1}$, are flattened on striking the substrate surface and adhere to it. For a given thickness, the coat will be more porous than dipped or electroplated coats, but thick layers are easily built up by repeated spraying. The process can be applied on-site after final erection of a structure. However, it is labour-intensive and requires grit blasting just before spraying. The coating metal is usually anodic to the substrate; aluminium or zinc is applied to steel by spray coating.

Clad coatings

Corrosion-resistant metal skins can be laminated with other metals. The other metals have the engineering properties needed in a structure, but lack the necessary corrosion resistance in the working environment. Skins have been applied by rolling, by explosive welding and by building up a welded coat on the substrate, a process known as **buttering**. Normally, diffusion across the interface between the metals produces an alloy-bonded layer which has very good adhesion. The main difficulty in obtaining a sound bond is the lack of diffusion in the presence of surface contaminants, mostly oxides. The pressure in the rolling mill must be sufficiently great to break up the oxide. Explosive welding scours the two metal surfaces at the moment they are forced together. Additional protection is required at cut edges, holes or other breaks where the substrate is exposed to the environment. The skin must be able to cope with the fabrication processes used to make the final product.

Diffusion coating

A number of processes exist for diffusing a coating metal, or even non-metals, into the surface layer of a substrate to form an alloy layer on a component. A very good bond is produced, but the process is limited to relatively small objects. Components to be treated are first degreased and cleaned. Then they are heated, either in contact with a powdered coating metal in an inert atmosphere (the solid route), or in a gaseous stream of a volatile compound of the coating metal (the gaseous route).

In the solid route, the components and the metal powder are mixed with sand and sealed in a drum. Next they are heated to just below the melting point of the coating metal and rotated for several hours. This technique is used to coat steel with zinc, **sheradising**, or with aluminium, **calorising**. The gaseous process uses halides to coat steel with chromium, **chromising**, or to produce silicon coatings.

14.10 BEHAVIOUR OF METALLIC COATINGS

The behaviour of metallic coatings is influenced by many factors. The nature of the electrolyte, oxygen concentrations, polarisation characteristics, relative areas of anode and cathode, and surface deposits on the coatings will all affect the performance of any corrosion cells which develop. It is not unknown for cell reversal to occur (see Section 5.1) or for insoluble corrosion products to block defects in coatings which might otherwise be expected to offer little corrosion protection.

If an anodic (sacrificial) coating is used the conductivity and continuity of the electrolyte will control the size of the surface defect that can be tolerated before corrosion occurs on the substrate. For a zinc coating on steel, a defect width as small as 3 mm will produce corrosion in distilled or soft water, whereas in sea-water the steel will be protected at a defect several decimetres across (but the rate of loss of zinc will be much higher). The following discussion relates to the performance of specific metals.

Zinc

Although zinc is more anodic than iron in the galvanic series, the zinc corrosion products, such as oxide, hydroxide and basic carbonates, form a protective film on the metal surface. The film reduces the corrosion rate of the zinc to levels well below those for iron or steel. A zinc coating has a long life, but provides sacrificial protection at any breaks in the coating. Zinc is applied by all the coating methods described in Section 14.9. The corrosion resistance of a zinc coat is independent of the method of application, although coatings containing 5.8% iron are said to be less prone to pitting attack than pure zinc. Post-coating heat treatment can be used to encourage diffusion across the interface between the metals to increase the iron content of the coating. Small additions of magnesium, aluminium and titanium improve the performance of galvanised coatings in flowing sea-water.

The useful life of a coating depends upon its thickness and the environment to which it is exposed. For most of its life, zinc simply forms a corrosion-resistant barrier between the steel substrate and the environment. However, owing to the sacrificial protection of the zinc, large areas of the coating can be lost before the steel is attacked, assuming the electrolyte has good electrical conductivity and extends across the bare substrate to the remaining coating. Under these conditions serious rusting will be delayed until as little as 10% of the original coating remains. In general, a coating of thickness 0.03 mm exposed to the atmosphere will last 11–12 years in rural areas or 8 years in marine locations, while sulphur oxide pollution in industrial areas will reduce the life to 4 years. Immersed in sea-water, each 0.03 mm of coating thickness lasts for approximately 1 year, but pollution, especially hydrogen sulphide arising from sewage contamination in estuaries, will increase the rate of deterioration of the coating. Figure 14.7 shows the life to first maintenance for specified minimum thickness of zinc coatings in different environments.

Case 14.8: Telford's suspension bridge over the Menai Straits in North Wales is exposed to an aggressive marine environment with high wind velocities and a wet climate. From 1938 to 1940 all the steelwork of the bridge was renewed, including the deck structure and the suspension chains. The deckwork was scratch-brushed and painted with two lead-based priming coats and one finishing coat. The tensile steel suspension chain-links were grit-blasted and a flame-sprayed zinc coat applied to a thickness of 125–150 μm. They were then primed with one coat of red

330

Galvanised coating, specified minimum weight (g m⁻²)

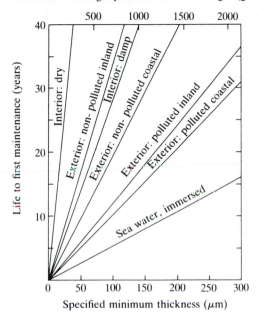

Fig. 14.7 Environment and thickness effects on the life of a zinc coating.

lead paint. After erection a further red lead primer coat and a finishing coat were applied to the chains. Zinc spray was used because the individual chain-links (approximately 6 m long) were too large to galvanise by dipping at that time. Furthermore, spraying, a cold process, would not affect the temper of the heat-treated steel links. In 1947 a detailed examination showed that the deck structure had corroded badly, especially on the undersurfaces. It has required a continuous repainting scheme since that date. On the other hand, the chains showed no sign of rust. The top-coat was thin in places, but the primer and underlying zinc-sprayed coat were intact. In 1950 the complete bridge was repainted. The chains required only a fresh topcoat to match the new colour scheme. A further examination in 1980, after 40 years' service, showed that there was still no sign of rusting on the chains, although areas of the deck needed replacement because of corrosion damage.

Cadmium

An examination of standard electrode potentials (Table 4.1) reveals that cadmium should be cathodic to iron, promoting its dissolution in a corrosion cell. However, it is found that cadmium behaves anodically to iron and steel and

331

protects them sacrificially. This is reflected in the practical galvanic series (Fig. 5.6). Again, for most of its life, a cadmium coating acts as a physical barrier between the substrate and the environment. The sacrificial role becomes operative only at breaks in the coating, or towards the end of its life when the coating becomes porous. Unlike zinc, the nature of the film which develops on the surface of cadmium is strongly dependent upon the atmospheric conditions. In rural or industrial locations soluble sulphates form part of the film; rainwater washes them away and the corrosion rate of the cadmium remains high. Insoluble carbonates and chlorides form in marine areas and the corrosion rate is much lower. For a given steel thickness, cadmium offers more protection than zinc in marine environments, but the situation is reversed at rural or industrial sites.

Cadmium is normally applied in thin coats by electroplating. It has a bright finish, but is more costly than zinc and is mainly used to coat small articles such as fasteners. Cadmium is preferred to zinc as a coating by the electronics industry because it is easier to solder. It is also useful as a coating on high strength steels, owing to the lower risk of hydrogen embrittlement from the plating bath (see Section 10.3). Unfortunately, cadmium salts are highly toxic and its use can form a significant pollution hazard in some instances. In the future, environmental regulations are likely to exert increased control over cadmium.

Aluminium

Coatings of aluminium are normally produced by spraying or hot dipping, although small articles are diffusion coated. Silicon is added to hot dips to retard the formation of brittle intermetallics at the interface between the coating and the substrate. Such intermetallics reduce the formability of the coated sheet and impose restrictions on the subsequent fabrication process. Sprayed coatings contain much higher levels of oxide than dipped coatings, and they are more porous. However, the corrosion product rapidly fills the microholes in the coating and forms an adherent, compact and impervious coating. Some corrosion of the substrate may occur during the sealing process, in which case unsightly rust stains may appear on the outer surface of the coat. This can be avoided by sealing the coat with a lacquer during the early stages of its exposure, although the rusting is not harmful. Lead-based paints must *never* be used on aluminium, but zinc chromate pigmented paints are suitable for use on aluminium coatings.

It has been shown that when aluminium coatings on steel are exposed to industrial atmospheres they give a longer life than zinc. In one test, a zinc coating of thickness 0.08 mm lasted about 7 years, whereas a similar aluminium coating gave a life of 12 years.

Both dipped and diffusion coatings of aluminium are used to protect steel exposed to temperatures up to 1,100°C in air. The coatings also offer good protection from attack by sulphur compounds and are used in the chemical industry, on some gas turbine blades and for car exhausts.

Nickel and chromium

Both of these metals are cathodic to steel and act as a barrier coating to protect the substrate surface. Once the coating is breached, the corrosion rate of the steel is faster than it would have been without the coating. There are many different types of coating arrangement involving nickel and chromium. Figure 14.8 shows two arrangements that will be discussed below. Some coatings have chemical additions to modify the different layers in the system. The addition of sulphur compounds to nickel produces a layer with a bright finish which is anodic to pure nickel. If pure nickel is plated on to steel and overcoated with the bright nickel, corrosion is initially restricted to the bright coat and penetration to the steel is delayed.

During prolonged exposure outdoors, nickel coatings tend to lose their surface brightness. To maintain a bright and therefore attractive surface, thin coatings of chromium are plated over the nickel. This is the traditional chromium plating used on articles intended for outdoor use.

Chromium coatings are highly stressed and it is difficult to deposit a coating that does not contain microcracking or porosity. When water enters the cracks, microcells are generated. The chromium is the cathode in the microcell; it has a large surface area compared to the anode (the nickel plate at the base of the

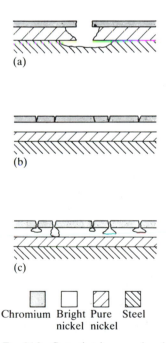

Chromium Bright Pure Steel
         nickel nickel

Fig. 14.8 Corrosion in normal and microcracked chromium plating: (a) normal chromium plate where steel is anodic to both coating layers; (b) usual arrangement of layers in microcracked chromium plating; and (c) microcracked chromium plating where the corrosion is confined for long periods to the bright nickel layer which is anodic to the chromium and the zinc in the nickel layer.

Fig. 14.9 A chromium-plated steel snap-hook showing coating failure all over, but mostly on the right-hand side (Case 14.9).

crack) and therefore the corrosion rate of the nickel is very high. The propensity of the chromium to crack is turned to advantage in the production of microcracked chromium plating. The steel is first flash-coated with copper to produce a smooth surface and facilitate buffing. This is followed by a pure nickel coating over which a bright nickel coat containing sulphur compounds is deposited. The process is completed with a thin overlay of chromium (Fig. 14.8(b)). The normal cracking of the chromium is enhanced by the sulphur compounds in the nickel coat. The bright nickel is anodic to both the pure nickel and the chromium, therefore corrosion proceeds sacrificially in this layer (Fig. 14.8(c)). However, the larger anode/cathode ratio produced by the increased cracking density reduces the tendency for pitting to occur. Gradually, flakes of the overlying chromium drop off, dulling the surface, but the appearance of unsightly rusting is postponed. Microcracked chromium coatings are used extensively for the bright trim on cars.

Case 14.9:   A steel snap-hook (Fig. 14.9) was supplied from a store as a 316 stainless steel component, but after exposure for only two years in a windswept coastal atmosphere, it was impossible to use the nut by hand. The component was actually chromium-plated with a copper alloy knurled nut. The correct underlayers to the chromium were not present. Thus, the microcracking of the chromium led to corrosion, worse near the nut where there was a galvanic effect. Corrosion of the thread just outside was worst of all, severe enough to prevent its use.

Inside the crevice between the nut and the good plating, the surface remained bright and uncorroded.

Tin

Large quantities of tin-plate are used for cans in the food industry. The coating is cathodic to steel on the outside of the container and promotes corrosion if damaged. However, cell reversal occurs with many organic acids, such as citric acid found in fruit juices, and sacrificial protection occurs on the inside of the can.

14.11 REFERENCES

1. Munger C G 1993 Marine and offshore corrosion control, past present and future. Paper 549 in *Corrosion 93 plenary and keynote lectures*, edited by R D Gundry, NACE International, Houston TX.
2. Porter F C 1978 *Comparative costs of protecting steel*, Zinc Development Association, London.
3. Duncan J 1984 Cost implication of offshore platform structural protection by coatings. In *Corrosion and marine growth on offshore structures*, edited by J R Lewis and A D Mercer, Ellis Horwood, Chichester UK.

14.12 BIBLIOGRAPHY

Anon 1983 *Solving rebar corrosion problems in concrete*, NACE International, Houston, TX.

Anon 1988 *Corrosion of metals in concrete*, NACE International, Houston TX.

Barry B T K 1984 The new tinplate industry, *The Metallurgist and Materials Technologist* **16**(12): 618–23.

Callcut V 1984 Surface finishes on copper and copper alloys, *The Metallurgist and Materials Technologist* **16**(11): 575–84.

Ledheiser H 1981 *Corrosion control by organic coatings*, NACE International, Houston TX.

Long D J 1994 'The state of the art' of painting weathered galvanized transmission towers and substation structures. Paper 103 in *Corrosion 94, Baltimore MD*, NACE International, Houston TX.

Munger C G 1985 *Corrosion prevention by protective coatings*, NACE International, Houston TX.

Sloan R N 1993 50 years of pipe coatings: we've come a long way. Paper 17 in *Corrosion 93 plenary and keynote lectures*, edited by R D Gundry, NACE International, Houston TX.

TPC Publication 1983 *A handbook of protective coatings for military and aerospace equipment*, NACE International, Houston TX.

15 THE CORROSION PROPERTIES OF SOME METALLIC MATERIALS

'Gold is for the mistress — silver for the maid —
Copper for the craftsman, cunning at his trade.'
'Good!' said the Baron, sitting in his hall,
'But Iron — Cold Iron — is master of them all.'
(Rudyard Kipling, *Cold Iron*)

It is clearly impossible to list materials to match every conceivable combination of environmental conditions; many factors are likely to restrict the range of materials from which a choice may be made. Outside the oil and chemical industries, most large structures are made from mild or low alloy steel, aluminium or concrete with steel reinforcing, owing to their cheapness, availability and strength. The choice is mainly decided by the pattern of stress within the structure, the fabrication and joining techniques to be used and the availability of labour with the skills to undertake the construction. The corrosion resistance of the material generally plays a minor role in the selection process; the engineer looks to a barrier coating or some other technique to control the loss of metal. Sophisticated corrosion-resistant alloys have traditionally been reserved for specific hazards, as may be found in the oil and chemical industries, or where reliability is of paramount importance. If future environmental legislation places new demands for leaner performance, reduced pollution and maximum lifetimes, this situation may change somewhat.

We have already seen how the expected life of a structure, and the environments in which it will be fabricated, stored and used, must be specified early in the design process. Certainly, these factors should be known before any decision over the choice of materials is made.

Case 15.1: The use of closed-loop dry cooling towers at an electricity generating station at Rugeley, Staffordshire, was expected to eliminate the need for 5700 m^3 of make-up water per day. The water-coolers in the towers, which at the time were the largest natural draught shells in the world, were made from louvred aluminium plate. The only other towers operating on this principle were in Hungary, where they had given several years satisfactory service. Within 18 months of installation, the Rugeley coolers had corroded and perforated. The problem was attributed to the difference in atmospheric environment between the Hungarian site and Rugeley. Rugeley's atmosphere was much more moist and contaminated with chloride ions. The aluminium coolers were the most expensive part of the plant and had been expected to last its lifetime. The problem was solved by modifying the design and applying an epoxy resin coat to the replacement units.

Many metal manufacturers supply data sheets on their products which include information on the behaviour of the metals in various environments. Often this information is derived from accelerated testing procedures. These data sheets should be treated with some caution, especially concerning the environmental and test conditions under which the information was obtained. We have already seen that changes in the environment can alter the corrosion characteristics of the metal and cause pitting, selective phase or localised attack, and environment-sensitive cracking; all shorten the useful life of the metal. Accelerated test data must be correlated with data from field trials in the expected environment. Field trials should include elements to represent the real structure, such as the relative anode to cathode area ratio, joint configurations and heat treatments, particularly those changes which will occur on welding.

Incorrect testing techniques can produce false results which can be financially disastrous.

Case 15.2: In a process plant, chlorides were stripped from a strongly acidic mixture with some residual chlorine and a very high salt content. The cold liquor was heated to 80°C in a glass-lined steel vessel to boil off the chlorides. The glass-lined vessels had a life of 1–2 years. In a search for a longer-life material, nickel, Inconel, Hastelloy, stainless steel and titanium were tested, but only titanium proved to be resistant to the process conditions. A complete titanium unit was installed but failed after 2 days' operation with a corrosion rate of 38 mm y^{-1}. In the laboratory tests, all the alloys had been placed in the same reaction chamber, where the iron (III) ions produced by some of the materials had proved an excellent corrosion inhibitor for titanium. The iron (III) ions were absent in the real plant. The use of individual test cells for the laboratory tests would have prevented the false result.

On the other hand, potentially useful materials may be missed.

Case 15.3: A large chemical scrubber was installed to operate in 40% sulphuric acid at 82°C. After some failures, a PVC-lined fibre-reinforced plastic was selected for the constructional material. It gave a life of 1 year. A stainless steel stack band had fallen into the scrubber and was observed, immersed in the acid, during an internal inspection following a PVC failure. Normally, it would be expected that stainless steel in hot 40% sulphuric acid would corrode badly, but in this case it was bright and shiny and appeared to be in pristine condition. Tests showed that the sulphuric acid actually contained sufficient nitric acid to passivate the stainless steel under the conditions in the scrubber. A replacement scrubber made from type 304 stainless steel gave more than 12 years' maintenance-free service.

This chapter is concerned with the general corrosion characteristics of a variety of metals. If there is any doubt about the performance of a metal in a given environment, *seek advice* from its manufacturer or run a test programme to evaluate its behaviour. The materials are grouped into the main alloy classes. Steels are broadly divided into three main groups: plain carbon and low alloy steels, using no more than about 2–3% of elements other than iron; stainless steels using much higher proportions of elements such as chromium, nickel, molybdenum and manganese; and cast irons, to which little or no elements are added. At the end of the chapter is a brief survey of materials for particular applications.

15.1   PLAIN CARBON AND LOW ALLOY STEELS

Corrosion in the atmosphere

Plain carbon steels corrode in most atmospheric environments when the relative humidity exceeds 60%. Once a moisture film has formed, the rate of corrosion depends upon many environmental factors. The most important are oxygen supply, pH and the presence of aggressive ions, particularly oxides of sulphur and chlorides.

The composition of the steel, its surface condition and the angle of exposure also affect the corrosion rate. Increases in carbon, manganese and silicon content tend to reduce the corrosion rate, although the presence of manganese sulphide in free-machining steels produces local galvanic cells (Section 7.2) and lower corrosion resistance. Copper additions have a beneficial effect, discussed later.

Rolled or pickled steels corrode faster than machined or polished samples. Mill scale, a layer of iron oxides produced during heat treatment, hot rolling and forging, forms a very effective cathode on steel surfaces if it is not removed before the metal is put into service. The layer consists of FeO adjacent to the metal with Fe_3O_4 between it and an outer skin of Fe_2O_3. Although mill scale may act as a barrier coat to reduce atmospheric corrosion in the short term, it cracks and spalls under shock, vibratory loading and differential thermal expansion, leaving bare areas of metal which corrode rapidly. Paint films applied over mill scale are useless as they spall with the scale.

Vertical plates are attacked more slowly than those at 45°, where more than half the metal loss is on the underside of the plate. Initially, the corrosion rate on bare metal is high, but it falls to fairly constant rates after the first year of exposure because of the formation of surface oxide films. Table 15.1 lists some typical corrosion rates for different locations and atmospheres.

Corrosion in aqueous liquids

The corrosion rate for steel immersed in water is influenced by the interaction of many factors which have already been discussed in Chapters 12, 13 and 14.

Table 15.1 Typical corrosion rates for mild steel plates freely exposed for one year in a vertical position in different locations and atmospheres

Atmosphere type	Location	Corrosion rate/mm y^{-1}
Rural	Khartoum (Sudan)	0.003
	Abisko (Sweden)	0.005
Suburban	Teddington (UK)	0.070
Marine	Brixham (UK)	0.053
	Singapore (Singapore)	0.015
	Sandy Hook NJ	0.084
Marine/industrial	Congella (RSA)	0.114
Industrial	Pittsburgh PA	0.108
	Derby (UK)	0.170
Marine/surf beach	Lagos (Nigeria)	0.615

Geometry and fabrication techniques may introduce poor design features and control the flow rates, and water quality also plays a major role. Increasing the flow rate increases the corrosion rate, especially when it produces turbulent flow.

In fresh water a corrosion rate of 0.05 mm y^{-1} is typical, although it may fall to 0.01 mm y^{-1} when calcareous deposits are formed. In sea-water the average rate is likely to be in the region 0.1–0.15 mm y^{-1}, but pitting attack may penetrate much deeper and faster. The presence of mill scale or exposure at half-tide locations, where a repeated wet/dry cycle occurs, increases these figures considerably. The fastest corrosion rate for mild steel in marine environments occurs in the splash zone, where there is an abundant oxygen supply. Here the rate of metal loss is likely to be four or five times faster than that for a fully immersed plate at the same location. A typical corrosion allowance on painted steel structures exposed in the North Sea is an extra 12 mm on plate thickness for a life of 15–20 years. The factors affecting the corrosion of carbon steel in sea-water are given in Table 15.2.

Corrosion in soils

Generalisations about corrosion rates for mild steels in soils are even more difficult. Soil quality is very variable, not only from one district to another, but with the seasons of the year. It is more profitable to consider the conditions required to maintain a sound barrier coat or cathodic protection system (Chapters 14 and 16). If a corrosion rate in a particular situation raises doubts about the safe operation of a system, trials should be made in that area.

Low alloy steels

Low alloy steels contain up to 3% of various alloying elements, notably chromium, nickel, copper, manganese, vanadium and molybdenum. These alloying additions improve the mechanical properties, but have marginal effects

Table 15.2   The effect of change in the environment upon the corrosion of carbon steel in sea-water

Factor in sea-water	Effect on iron and steel
Chloride ion	Highly corrosive to ferrous metals; carbon steel and common ferrous metals cannot be passivated (sea-salt is >55% chloride ion).
Electrical conductivity	High conductivity makes it possible for anodes and cathodes to operate over long distances, thus corrosion possibilities are increased and the total attack may be much greater than that for the same structure in fresh water.
Oxygen	Steel corrosion is cathodically controlled for the most part. Oxygen, by depolarising the cathode, facilitates the attack, thus a high oxygen content increases activity.
Velocity	Corrosion rate is increased, especially in turbulent flow. Moving sea-water may (1) destroy the rust barrier and (2) provide more oxygen. Impingement attack tends to promote rapid penetration. Cavitation damage exposes fresh steel surfaces to further corrosion.
Temperature	Increasing ambient temperature tends to accelerate attack, but heated sea-water may deposit protective scale or lose its oxygen. Either or both actions tend to reduce attack.
Biofouling	Hard shell animal fouling tends to reduce attack by restricting access of oxygen. Isolated hard-shell animals promote localized pitting by differential aeration under the shell. Bacteria sometimes take part in corrosion reactions.
Stress	Cyclic stress sometimes accelerates failure of a corroding steel member. Tensile stresses near yield also promote failure in special situations.
Pollution	Sulphides which are normally present in polluted sea-water greatly accelerate attack on steel. However, the low oxygen content of polluted waters could favour reduced corrosion.
Silt and suspended sediment	Erosion of the steel surface by suspended matter in the flowing sea-water greatly increases the tendency to corrode.
Film formation	A coating of rust and mineral scale (calcium and magnesium salts) will interfere with the diffusion of oxygen to the cathode surface thus slowing attack.

on the corrosion rate of immersed or buried components, where mild steel, low alloy steel and wrought irons corrode at approximately the same rate. Slightly larger additions of chromium can produce marked improvements in corrosion behaviour, and small amounts of copper, up to 0.3%, are said to reduce the pitting corrosion of steels in boilers, giving rise to lighter general corrosion instead.

The commercial pressure to develop new oil and gas fields has led to a significant increase in the number of sour wells (i.e. wells containing hydrogen sulphide). Low alloy steels used in these drilling rigs and production wells have suffered higher rates of general corrosion, together with sulphide-promoted stress-corrosion cracking problems. Of the elements listed above, only copper in amounts up to 0.3% reduces the corrosion rate in hydrogen sulphide, but vanadium and molybdenum appear to be beneficial in resisting stress-corrosion cracking.

Where components are exposed to the atmosphere, up to 0.2% copper is very beneficial, yielding a dense compact corrosion product which inhibits further attack. However, the conditions of exposure are important. The improvement is lost unless the steel is exposed in the open. In sheltered conditions where the steel is protected from rain, the corrosion rate is little different from mild steel. Small additions of chromium also improve the corrosion resistance of steels.

Despite all the limitations imposed by the corrosion behaviour, mild and low alloy steels have well-known engineering properties and with a sound, properly maintained protection system in a well-designed structure they will give a long life. They remain very economical to produce because of the abundance of iron ore in the Earth's crust and their production probably consumes less energy than many other materials because of the economies of scale. They do not generate toxic materials when they corrode and are therefore quite environmentally friendly.

15.2 STAINLESS STEELS

Stainless steels are iron alloys which depend upon a very thin transparent passive surface film of chromium oxide to resist corrosion. A supply of oxygen and a minimum level of 11% chromium dissolved in the matrix material are necessary to maintain the surface film, which is self-healing in air at room temperature. If the film is not repaired after damage, the corrosion of stainless steel will be very rapid in most environments.

Case 15.4: In the 1930s, a US warship was fitted for the first time with 18/8 stainless steel pipes for its sea-water systems, but the material pitted extremely badly within 18 months and a directive was issued in 1934 banning the use of stainless steel in warships for any purpose whatsoever. At the time, Herb Uhlig was a contract researcher at MIT. He was assigned to investigate the problem, which involved the breakdown of the passive film in sea-water. Uhlig went on to become one of the great names of corrosion science. [1]

During the latter half of the nineteenth century, a number of commercial uses for steels containing a few per cent chromium were made, but the steels were not stainless. Julius Baur opened the Chromium Steelworks in Brooklyn, New York, in 1869, and, besides being used in armaments, chromium steels were used in turbine generators at Niagara Falls in 1893, for a crankshaft in the liner, *Deutschland*, in 1899, and for early aircraft parts in the Wright brothers' Kitty Hawk, and Lindbergh's Spirit of St. Louis. [2]

There is some debate about the inventor of stainless steel. Americans like to think the first genuinely stainless alloys were of 12% chromium and 0.3%

carbon, developed by Elwood Haynes in the United States, whereas the British believe Harry Brearley made the first stainless cutlery in 1915 in Sheffield.

The main corrosion problems associated with stainless steels are environment-sensitive cracking (Chapter 10), and crevice and pitting corrosion (Chapter 7). These processes are aggravated by chloride ions, which are aggressive to the chromium oxide film and which often enable the achievement of the optimum pit chemistry for the localised corrosion in pits and crevices. Too high a carbon content (>0.03%) can be detrimental to the corrosion behaviour of stainless steels, owing to the precipitation of chromium carbide which leaves some areas of the matrix with chromium levels below the critical minimum to maintain the oxide film. The depleted areas become the anodes in a cell and the remaining oxide-coated metal forms the cathode (Section 6.2). Although the carbon content of the steel may be low, additional carbon may dissolve in the steel during any heating process which carbonises oil, paints or greases on the metal surface, or even from heating where gas flames are used. All surface coats that may provide a source of carbon should be removed before stainless steels are welded, heated for forging or pressing, or for general heat treatment. Do not forget to remove colour code paints!

In the 1970s, stainless steelmaking was revolutionised by the introduction of argon oxygen decarburisation (AOD). This allowed carbon to be reduced to very low levels, while the amount of other alloying elements could be controlled within very narrow limits. It became practical to specify nitrogen as an alloying element at levels of about 0.1%, while keeping the carbon content at 0.03%. The amounts of elements such as sulphur, bismuth, tin and lead, which affect the fabrication and rolling properties, could also be controlled. These improvements in production techniques allowed much tighter control on the specifications for existing steels and the introduction of a whole new range of higher alloyed steels with improved corrosion resistance. The content of chromium and molybdenum is very important. Molybdenum additions for good corrosion resistance, and relative freedom from crevice corrosion in low velocity sea-water, depend on the level of chromium in the steel. Austenitic steels with 20–22% chromium require 6% molybdenum; ferritic steels with 25–28% chromium require 3% molybdenum. This chromium/molybdenum content, together with the influence of nitrogen content, has led to the increased use of so-called **pitting indices**. The most popular index, often referred to as the **pitting resistance equivalent number (PREN)**, is obtained by the formula

$$\text{PREN} = \%\text{Cr} + 3.3 \times \%\text{Mo} + 16 \times \%\text{N} \qquad (15.1)$$

with larger values of PREN indicating superior properties. This index obviously takes no account of the other alloying elements, such as nickel and copper, which must have an influence. Empirical though it may be, PREN is finding considerable use as a method of ranking the corrosion performance of many stainless steels. A second valuable indicator of performance is claimed to be the critical pitting temperature, which involves exposing samples to 6% iron (III) chloride solutions and raising the temperature by incremental amounts until the onset of pitting. The minimum accepted temperature for North Sea offshore applications is 40°C, whereas in the pulp and paper bleaching environments this would typically be 50°C.

Stainless steels are often sold under trade names, and some typical compositions are given in Table 15.3. But there are many stainless steels, and only the broadest discussion can be given here. Thus we summarise separately the general features of each group. The generic type of stainless steel is dependent on the relative amounts of the main phases which can be present, i.e. ferrite, austenite and martensite.

Martensitic stainless steel

Many martensitic steels have a chromium content lower than the 11% minimum and so are not really stainless. However, most of the martensitic stainless steels have 11.5–13.5% chromium and they can be hardened by heat treatment, just as ordinary steels. Grades 410 to 420 are the basic 13% chromium type with varied carbon content. They are used where the corrosion resistance need not be great, such as steam and water valves, pump and steam turbine components and shafting. Because of the hardness of martensite, 13% chromium steels are used for corrosion and abrasion-resistant cutting edges.

Ferritic stainless steels

With increasing chromium content, mixed martensite–ferrite alloys are obtained first, then fully ferritic alloys in the range 16–30% chromium. Although they have better corrosion resistance than the martensitic steels because of their higher chromium content, they are not heat treatable. Relative to martensite, ferrite is soft but also brittle, so ferritic steels are used mainly in small section. Many of them contain significant amounts of martensite.

In contrast to the relatively high solubility of carbon in austenite, the very low solubilities of carbon and nitrogen in ferrite cause the ferritic stainless steels to be easily sensitised to intergranular corrosion by the precipitation of carbides and nitrides. These stainless steels suffer both weld decay and reduced toughness unless the levels of carbon and nitrogen are kept very low. The new AOD and its extension to vacuum oxygen decarburising (VOD) allow these elements to be controlled while producing a wide range of chromium and molybdenum levels. Additions of titanium and niobium also help to minimise chromium depletion by grain boundary precipitation in the ferritic stainless steels.

Ferritic stainless steels are finding a wide variety of uses in industry. Type 409 has proved successful in trials for catalytic converters to be fitted to car exhausts, while small additions of nickel, molybdenum and aluminium to 409 make it a financially viable alternative to mild steel for the exhausts themselves. If the carbon and nitrogen contents are kept to suitably low levels, so the material can be welded and fabricated without sensitisation to weld decay, type 409 can be used where abrasion would quickly damage normal barrier coatings, for example, in chutes, waggons, ducting and shields. Although the initial cost is higher than for mild steel, the total life cost may be lower. When fresh water is

Table 15.3   Some stainless steels and their approximate compositions

Material type (AISI no.) *Approximate percent composition (Fe remainder)*

	Cr	Ni	Mo	C	N	Other
Austenitic						
201	16–18	3.5–5.5	–	0.15	0.25	5.5–7.5 Mn
202	17–19	4–6	–	0.15	0.25	7.5–10 Mn
205	16–18	1.00–1.75	1.0–1.7	0.25	0.4	14.0-15.5 Mn
304	18-20	8.0–10.5	–	0.08	0.10	–
304L	18–20	8–12	–	0.03	0.10	–
304N	18–20	8–10.5	–	0.08	0.10–0.16	–
304LN	18–20	8–12	–	0.03	0.10–0.16	–
316	16–18	10–14	2–3	0.08	0.1	–
316L	16–18	10–14	2–3	0.03	0.1	–
316N	16–18	10–14	2–3	0.08	0.1	–
316LN	16–18	10–14	2–3	0.03	0.1	–
317	18–20	11–15	3–4	0.08	–	
317L	18–20	11–15	3–4	0.03	–	
321	17–19	9–12	–	0.08	–	<0.4 Ti
254 SMO*	20	17.5–18.5	6.0–6.5	0.02	0.18–0.22	0.5–1.0 Cu
AL-6XN	21	25	6.5	0.02	0.2	–
25-6MO	19–21	24–26	6–7	0.02	0.15–0.25	<2 Mn
20 MO-6	22-26	33–37	5.0–6.7	0.03	–	2–4 Cu
Martensitic						
410	11–14	1	–	0.15	–	1 Mn, 1 Si
420	11-14	1	–	0.2	–	1 Mn, 1 Si
Ferritic						
403	12–14	1	–	0.08	–	1 Mn
409	11	1	–	0.08	–	<1 Ti
430	16–18	1	–	0.08	–	–
434	16–18	–	0.90–1.25	0.08	–	–
440A	16–18	–	0.75	0.7	–	1 Mn
444	17.5	19.5	1.75–2.50	0.025	0.035	Ti + Nb
446	23–27	–	–	0.20	0.25	–
AL 29-4-2	28–30	2.0–2.5	3.5–4.2	0.01	0.02	–
Duplex						
3 RE 60	18–19	4.25–5.25	2.5–3.0	0.03	–	–
SAF 2205	21–23	4.5–6.5	2.5–3.5	0.03	0.08–0.2	–
329	23–28	2.5–5.0	1–2	0.08	–	–
Ferralium† 255	25.3	6	3	0.04	0.17	2 Cu
Ferralium† SD40	26	6.4	3.3	0.03	0.26	1.6 Cu, 1 Mn
DP-3	24–26	5.5–7.5	2.5–3.5	0.03	0.1–0.2	0.2–0.8 Cu 0.2-0.5 W

*254 SMO is a registered trade mark of Avesta Jernverks Aktiebolag.
†Ferralium is a trade mark of Langley Alloys Ltd.

used in condensers and heat exchangers in the food industry, chemical works and refining plants, the higher levels of chromium, with some molybdenum, make type 444 a satisfactory material, combining good corrosion resistance with resistance to stress-corrosion cracking.

The very highly alloyed ferritic steels are resistant to severely corrosive environments, AL 29-4-2 can withstand boiling 10% sulphuric acid. Highly alloyed ferritic steels have been used extensively in caustic evaporators and in heat exchangers in the chemical and petrochemical industries. In some cases, they offer satisfactory replacements for more expensive nickel, nickel alloys and titanium.

Since 1973, both ferritic and austenitic stainless steels have found increasing use in sea-water-cooled condensers at electricity generating stations. Initially, they replaced copper alloy tubes which had suffered sulphide attack owing to pollution of the water.

Case 15.5: The first proprietary superferritic stainless steel was a 26% Cr, 1% Mo alloy and unusually low carbon and nitrogen content. In laboratory tests, it showed excellent resistance to chloride pitting and was used to replace cupronickel sea-water condenser tubes which had failed after several years' service. The superferritic tubes failed by severe pitting in less than six months.

In power station cooling systems ferritic stainless steels are operated below the temperature at which stress-corrosion cracking becomes a problem. And practical flow rates are below the maximum critical water velocities.

Austenitic stainless steels

Owing to the high solubility of carbon in austenite, the very low carbon contents of the L grade austenitic steels permit welding without the danger of sensitising the metal to weld decay (see Section 6.2).

The higher alloyed stainless steels are more resistant to crevice and pitting corrosion, especially in sea-water. In the early 1970s, it was shown that type 316 could be used in fast-flowing sea-water when crevices were absent from the structure; it could not be used in stagnant water, or where differential-aeration cells might occur. Intermediate grades of stainless steel with 4–5% molybdenum gave improved performance, but were still subject to corrosion in crevices or beneath biofouling. More recently, 6–7% molybdenum steels, such as Inco alloy G-3 and 25-6MO, have proved to be resistant to crevice corrosion, even under adverse conditions. Highly alloyed steels containing molybdenum are more resistant to stress-corrosion cracking.

Nitrogen can be substituted for nickel in these steels as an austenite stabiliser. It reduces the possibility of precipitating secondary phases (i.e. σ phase) which are harmful to the general corrosion behaviour. It also improves the pitting resistance of molybdenum-containing austenitic steels, but adversely affects their stress-corrosion cracking performance. Austenitic steels which include nitrogen as an alloying element usually have higher molybdenum contents to offset the deleterious effects of the nitrogen on stress-corrosion cracking behaviour.

Molybdenum-containing austenitic stainless steels have been widely used in flue gas scrubbers attached to North American coal-fired power stations. The scrubbers utilise limestone slurries to neutralise the sulphurous and sulphuric acids in the flue gases, which may also contain chlorides and fluorides. In tests to simulate flue gas conditions, type 316 stainless steel developed pits 0.26 mm deep, type 317 pits 0.017 mm deep, while 254 SMO had no visible corrosion or pitting after the test. The introduction of stricter controls on effluent pollution from the paper manufacturing industry has also led to the use of highly alloyed stainless steels for the construction of waste processing systems. Although the use of highly alloyed austenitic steels greatly increases the capital cost in such systems, the reliable long-term performance of these materials has justified the initial investment.

Austenitic stainless steels have found very wide usage in industry. We have already mentioned their uses (and problems) in the nuclear industry. In architecture they are used for shop-fronts, door- and window-frames, fittings and external cladding; type 316 is particularly well used. Austenitic stainless steel masonry fixings are now common. In the marine world, they are used for propeller shafts, and where high torque is present, high yield and precipitation hardening steels are used. Wire ropes, rigging and deck fittings are frequently of these materials. They are frequently used in domestic and food applications, for sinks, drainers, cooking vessels, pans and many large containers. In medicine, they have been much used for surgical instruments, theatre equipment and implants, although implants have created some problems and much research is under way at present for improvements in performance. Finally, there has been extensive use of austenitic stainless steels in the chemical industries, particularly for acids production, storage and transit.

Duplex stainless steels

Duplex stainless steels usually have ferritic/austenitic microstructures, even though duplex suggests any two phases. Broadly they combine the toughness and weldability of the austenitic steels with the strength, corrosion resistance and stress-corrosion cracking resistance of the ferritic steels. The solubility of the various alloying elements differs for the two phases, and this can produce variations in properties and corrosion resistance following the rapid temperature changes which occur during welding. However, they are readily produced in heavy sections, and possess excellent toughness and corrosion resistance. A small increase in chromium and molybdenum content to ensure adequate concentrations of each element in both phases is a slight cost to pay for their excellent properties. Although a 'standard' duplex stainless steel would have the composition, 22Cr–5.4Ni–2.9Mo–0.02C–0.19N (PREN 34), other proprietary alloys such as Ferralium 255 or the more recent SD40 are reported to have even better corrosion performance, with PRENs of 36 and 40 respectively. Duplex stainless steels have also proved very resistant to the general corrosion and stress-corrosion conditions which are found in sour gas wells.

15.3 CAST IRONS

The corrosion resistance of grey cast iron immersed in water is relatively good when compared with that of mild steel. A slight improvement in corrosion resistance and an increase in strength is obtained by adding up to 3% nickel. Further improvements in the impact resistance of the iron can be obtained by modifying the casting procedure to produce graphite spheroids instead of the normal flakes.

The graphite flakes in cast irons are interconnected and noble to the surrounding matrix. When an iron corrodes, the graphite flakes are often left standing proud of the surface, gradually forming a noble carbon-rich layer on the metal. This noble layer, acting as a cathode, promotes the corrosion of many metals when they are connected to iron, although the galvanic series leads one to expect that iron would sacrificially protect the second metal. This reversal of the role of the iron is called graphitisation. It is less of a problem in spheroidal graphite irons where the spheroids are not interconnected.

Additions of nickel in the range 13–36% yield an austenitic matrix in the iron to produce the NiResist series of cast irons. These irons have greatly increased corrosion resistance when compared with grey irons. For example, in flowing sea-water, the corrosion rate for NiResist is only one-tenth that of a grey or low alloy iron. Austenitic irons are used for valve and pump bodies, and for condenser water boxes in refinery, marine and electricity generating systems. They are almost immune to graphitisation.

In concentrated sulphuric acid (greater than 65%) a layer of insoluble iron (II) sulphate forms on the surface of grey iron to protect it from further attack. However, anything which destabilises the film, including increased flow rates, fluctuations in temperature or the presence of abrasive particles, can lead to pitting attack on the surface. Dilute sulphuric acid and other acids readily attack unalloyed irons.

The corrosion rate of cast iron in dilute caustic solutions, below 30% concentration, is negligible. Above this level, the addition of 1–3% nickel is said to be beneficial in controlling the rate of attack. Nevertheless, cast iron is used for caustic fusion pots, where the thick walls compensate for the initial high corrosion rate which falls to acceptable levels after a few days.

The austenitic irons have better corrosion resistance than the unalloyed or low alloy irons in dilute sulphuric acid and many other acids, as well as in caustic solutions, although they show poor resistance to solutions of nitric acid.

15.4 ALUMINIUM AND ITS ALLOYS

Aluminium is a very active metal. Exposed to a source of oxygen, it reacts to form a thin transparent oxide film over the whole of the exposed metal surface. This film controls the rate of corrosion and protects the substrate metal, allowing the production of long-life components in aluminium and its alloys. If the film is damaged and cannot be repaired, corrosion of the substrate is very rapid, as shown by Expt 3.14.

Case 15.6: A series of 18 mercury-in-glass manometers were used to check pressures in an experimental condenser system. Just before the end of the trial, one manometer was broken and mercury spilled on to the aluminium alloy platform which surrounded the rig. To save an expensive test programme, the trial continued. Fifteen minutes later, when it was finished, the technicians were surprised to find that large areas on the platform had grown long filaments of white powdery oxide where their shoes had ground mercury into the aluminium surface. The facility was closed for 48 hours for cleaning-up operations.

Aluminium alloys can be divided into two groups, casting alloys and wrought alloys. The wrought alloys can be subdivided into those which can be hardened and strengthened only by mechanical working, and those in which the properties can be varied between given limits by heat treatment alone. They are referred to respectively as non-heat-treatable (N) and heat-treatable (H). Typical compositions of aluminium alloys are shown in Table 15.4.

The corrosion performance of wrought aluminium alloys can be improved by cladding the material with thin sheets of pure aluminium or an aluminium alloy, often an Al–1Zn alloy. One such material is Alclad. The cladding is pressure welded to both sides of the material by rolling. The core material provides the desired mechanical properties for the components being manufactured, while the cladding is chosen to be inherently more corrosion resistant than the core, but anodic to it, so that sacrificial protection is given at any break in the cladding. Aluminium alloy tube clad on the bore has been used as water piping for several irrigation systems on farmland. If the coating is penetrated, the corrosion is confined to the cladding layer. This protection is particularly important at cut edges or drilled holes where the core is exposed. Among the factors which control the life and performance of the cladding are the following:

Cladding thickness, which is directly proportional to its life.

Electrolyte conductivity, which dictates the size of the cladding defect before corrosion of the substrate occurs; the better the conductivity, the larger the defect.

The oxygen supply, which maintains the corrosion resistance of the cladding.

Table 15.4   Some aluminium alloys and their approximate compositions

Specification		Approximate percent composition					
UK	USA	Cu	Mg	Si	Mn	Cr	Other
Wrought alloys							
	2011	5.5	–	0.4 max	–	–	0.5 Pb 0.5 Bi
H15	2014	4.4	0.4	0.8	0.8	0.1 max	–
L97	2024	4.5	1.5	–	0.6	–	–
	2218	4.0	1.5	0.9 max	–	–	2.0 Ni
N3	3003	–	–	0.6 max	1.3	–	–
	4032	0.9	1.0	12.5	–	–	0.9 Ni
	5050	–	1.2	0.4 max	–	–	–
N4	5052	–	2.5	0.4 max	–	0.25	–
N8	5083	–	4.5	–	0.7	–	–
	5086	–	4.0	–	0.5	–	–
N5	5154	–	3.5	–	–	0.25	–
H20	6061	0.25	1.0	0.6	–	0.25	–
	6063	–	0.7	0.4	–	–	–
	6101	–	0.5	0.5	–	–	–
	6151	–	0.6	1.0	–	0.25	–
DTD5074	7075	1.5	2.5	–	–	0.3	5.5 Zn
	7079	0.6	3.3	–	0.2	0.2	4.3 Zn
	7178	2.0	2.7	–	–	0.3	6.8 Zn
Cast alloys							
	13	–	–	12.0	–	–	–
	43	–	–	5.0	–	–	–
	108	4.0	–	3.0	–	–	–
	A108	4.5	–	5.5	–	–	–
	D132	3.5	0.8	9.0	–	–	0.8 Ni
	319	3.5	–	6.3	–	–	–
	356	–	0.3	7.0	–	–	–
	380	3.5	–	8.5	–	–	–

Anodising may also be used to enhance the corrosion resistance of aluminium and its alloys. An electrolytic process is used to thicken the surface film and produce a hard, compact, strong and tightly adherent layer. The process can be applied to complicated shapes. Various finishes are available and the film may be dyed to give attractive colours to the finished object. (Details of the anodising of titanium to yield an attractively coloured finish are given in section 15.6.) The final process in anodising is to seal the surface by boiling in water or exposing the metal to steam.

Both anodising and cladding are frequently applied to the copper-containing 2000 series of alloys, which have poorer corrosion resistance than the other alloys. Without some form of additional protection, the 2000 series should not be exposed outdoors, immersed in water or buried in soil.

Care is needed in the heat treatment of aluminium alloys. Intermetallic phases are easily precipitated if the temperature or duration of treatment is outside those ranges specified for the alloy. Some phases, for example $CuAl_2$, are cathodic to the matrix and produce corrosion damage in the areas around the precipitate particles, whereas others, such as Mg_2Al_3, are anodic to the matrix. Continuous grain boundary films of Mg_2Al_3 are particularly dangerous because the large cathode size compared to the anode causes corrosion to penetrate deeply into the cross-section of the material for a relatively slight loss of weight.

Stress-corrosion cracking (Chapter 10) affects some alloys in specific heat treatment conditions. Those in the 2000 and 7000 series appear to be particularly at risk, especially in thick cross-sections. The heat treatment and environmental conditions which produce stress-corrosion cracking are complex; consult a specialist text should problems of this nature occur (see the references and bibliography of Chapter 10). One method of reducing SCC in high-strength aluminium alloys is to use cladding of pure aluminium.

Most aluminium alloys suffer a higher rate of attack during the first year or so of exposure to the open atmosphere, but once they have developed a weathered surface, the rate falls, except in the case of the 2000 series. Rainwashing can be beneficial to alloys exposed to the atmosphere, especially in marine locations where aggressive ions and deposits are washed away. Alloy 3003 is often used for building and constructional work outdoors. It may be clad when sulphur oxides and condensation are likely to be present, to prevent the formation of loose voluminous corrosion products. Alclad is also beneficial in controlling differential-aeration corrosion in gutters, channels and roofing panels where leaves and other debris may collect. The 5000 series is also resistant to atmospheric corrosion, especially in marine locations or splash zones.

When immersed in water, including sea-water, most aluminium alloys exhibit reasonable corrosion resistance. The major exceptions are the 2000 series, and some of the 7000 alloys which must be in the form of Alclad or protected with a barrier coating. Polluted water can promote pitting corrosion on many aluminium alloys. Copper and its ions, especially in the presence of chloride and carbonate, are particularly troublesome in this respect. The copper plates on to the surface and causes severe pitting attack (see Case 12.7). The magnesium-containing alloys of the 5000 series offer a good combination of strength and corrosion resistance when immersed in water. Higher strength, at the expense of some corrosion resistance, may be obtained by using alloy 6061. When water temperatures are above 50°C, the increased rate of oxide formation tends to prevent the initiation of pits, but in cold static water, most alloys benefit from cladding.

Bimetallic couples involving aluminium should be treated with great caution and insulated where possible. Owing to the ease of ion exchange processes, copper and its alloys should not be used in the same system as aluminium, even where there is no direct contact between the two metals. The severe damage which results from ion exchange also rules out the use of copper and mercury antifouling paints on aluminium hulls.

Although aluminium and stainless steels are widely separated in the galvanic series, the oxide films present on both metal surfaces can act as a natural insulator and reduce corrosion to acceptable levels when the metals are coupled. However, trials should always be undertaken before such couples are put into service.

Aluminium alloys are not suitable for handling inorganic acids, except concentrated nitric and sulphuric acids, although they do have a satisfactory performance with many organic acids. Alkalis attack aluminium, but dilute alkali solutions can be inhibited with silicates to reduce the rate of attack to acceptable levels.

A new class of alloys using lithium as the major addition are under development. Although additions of up to 12% lithium are being considered, current interest seems to be centred on 3% lithium with smaller amounts of zinc and magnesium. It has been predicted that these alloys will supplant both 2024 and 7075 (see Table 15.4) in the construction of aircraft wing skins and fuselages over the next 10 years. The alloys offer a weight saving of up to 10% and the corrosion resistance appears to be satisfactory. However, there is some loss of ductility when compared to 2024 or 7075.

The cast alloys are very similar in performance to the wrought alloys. Again, the presence of copper in an alloy reduces corrosion resistance, especially in marine or other aggressive environments. However, the selection of a casting alloy will depend upon many factors, including the casting process (sand, die or permanent mould), the fluidity required to produce complicated shapes, weldability when sections have to be joined, as well as the characteristics required in the final product (good wear resistance, heat transfer, thermal expansion and mechanical properties). In general, the alloys containing silicon and magnesium, or combinations of both elements, have good castability and excellent corrosion resistance.

15.5 COPPER AND ITS ALLOYS

Copper and some of its alloys are the oldest toolmaking metals known to man. They have given excellent service in a wide variety of environments and the artefacts to be seen in many museums testify to the long-lasting qualities of these materials.

Pure copper is very soft and malleable. It is alloyed with small quantities of metals such as beryllium, tellurium, silver, cadmium, arsenic and chromium to modify the properties for particular applications, while retaining many of the characteristics of the pure metal. Larger alloying additions of zinc, tin and nickel are made to improve the mechanical properties of the metal, and to retain its excellent corrosion resistance under more arduous service conditions. Nickel permits increased flow rates in water systems, whereas zinc gives increased resistance to sulphide attack. Typical compositions of some copper alloys, available in wrought or cast forms, are given in Table 15.5.

Table 15.5 Some copper alloys and their approximate compositions

Alloy	ISO designation	BS	Approximate percent composition						
			Cu	*Zn*	*Sn*	*Ni*	*Al*	*Fe*	*Other*
Wrought Alloys									
Arsenical copper			99.6	–	–	–	–	–	0.4 As
Aluminium bronze		AB1	85	0.5	0.2	1	10	2.5	0.1 Pb, 1 Mn
		AB2	77	0.5	0.1	5	10	5	0.03 Pb, 3 Mn
		CMA1	71	1.0	0.5	3.5	8	3	0.05 Pb, 13 Mn
90/10 cupronickel			88.5	–	–	10	–	1.5	
Aluminium brass (inhibited)	CuZn20Al2	CZ110	75.9	22	–	–	2	–	0.04 As, <0.07 Pb
70/30 cupronickel		CN2	67.5	0.5	–	30	–	1	0.01 Pb, 1 Mn
Admiralty brass (inhibited)	CuZn28Sn1	CZ111	71	29	1.25	–	–	–	0.04 As, <1 Pb
Manganese bronze			65.5	23.3	–	–	4.5	3	3.7 Mn
Muntz metal	CuZn40	CZ109	60	39	–	–	–	–	<1 Pb
Naval brass (inhibited)	CuZn38Sn1	CZ113	58	38	1	–	–	–	0.04 As, <3 Pb
Cast Alloys									
Phosphor bronze			89.5	–	10	–	–	–	0.5 P
Aluminium bronze			88	–	–	–	9	3	–
Gunmetal		LG4	85.5	2.5	7	2	–	–	3 Pb, 5 Pb
		LG1	82	8	3	2	–	–	–
Nickel aluminium bronze			57.9	40	–	1	1	–	0.1 Mn
High tensile brass	CuZn37Al4Mn2	CZ116	66	27	–	–	4.5	1.0	0.3–2.0 Mn

Atmospheric corrosion

In the open air, copper forms a green patina which, in its most stable form, consists of basic copper sulphate, $CuSO_4 \cdot 3Cu(OH)_2$, although it may contain chloride in marine atmospheres or carbonate in industrial areas. This decorative long-lasting coating makes copper an ideal material for low maintenance roof coverings, for gutters and for channels. A small amount of copper dissolves in water runing over the metal surface. This can precipitate on to other less noble metals downstream in the water cycle, leading to galvanic corrosion. Cast iron gutters and pipes used in conjunction with copper roofs benefit from a bituminous or other impervious coating to reduce the possibility of galvanic corrosion. General corrosion or stress-corrosion cracking may become a problem in industrial areas if ammonia or ammonium compounds are present in the atmosphere (see Case 13.5). This is frequently associated with animal or bird droppings and many copper alloy components in store have later been found to be cracked.

Copper suffers general corrosion which thins the cross-section in sulphurous atmospheres.

Case 15.7: When British Rail converted from steam to electric traction, copper overhead cables were installed over all the running rails and sidings still required under the new system. During the transition, steam engines always stopped at a particular spot in a station. In a matter of months, the overhead cable at that one spot was reduced to half its original section by the sulphur-based gases in the steam engine exhaust.

Use in soil

Copper and its alloys can safely be buried in most soils, although high corrosion rates have been experienced in those containing cinders or acid peat. If it is expected that corrosion will be a life-limiting factor in the use of the material, it can be protected with bituminous, plastic or paint coatings. Dezincification can be a problem in brasses, so it is best to avoid them unless they are specifically required to counter the difficulties which may result from high sulphide levels in the soil.

Corrosion behaviour in aqueous media

The same forms of corrosion are found in freshwater and sea-water environments, although the change from pure water to salt-water with varying degrees of pollution should be regarded as gradually increasing in aggressiveness, aggravated by increases in flow rates and changes in the temperature and oxygen content of the water. Long-term steady-state corrosion rates for most copper alloys in sea-water range from 0.01 mm y^{-1} to 0.025 mm y^{-1}, while the best copper–nickel alloys have rates of about 0.001 mm y^{-1}.

The main problems which beset copper alloys in water systems are differential-aeration corrosion, erosion corrosion, stress-corrosion cracking and selective metal leaching. Differential-aeration corrosion is mainly a design problem, although pitting may occur with very slow flow which starves the metal surface of oxygen.

Ammonia and its salts, together with mercury-based compounds, are the prime cause of stress-corrosion cracking (Chapter 10). Aluminium bronzes are more resistant by an order of magnitude, but will still crack in hot water or steam; sometimes a small (0.3%) addition of tin is made to reduce the effect. Aluminium bronze is rendered susceptible to stress-corrosion cracking when welded, and the solution is to weld with a filler metal of 10% aluminium content, the final pass being of nickel aluminium bronze (NAB) to minimise the risk of galvanic effects between body and filler metals.

A protective film of carbonate may be deposited on the metal surface from water containing carbon dioxide and oxygen (see Section 13.4.) However, residual carbon deposited during manufacture must be avoided or cleaned from the internal surfaces of pipes, otherwise pitting can be initiated much later in the life of the system (remember Case 7.3.)

Dealloying

Dealloying affects many of the alloys. The commonest is dezincification of brasses containing more than 15% zinc, although dealuminification and denickelification have been reported for aluminium bronzes and cupro-nickels. Dealuminification is most prevalent in aluminium bronzes containing the γ-2 phase in the microstructure, and is most serious when the γ-2 forms a continuous grain boundary network. Dealuminification is controlled by rapid cooling from above 600°C, by addition of 1–2% iron or by addition of >4.5% nickel, but microstructural changes which occur during welding can still lead to corrosion problems in the heat-affected zone around the weld. Reheat treatment to remove unsatisfactory post-weld microstructures can pose serious problems to the engineer. Heating, handling and quenching large units, to achieve uniform properties throughout the material, may be impossible without introducing distortion.

Fittings made from brass may suffer dezincification (Section 6.3), especially when the β phase is present. The loss of zinc is accelerated by high temperature, by increased chloride content and in stagnant or slow flows, where differential-aeration occurs. Additions of 1% tin and about 0.04% arsenic, phosphorus or antimony inhibit dezincification. However, phosphorus can lead to inter-granular corrosion and most manufacturers concentrate on arsenic as an inhibitor in brasses. Inhibited α brasses are largely immune to dezincification in most waters, but the effect of tin and arsenic additions to α/β brasses is not so reliable or predictable in controlling the problem. There have been many cases of dezincification in the duplex brasses in both fresh water and sea-water. In potable water distribution systems, some duplex fittings which have given many years' service suddenly begin to lose zinc after only a slight change in water chemistry.

Performance under flowing conditions

Erosion corrosion is a function of flow rate. Pressure changes in a liquid on passing through valves and pumps give rise to cavitation damage, whereas entrained air or abrasive particles disrupt protective surface films to produce shallow horseshoe-shaped pits. The deterioration can be very rapid (see Chapter 8).

As the flow rate increases, copper and brass tubes become prone to impingement attack. Aluminium brass and cupronickel offer a greater resistance to higher flow rates, but both have maximum limits which must not be exceeded or the surface film on the metal will be destroyed. While the maximum velocity for inhibited Admiralty brass and aluminium brass is lower than that for cupronickels, they both give better service should sulphide be present in the water, either as a pollutant in rivers or estuaries, or in chemical and oil production plants.

Case 15.8 At flow rates above 2 m s^{-1} the Shell Oil Company experienced poor performance with aluminium brass condenser tubes in bulk carriers. The problems were eliminated by addition of ferrous sulphate to the cooling water (see Case 8.6). In contrast, use of alloy CN108 (Cu–30Ni–2Fe–2Mn) gave 14 years' trouble-free service in the velocity range $0.25–4 \text{ m s}^{-1}$. However, in some cases where carbon deposits were created during manufacture the tubes failed within a year (Case 7.4).

Tin bronzes and phosphor bronzes have good resistance to flowing sea-water. The alloys containing 8–12% tin are less susceptible to stress-corrosion cracking than brasses and have excellent resistance to impingement attack and to attack in acid waters.

The aluminium oxide film on both aluminium brass and aluminium bronze, which confers the additional corrosion protection, reduces the dissolution rate of copper ions from the alloy and makes them less effective in resisting biofouling.

The resistance of some copper alloys to erosion corrosion is improved when small quantities of iron are present in the alloy or the water. The iron apparently produces a tougher surface film. This has led to the use of iron sacrificial pieces, in preference to the normal zinc sacrificial anodes, in the water boxes of condensers and heat exchangers which use copper-based tubes and tube plates. The iron ions are absorbed into the film from the water and confer the additional resistance to impingement attack. Zinc ions do not have this beneficial effect.

Performance in chemical environments

A very clear distinction can be made between acids which can be safely handled in copper-based equipment and those which cause catastrophic attack. Non-oxidising acids such as acetic, phosphoric, dilute sulphuric and hydrochloric can be safely handled providing the concentrations of oxidising agents such as entrained or dissolved air, chromates and iron (III) ions are kept very low. Oxidising acids, nitric or concentrated sulphuric, and acids containing oxidising agents must not be handled in copper-based systems. Small additions of oxidising agents are particularly dangerous in hydrochloric acid, causing a dramatic increase in the rate of metal loss. Before using copper alloys in acid systems, tests should always be made with the particular liquid to be processed, reproducing the actual plant conditions as closely as possible.

In general, copper and its alloys are resistant to attack by alkalis, except ammonium hydroxide and alkalis containing ammonium or cyanide ions. Iron (III) and tin (IV) salts are aggressive to copper alloys.

Biofouling

Biological organisms abound in the open sea, coastal estuaries and rivers. They are found on piers, pilings, boat hulls, offshore platforms and all marine structures, as well as inside pipes and condensers. Copper and its alloys are more resistant to the attachment of biofouling organisms than other metals, a property which was used to great advantage in the copper cladding of ship hulls (Chapter 1). In the absence of wave action above 0.5 m s^{-1}, slime gradually thickens on copper alloys until, after about 18 months, organisms attach to the slime. At higher velocities, biofouling is much less. But when the copper is galvanically coupled so that release of copper ions is greatly reduced, its biofouling becomes as great as on other metals.

Applications of copper and its alloys

The applications of copper and its alloys are numerous. They are used for structures open to the atmosphere, such as architecture and sculpture. Copper and its alloys are used in freshwater- and sea-water-cooled heat exchangers and condensers. They occur in chemical, industrial and power-generating plant, and buried in the earth for water distribution systems. Many components in valves, pumps and taps, as well as pipes and pipe fittings, are made from copper alloys.

15.6 TITANIUM

Titanium has excellent resistance to atmospheric corrosion in both marine and industrial environments, and to corrosion and erosion corrosion in fresh water and sea-water at normal ambient temperatures. Indeed, it could be said that titanium and titanium alloys offer the best corrosion resistance of all alloys in many everyday environments, natural waters in particular. Coupled with medium to high strength and low density, titanium's corrosion resistance makes it especially valuable in engineering. After extensive use in the Mercury and Apollo spacecraft, it is still used widely by NASA in space applications for solid

solid rocket booster cases and other pressure cases. The SR71 Blackbird still holds the altitude and speed records for any aircraft. The Blackbird's fuselage frames, wing-beams and landing-gear were made with large amounts of thick secton Ti–13V–11Cr–3Al alloy, having the same strength as stainless steel and half the density. Titanium alloys are also very common in passenger aircraft components.

Titanium is one of the most abundant metallic elements in the earth's crust, but it requires a large amount of energy to separate it from its ore. This makes it expensive, as anyone who has ever bought titanium jewellery will know.

Surprisingly, it was the very good reputation which titanium enjoyed as a corrosion-resistant material that led to early unsatisfactory performance and sometimes failure. The problems stemmed from inadequate preservice testing when the alloys were first introduced. It was incorrectly assumed that the metal would be unaffected by the environment. In fact, titanium is a highly active metal, but its excellent corrosion resistance over a wide temperature range in a variety of environments is conferred by the formation and maintenance of a thin adherent oxide film. Even trace amounts of water during film formation influence its ability to protect the substrate metal. Those films developed in the absence of water in strongly oxidising atmospheres are often non-protective and may even result in pyrophoric reactions. Dry chlorine attacks titanium, but the metal is quite resistant to chlorine containing more than 0.01% water.

The oxide film, which is almost always present on titanium and its alloys in air, has the most exciting property of colour. Oxide films of different thicknesses can be made which have a wide range of attractive colours. This property has recently been used to great advantage in titanium jewellery, and works of art.

Experiment 15.1

If you have access to some titanium (any convenient size that will fit into a medium size beaker), abrade it using wet and dry papers so that a good surface finish is obtained. Then using a dc power supply capable of supplying up to 100 V, but no more, and not more than 1 A, connect the polished specimen to the positive terminal of the power supply, and another unpolished piece to the negative terminal as cathode. Wearing rubber gloves, set the power supply to 40 V, immerse both electrodes in a solution of 10% ammonium sulphate and switch on the power for a few seconds. Gas (oxygen) will be evolved and a beautiful colour will develop on the immersed part of the anode. By experimenting with different times and potentials, fascinating results can be achieved. A used specimen can be abraded and reused.

Try soldering an electrical wire to the metal part of an artist's ordinary soft paintbrush. Connect the other end of the wire to the negative terminal of the power supply; then connect the freshly abraded specimen of sheet titanium to the positive terminal. Wear rubber

357

gloves; dip the brush in the electrolyte and switch on the power, setting the supply to 40 V. 'Painting' with the brush on the metal will produce startlingly beautiful effects. Different colours will be obtained depending on the length of time the brush is in contact with the metal and the value of the potential. With practice, all sorts of *objets d'art* can be made. Always switch off the power before removing the rubber gloves.

Alloys

Titanium can exist in two allotropic forms, α and β, with different crystal structures and therefore the alloys are of three kinds — α, β and α/β. Alloying additions, which tend to stabilise one or other of the two phases at room temperature, are made principally to modify the mechanical and physical properties of the alloy, but the structural change also modifies the corrosion behaviour. Aluminium and tin are α stabilisers, whereas vanadium, molybdenum, chromium and copper are β stabilisers. The α alloys are non-heat treatable but have good weldability and medium strength. The more highly alloyed α alloys offer the best high temperature creep strength and oxidation resistance. Normally, the α/β and β alloys are heat-treatable, weldable and have superior mechanical properties but slightly inferior general corrosion resistance compared to the α alloys. A summary of titanium alloys and their compositions appears in Table 15.6.

Resistance to waters

Turning our discussion to specific environments, titanium resists all forms of corrosive attack in fresh water and steam to temperatures in excess of 310°C, although it does tarnish slightly in hot water and steam. In sea-water, titanium is usually totally free of crevice and pitting attack, even if marine deposits form; in splash zones or sulphide-containing waters, titanium is also inert. Velocities as high as 35 m s^{-1} are reported to cause only a slight increase in erosion corrosion, even in the presence of abrasive sand particles, which are particularly damaging to copper alloys. Titanium tubes in sea-water heat exchangers have given over 20 years' trouble-free service and are considered not to need replacement in most designs. Although they are not resistant to marine fouling, chlorination reduces the fouling and does not affect the titanium.

In hot aqueous chloride media, most titanium alloys will suffer crevice corrosion, but those rich in molybdenum such as 3Al–8V–6Cr–4Zr–4Mo have excellent resistance and are used for high temperature sour well and geothermal brine applications.

Table 15.6 Some titanium alloys and their properties

Composition	Properties	Description
Ti	Strengths in range 250–540 MPa; good general corrosion resistance for marine and chemical industry	ASTM grades 1–4 (commercial purity, cp)
Ti–0.2Pd	Strength 340 MPa	ASTM grade 7 (cp)
Ti–0.3Mo–0.7Ni	For mildly reducing environments or variable oxidising/reducing conditions	ASTM grade 12 (cp)
Ti–5Al–2.5Sn	Strength 810 MPa; airframe and jet engine components requiring good weldability and strength at elevated temperatures; also for cryogenic service	(α)
Ti–6Al–4V	Medium strength (880 MPa); used for jet engine components, air frame forgings, aerospace sheet components	ASTM grade 5 (α/β)
Ti–3Al–2.5Sn	Aircraft hydraulic and fuel lines	ASTM grade 9 (α/β)
Ti–2.5Cu	Ductile, medium strength, weldable, age-hardenable	(β)
Ti–4Al–4Sn–4Mo	Very high strength	
Ti–11Sn–5Zr–2.5Al–1Mo	Creep resistance to 450°C	
Ti–6Al–5Zr–0.5Mo	Medium strength (880 MPa), weldable, creep resistance to 550°C	
Ti–6Al–7Nb	Medium strength for surgical implants	
Ti–6Al–2Sn–4Zr–6Mo	High strength (1,150 MPa); jet engine components; intermediate creep resistance	(α/β)
Ti–15Mo–3Nb–3Al	High strength, oxidation resistant	
Ti–3Al–8V–6Cr–4Mo–4Zr	High strength (1,250 MPa), deep hardenable; for fasteners, rivets and springs in aerospace industry and for pipe for oil and geothermal applications	
Ti–13V–11Cr–3Al	High strength, thick-section components for SR71 aircraft	(β)

Resistance to chemicals

The key to titanium's supreme corrosion resistance is oxidisability. Thus, titanium is very resistant to attack in oxidising solutions, providing they are not dry. For example, as little as 1% water in chlorine gas is sufficient to alleviate the rapid reaction with dry chlorine. Resistance to chlorides is extremely good, a rare property among metals. Although there may be an enhanced corrosion rate in the first few hours of exposure in normal nitric acid solutions, this quickly disappears and the metal can be safely used to handle nitric acid in all strengths and at temperatures up to 250°C. Additions of silica compounds and silicone greases appear to inhibit the initial enhanced corrosion rate in nitric acid and mixtures of nitric and hydrochloric acids, even at temperatures as high as 200 °C. Other oxidising acids, such as chromic and hypochlorous acids, do not affect the metal significantly.

Reducing acids do attack the metal; fluorine is particularly aggressive if there is any possibility of formation of hydrofluoric acid. Aqueous solutions of acidic fluorides also produce rapid general corrosion of titanium and its alloys. Depending upon the alloy used, hydrochloric, phosphoric and certain concentrations of sulphuric acid may attack titanium.

Four techniques are generally used for controlling corrosion in reducing acids other than hydrofluoric acid. **Oxidising inhibitors** can be used. Nitric acid or oxidising ions, such as Fe^{3+}, Cu^{2+} or Ti^{3+}, are added to hydrochloric or sulphuric acid. Small quantities of inhibiting ions are often present in electrolytes used on production flow lines, but these impurities are not present in laboratory reagents used in test or monitoring programmes. Therefore, as far as possible, any testing to determine the reliability of titanium as a constructional material should be carried out in the process liquors, not in artificial solutions.

Aeration of the solution has been used to maintain the passive film on titanium, particularly in organic acids. Aerated hot formic acid, for example, does not attack titanium, but de-aerated solutions attack the metal quite readily.

Alloying additions can be made. Some alloys are particularly resistant to attack in reducing acids, particularly the alloy containing 0.2–0.5% Pd and the one containing 0.3% Mo. The high molybdenum alloys are said to resist hot concentrated hydrochloric acid, sulphuric acid or phosphoric acid. An explanation for the corrosion resistance of the alloy with palladium can be found in Section 5.2.

Anodic impressed current protection (Section 16.6) can be used to stabilise the oxide film on the metal surface. It is claimed that the corrosion rate in 40% sulphuric acid at 60°C is reduced by a factor of 33,000 after application of an anodic impressed current system.

Although it is attacked by hot concentrated alkalis, titanium is resistant to attack by dilute alkalis. It is also very resistant to crevice corrosion, although this has occurred in solutions containing chlorides, iodides and bromides at temperatures above 60°C. The problem is most likely to be found in condensers, and an upper limit of 130°C has been suggested for condensers operating in strong aqueous solutions of these ions.

Environment-sensitive cracking resistance

In specific environments, particularly chloride in some cases, titanium alloys may suffer stress-corrosion cracking, although the commercially pure material is much more resistant. The damage occurs when the oxide film is disrupted and the crack growth rate exceeds the rate at which the film is repaired. Thus, the active metal at the base of the crack is continuously exposed to the local environment. At faster strain rates, normal ductile failure ensues. In most cases a pre-existing crack is necessary to initiate the SCC failure mechanism, but in certain environments, notably methanolic chloride solutions, pitting attack by the electrolyte is the initiating mechanism. Pre-existing cracks are often developed by fatigue. There can be a long delay while a fatigue crack initiates, but propagation is much faster, assisted by the environment.

Hydrogen embrittlement can also be a problem because of the propensity of titanium to form brittle hydrides. The problem is greatest when titanium components are cathodic and hydrogen is generated directly on the surface. Investigations have shown the following conditions must exist simultaneously: pH <3 or >12, with abrasion and potential more negative than -0.7 V SCE; high tensile stress or temperature $>80°C$.

15.7 NICKEL AND ITS ALLOYS

Compared with many of the materials we have considered so far, nickel and its alloys are better at resisting corrosion, even in the aggressive environments of the oil and chemical industries. But nickel alloys are very expensive; only when a cheaper alternative has failed to produce an economic life will a nickel alloy be selected.

Most of the alloys are sold under trade names, which tend to run in a series differentiated only by a number or type letter. A limited selection of commercial nickel alloys and their typical compositions is given in Table 15.7. The most important alloying additions are copper, molybdenum, chromium and iron; each one confers a significant corrosion resistance in a particular environment. The response to an environment may be quite different for two alloys in a series (remember Case 13.11), so great care should be taken over their selection. Information on specific alloys is given in the bibliography, or should be sought from the trade literature and the manufacturers.

Commercially pure nickel is resistant to attack in air at normal temperatures. However, if sulphur dioxide is present and the relative humidity exceeds 70%, very common in industrial or urban areas, fogging of the metal surface occurs. Fogging is a tarnishing process in which the nickel acts as a catalyst in the conversion of sulphur dioxide to sulphuric acid, and finally to a surface film of basic nickel sulphate. Fogging spoils the aesthetic appearance of the metal. To maintain a bright finish, nickel is frequently coated with a thin layer of chromium (see Section 14.10).

Nickel is resistant to attack by hot or cold alkaline solutions and fused (molten) sodium hydroxide. It is slowly attacked by reducing acids, but dissolved oxygen increases the rate of attack. Oxidising acids and ions such as HNO_3, Cu^{2+}, and Fe^{3+} rapidly attack the metal; pitting may occur in sea-water.

Table 15.7 Some commercial nickel-containing alloys and their approximate compositions

Alloy	Approximate percent composition												C_{max}
	Ni	Cu	Cr	Fe	Mo	Ti	Mn	Co	Al	W	Si	Nb + Ta	
Alloy 201	99	—	—	—	—	—	—	—	—	—	—	—	0.15
Inconel 600	77	—	15	8	—	—	—	1	—	—	—	—	0.15
Hastelloy* B2	68	—	—	2	28	—	1	1	—	—	—	0.015	0.01
Hastelloy B3	68.5	—	1.5	1.5	28.5	—	—	—	—	—	—	—	—
Hastelloy C	64	—	16	—	16	—	—	—	—	4	—	—	0.005
Hastelloy C4	68	—	16	3	16	0.7	—	2	—	3	—	—	0.01
Hastelloy C22*	56	2	22	3	13	—	0.5	2.5	—	3	—	—	0.03
Hastelloy D205	64.5	2	20	6	2.5	—	—	—	—	1	5	—	0.03
Hastelloy G30	58	2	30	15	5	—	2	5	—	1	0.8	0.3–1.5	0.15
Monel 400 (CM400)	66	30	—	—	—	—	2	1	—	—	0.5	—	0.15
Monel K500	66	30	—	—	—	0.5	—	—	3	—	—	—	0.1
Inconel† 625	63	—	21.5	2.5	9	0.2	—	2.5	0.2	—	—	3.65	0.1
Hastelloy C276	57	—	16	5.5	16	—	—	2.5	—	4	—	—	0.005
Incoloy 825	40	2	21	33	3	1	—	—	—	—	—	—	0.05
Incoloy 800	33	—	20	46	—	0.6	—	—	0.6	—	—	—	0.1
Carpenter 20 Cb	30	3.3	20	44	2.5	—	—	—	—	—	—	—	—

*Hastelloy and C22 are trade marks of Haynes International Inc.
†Inconel is a trade mark of the Inco family of companies.

When sulphur is present in the atmosphere, the metal fails by grain boundary attack at temperatures above 320°C, but in sulphur-free atmospheres it gives satisfactory service to 850°C and is often used at even higher temperatures. The sulphur embrittles the grain boundaries at lower temperatures, with the greatest rate of penetration occurring in the range 550–650°C. Above 645°C a low melting point eutectic forms in the grain boundary and results in catastrophic failure. The maximum solubility of sulphur in nickel is 0.005% and any larger amount will form the eutectic. Trace amounts of sulphur acquired from oil and grease residues, marking paints and even fingerprints have caused failure in nickel alloy components at 650°C. Additions of chromium and iron produce alloys which are resistant to sulphur attack at temperatures up to 1,200°C and these alloys are used in furnaces and chemical processing plant. The wet corrosion resistance of the metal is also improved by alloying additions, notably copper, chromium, molybdenum and iron. Monel is a nickel alloy containing 30% copper. At room temperature the metal fogs in sulphur-polluted industrial atmospheres, and it is attacked by oxdising acids and ions, which destroy the passive surface film. Compared to the commercially pure metal, Monel is more resistant to reducing acids and alkalis, although it is attacked by hot concentrated caustic and aerated ammonium chloride. In stagnant sea-water the alloy is less prone to pitting or crevice attack than commercially pure nickel. Monel gives excellent corrosion and erosion resistance in fast-flowing (i.e. >2 m s^{-1}), well-aerated sea-water. It is often used to make valve trims and pump components for flowing sea-water systems in land-based power stations and chemical plants, as well as ships and offshore platforms. De-aerated hydrofluoric acid does not attack Monel, even in boiling solutions, but aerated acid will rapidly attack the metal. The rate of attack in 35% HF at 120°C has been reported as 0.025 mm y^{-1} when saturated with nitrogen and 3.75 mm y^{-1} when saturated with air.

Several alloys contain chromium. A minimum of 14% chromium must be added to reduce the risk of attack by nitric acid and oxidising ions. The Hastelloy C alloys contain 16–22% chromium. In 1955 alloy C was used in Bahrain in the first large-scale desalination plant. In 1964 an atomic-powered weather station entered service in the Gulf of Mexico using alloy C in double layers to encase the fuel and guard against corrosion. Its design lifetime was 500 years! In stagnant sea-water Inconel tends to pit, but the addition of molybdenum to Hastelloy C in alloy C-276 confers outstanding resistance to pitting and crevice attack in sea-water (see Section 7.4 and Fig. 7.8(b)). Heat exchangers of C-276 are in use to cool sour gas with sea-water in an offshore platform; other nickel alloy tubes in this application lasted only months.

On the other hand, Hastelloy B is not resistant to attack by oxidising acids or ions. However, it withstands attack from hydrochloric acid over a wide composition and temperature range; the corrosion rate is 0.26 mm y^{-1} in boiling 10% HCl and 0.05 mm y^{-1} in 35% HCl at 65°C. Hastelloy B is also resistant to de-aerated sulphuric and phosphoric acids (see Case 13.11).

Hastelloy C-22 is being used as a thin sheet liner in flue gas desulphurisation (FGD) equipment and for critical parts in a deep-well hazardous waste disposal system handling oxidising and reducing acids and metal-containing alkali solutions. Vessels of C-22 have been used for concentrating nuclear waste prior to glassification as well as for a scrubber for a nuclear waste incinerator at Los Alamos.

15.8 APPLICATIONS IN WATER SYSTEMS

Pipes and valves

Pipe systems and valves have to contend with water velocities which may vary from prolonged stagnation in dead legs, to 3–5 m s^{-1}, or higher, in pumped flow lines. Within a valve or on bends, the local velocity may be much higher than the design velocity of the system. The turbulence which results from this increased flow rate can disrupt the surface films on some metals and lead to premature impingement failure. Such metals have maximum flow rates which must not be exceeded. Other materials, notably stainless steels, require a supply of oxygen to maintain the surface film which protects the metal from corrosion. For these materials there are minimum flow rates which ensure an adequate oxygen supply.

Case 15.9: It took about 60 minutes to transfer 98% sulphuric acid through a pipeline. In order to reduce the transfer time to a more acceptable 15 minutes, a stronger pump was installed. This also increased the flow rate from 0.7 m s^{-1} to over the 1.6 m s^{-1} critical velocity for the pipe material in sulphuric acid. The pipe failed less than a week after the modification. [3]

All bends, transitions to smaller bore pipe and the internal surfaces of welds should be as smooth as possible.

Valve bodies are normally made anodic to the trim to give sacrificial protection to the internal components. Lists of typical pipe and valve materials are given in Tables 15.8 and 15.9.

Condensers and heat exchangers

Condensers and heat exchangers are often the critical components for processing plant in the electricity generating, oil and chemical industries. Although a valve change can be a great inconvenience, the downtime involved in retubing or changing a condenser can impose severe financial penalties on the user.

Plastic or metal pipe inserts are frequently placed in the ends of condenser tubes where impingement damage is most severe. These can be cheaply replaced during routine maintenance shutdowns.

Although sacrificial zinc or iron pieces may be incorporated in the water-box to protect the whole unit, the box is normally made from a metal which is

Table 15.8   Material characteristics for pipe systems

Material	Characteristics	Limiting velocity/m s^{-1}
Cast iron	Cheap; gives good service at low flow rates; suitable for water distribution systems; thickness of pipe often controls life	Depends upon the oxide which forms on the surface; 6 or above with adherent impermeable film
Galvanised steel	Cheap; must protect welded joints; life depends on coating thickness; use low flow rates; poor resistance to turbulence; contact to copper fittings or copper ions in water can cause premature failure	<1.0
Copper	Poor resistance to turbulence; toxic to biofouling organisms	1.0
Inhibited Admiralty brass	Resists dezincification	2.0
Aluminium brass	Better resistance to turbulence than copper; better resistance to sulphide pollution than cupronickel; less resistant to biofouling	4.0
90/10 cupronickel + 1.5% iron	Very good overall performance; resists stress corrosion cracking	3.6
70/30 cupronickel + 0.5% iron	Best corrosion resistance of any copper alloy; resistance to sulphide pollution inferior but improved by iron and manganese additions	4.6
Highly alloyed stainless steel	High capital cost (see Section 15.2)	no limit
Titanium	High capital cost; requires more support owing to lower Young's modulus	no limit

Table 15.9   Materials for valves in water systems

Body	Trim	Pipe material	Notes
Cast iron	Brass	Cast iron/steel	Old systems; requires constant maintenance
SG iron	Brass	Cast iron/steel	Reduces cracking in body
Leaded tin bronze	Alloy 400 Ni/Cu	Copper base	Good performance; some limits on velocity
Nickel aluminium bronze	Alloy 400 Ni/Cu	Copper base	Tolerates high flow rates

anodic to the remainder of the system, especially the tubes. Cast iron and steel are often used for water-boxes, although graphitisation of cast iron (see Section 15.3) may ultimately make the metal cathodic to the tubes, hence the sacrificial anodes. Internal nylon coatings are applied to ferrous water boxes to reduce metal loss. However, this reduces the sacrificial protection available to the tubes and generates high corrosion rates at any small break in the coating. NiResist cast iron, aluminium brass, aluminium bronze and cupronickels have also been used for water-boxes, as well as some non-metallic materials such as glass-reinforced plastics and thermoplastics. The tube-plate is normally made from the same metal as the tubes to avoid galvanic corrosion; in some areas it may be made anodic, but never cathodic, to the tubes. Common metals employed for

Table 15.10 Condenser tube materials

Material	Critical velocity/m s^{-1}	Notes
Aluminium	Min 0.3	Resistant to attack from CO_2 and O_2 in steam condensate; limited to maximum temperature 150°C; very low levels of copper ions promote pitting (<0.1 ppm)
Copper	Max 1.0	Fair performance
Admiralty brass	Max 1.8	Good performance; inhibited against dezincification
Aluminium brass	Max 2.5	Good performance; not so resistant to biofouling; better than 70/30 Cu/Ni if sulphide present
90/10 cupronickel + 1.5% iron	Max 3.0	Better than aluminium brass in waters polluted with ammonium ions
70/30 cupronickel + 0.5% iron	Max 3.6	Best flow rate of copper alloys; prone to failure when sulphide present
316 stainless steel	Min 1.5; no max limit	Very good in polluted (H_2S) environments but suffers crevice corrosion at low flow rates
High alloy stainless steel	No limits	Replacing 70/30 Cu/Ni for sea and estuarine water-cooled systems; very good in sulphide-polluted water
Inconel 625	No limits	Excellent performance; tolerates sulphide pollution; expensive
Titanium	No limits	Expensive; very reliable; tolerates a variety of pollutants; requires additional support owing to lower Young's modulus.

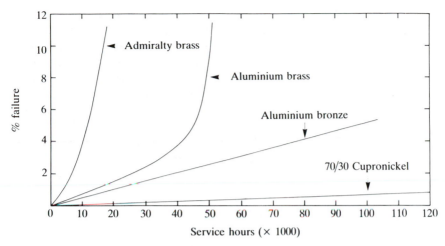

Fig. 15.1 Comparison of the performance of copper-based alloys used for condenser tubes operating in sea-water at the Harbourstream plant.

tube-plates include Admiralty brass, aluminium brass, aluminium bronze and cupronickels. When titanium tubes are used for maximum reliability, the tube-plate may also be made from this metal (Table 15.10). A comparison of the performance of different copper-based alloy condenser tubes operating in a marine environment is shown in Fig.15.1.

Pumps

Pump materials (Table 15.11) are subject to a wide variety of corrosion problems. It is common practice to make the body of the pump anodic to the critical components — impeller, shaft and seals — to ensure they are sacrificially protected. Both the impeller and shaft undergo crevice attack, whereas the impeller suffers impingement and cavitation damage, especially on the blade tips. Again, the liquid velocity in some parts of the pump may be considerably higher than the output velocity of the pump.

Stainless steels perform best as velocity increases, but are vulnerable to localized attack in slow-moving or stagnant waters, under O-rings and gaskets. Stainless steels are ideal for impellers and other internals that give galvanic protection during stand-by and shutdown periods. If the molybdenum content exceeds 5%, galvanic protection is not essential. Brines for oilfield injection are often de-aerated, which improves the performance because the differential oxygen effect with a crevice is reduced. However, when waters have low pH, low oxygen, high chloride and high sulphide, highly alloyed molybdenum-containing steels are necessary.

In the absence of sulphides, copper alloys have good performance up to a characteristic maximum velocity. Above this critical velocity, protection from the film is lost by erosion effects.

Table 15.11 Pump materials

Component	Characteristics	Low cost	Preferred	Maximum Life
Body	Should be anodic to other components Resistant to high velocity flow and cavitation	Cast iron G or M bronze Cement or epoxy-linked cast iron 2.5% Ni cast iron	NiResist iron Nickel aluminium bronze 316 stainless steel	NiResist iron 316 stainless steel
Impellers	Resistant to high flow rates and cavitation damage	2.5% Ni cast iron G or M bronze	Nickel aluminium bronze 316 stainless steel Alloy 400 Ni/Cu	316 stainless steel Stainless steel 29Cr–11Ni–Mo
Shafts	Resistant to pitting and crevice corrosion	410 stainless steel Mild steel	316 stainless steel Alloy 400 Ni/Cu	Alloy 400 Ni/Cu Alloy 20
Wear rings	Must be cathodic to other components owing to small size	G or M bronze	316 stainless steel	Alloy 506 Ni/Cu/Si Alloy 20
Bolts	Must be cathodic to pump body	Silicon bronze Carbon steel	Alloy 400 Ni/Cu	Alloy 400 Ni/Cu

Beware of graphitisation in cast irons.
Materials should be matched to flow rates, temperatures and pollutants.
Stainless steels require thorough flushing to avoid crevice corrosion on shutdown.

(a)

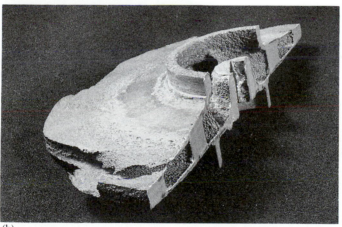

(b)

Fig. 15.2   A sea-water pump which failed after only a short time in service (Case 15.10):
(a) with the cover removed, the impeller appears very badly eroded; and (b) in section, the
internal erosion effects are visible.

Case 15.10: A sea-water pump failed after only a short time in service in a
coastal craft. When the cover was removed (Fig. 15.2(a)), the impeller
was found to be suffering from a severe case of erosion corrosion, both
on the edges, where it had lost much of its material, and internally,
where it showed the pitted surfaces typical of this form of attack (Fig.
15.2(b)). The material selected for the impeller was manganese bronze,
the cheapest choice, and plainly inadequate for the design velocity of
the pump.

369

Copper alloys are thus best for general medium speed performance when long periods of shutdown are expected. When sulphide is present, as is the case in polluted waters or those containing long-standing sediments, NiResist is preferable. Monel has a velocity resistance approaching that of stainless steel and is rather better with regard to crevice attack. It also performs better than copper alloys in the presence of sulphides.

The galvanic coupling of materials in pumps is successful on the large anode/small cathode principle. Pump cases should always be the anodic component, often providing galvanic protection to the smaller shafts and impellers. The combination of NiResist case/stainless steel impeller and shaft has been particularly successful, whereas a copper alloy case would lead to crevice attack of the steel. Copper alloy pumps are always matched to other copper alloy components. Tin bronze case/Monel impeller pumps have seen many years of service on ships. More recently, NAB impellers are used in 70/30 Cu/Ni or aluminium bronze cases, which again provide protection.

Graphite in pump packing and gaskets, and carbon fillers in rubber and Teflon, are to be avoided because the strong galvanic couple has led to many rapid failures.

Propellers

When ships acquired engines in the nineteenth century, the first propellers were made of cast iron. These could be supplied at low cost, but needed continual replacement because of low resistance to general corrosion. Additions of nickel were found to improve the corrosion resistance. In the 1880s Parsons discovered a high tensile brass which offered significant improvements over cast iron propellers. The material was manganese bronze, a 60/40 duplex brass with about 1% aluminium and 0.1% manganese. Propellers of this material have seen considerable service, though there are many classic examples of serious damage resulting from erosion and cavitation.

Since the early days, propellers have been made almost exclusively from various types of bronze, with the exception of some applications of stainless steels. The addition of aluminium to copper gives aluminium bronze, but both it and manganese bronze suffer from erosion corrosion and cavitation erosion and their use today as propeller materials is declining because of the availability of superior materials such as nickel aluminium bronze (NAB), developed in the 1930s and providing greater hardness and cavitation resistance. In the early 1950s, considerable research and development was carried out to find even better materials. A number of high manganese bronze materials were developed, notably Superston 70, a trade name of Stone Manganese Marine Ltd, its UK manufacturers. This was typically Cu–15Mn–6Al–4Fe–2Ni. The Superston range was found to have superior mechanical properties, but slightly inferior corrosion performance. Its main advantage was in providing a lower level of acoustic noise during operation, an obvious help to warships. Superston 70 was universally adopted in the Royal Navy for controllable-pitch propeller blades until a series of blade failures occurred in the 1970s, ascribed to corrosion

Table 15.12   Materials for marine propellers

Material	Comment
Cast iron	Low initial cost but requires frequent replacement; used on slow vessels; additions of nickel improve corrosion resistance
Manganese bronze	Medium cost; suffers dezincification, cavitation and crevice corrosion; suitable for low powered ships
Nickel manganese bronze	Higher cost but better performance than manganese bronze
Nickel aluminium bronze	Very durable; resists cavitation and erosion damage; aluminium content limited to 8.4–9.0% to gain maximum benefit without forming corrosion-susceptible phase; welding or hot repairs require caution; used for high performance propellers
Stainless steel	Used for harbour vessels; may suffer differential-aeration attack if low oxygen conditions exist, as when buried in mud or stationary for long periods in oxygen-shielded environments

fatigue originating at points of poor machining. Superston was discarded in favour of more tightly controlled NAB.

Some commercial ships, ferries in particular, are fitted with stainless steel propellers; in Europe they are typically 12% chromium steel. In instances where stainless steels have been used for bow thrusters, serious corrosion problems have been observed, whereas stern propellers of the same material suffer no serious corrosion. This has been attributed to the crevice-type low oxygen conditions existing at the relatively shielded location of the bow thruster; the propellers are in open water. Operation in low salinity coastal waters may also be significant. A summary of propeller materials is given in Table 15.12.

15.9   SELECTION OF MATERIALS

When choosing the material for a component or structure, in theory the designer begins by considering a large number of factors other than corrosion resistance. Figure 15.3 shows in schematic format some of the many decisions which must be made. The cost and availability of materials are as important as their properties; a gold-plated car would have excellent corrosion resistance, but would be rather expensive and hard to procure. Engineering design must always be concerned with compromise so that the best match of design criteria can be converted into an economic, readily available and durable product.

Corrosion resistance is an essential part of materials selection, but is often given scant consideration relative to other mechanical properties which might, at first, seem more important. Poor corrosion resistance will always lead to premature failure no matter how well the component has been designed in other

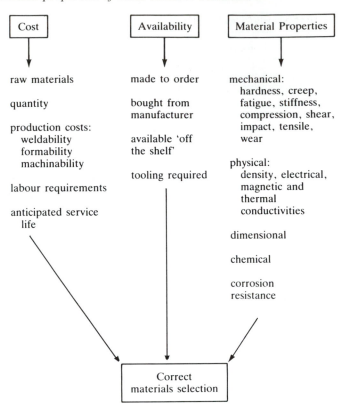

Fig. 15.3 The contribution of corrosion resistance to the process of materials selection.

respects. The previous sections have provided little more than the most general coverage of five major groups of alloys. Nevertheless, they will have achieved something if you have begun to appreciate the enormous number of materials available and the complexity of their behaviour in different environments.

Many designs evolve. Petroski makes this point in his study of engineering failures. Viewed broadly, the whole of engineering history has been an evolution of design, capitalising on the successes, but carefully studying the failures. Petroski believes it is the failures which teach us how to proceed to success — how to tune the designs to give even better performance than before. Materials are rarely selected using a blank sheet of paper. Most often choices are made by studying the precedent, looking carefully at alternatives and choosing according to the best available evidence at the time.

Even so, the decision is not easy; a good source of reliable information is necessary. As shown by many of the case histories, no solution has yet been found to the problems of selecting the correct material for a particular corrosive environment. At present, the best available method is to use a large-scale database of material properties, such as can be obtained in the Fulmer Materials Optimizer, the Cambridge Materials Selector or the collected works entitled *Engineering Science Data Units*. The complete design process remains arduous and littered with pitfalls for the inexperienced. However, at a more basic level, we suggest that elementary errors are the cause of the majority of corrosion problems, and these errors can be eliminated by the application of relatively simple ideas, combined with a general knowledge of material properties.

The best hope for the future is in the development of an entirely computer-aided materials selection process. The vast amount of data storage and high speed computing required for such a task is only now becoming available, though the first steps along the road towards achieving optimum selection have already been taken. In many areas of engineering, computers are being used as databases for materials selection in restricted applications and environments. The setting up of national facilities for on-line information retrieval has also been successful and will undoubtedly be considerably extended in the future.

15.10 REFERENCES

1. Uhlig H H 1981 Corrosion trails, passing scenery and distant views, *Corrosion Science* **21**:159–76.
2. Hooper R A E 1986 Stainless steels — past, present and future, *Metals and Materials* **2**:10–12.
3. Seibert O W 1983 Classic blunders in corrosion protection revisited, *Materials Performance* **Oct**:9–12.

15.11 BIBLIOGRAPHY

Anon 1989 *Metals and alloys in the unified numbering system,* 6th edn. ASTM, Philadelphia PA.

Anon 1990 Stainless steel, *Materials Performance* **29**(3):64–68.

Bogaerts W F and Agema K S 1992 *Active library on corrosion,* CD-ROM for DOS-PC, Elsevier, Amsterdam.

Callcut V A 1989 Aluminium bronzes for industrial use, *Metals and materials* **5**(3):128–32.

Davison R M and Redmond J D 1990 Practical guide to using duplex stainless steels, *Materials Performance* **29**(1):57–62.

Dillon C P 1987 *Performance of tubular alloy heat exchangers in seawater service in the chemical process industries,* MTI publication 26, MTI, St Louis MO. (Also available from the Nickel Development Institute.)

Gladman T 1988 Developments in stainless steels, *Metals and Materials* **4**:351–5.

Graver D L 1985 *Corrosion Data Survey — Metals Section,* 6th edn. NACE International, Houston TX.

Hooper R A E 1986 Stainless steels — past, present and future, *Metals and Materials,* **2**:10–12.

Lacombe P, Baroux B and Beranger G 1990 *Les aciers inoxydables,* Les Editions de Physique, Les Ulis, France.

Mantle E C 1986 Copper alloy castings for marine applications, *Marine Engineering Review* **July**:19–21.

Ross R W 1993 The evolution of FGD materials technology. Paper 414 in *Corrosion 93 plenary and keynote lectures,* edited by R D Gundry, NACE International, Houston TX.

Sedriks J 1993 Advanced materials in marine environments. Paper 505 in *Corrosion 93 plenary and keynote lectures,* edited by R D Gundry, NACE International, Houston TX.

Shone E B and Grim G C 1986 25 years experience with seawater-cooled heat-transfer equipment in the Shell fleets, *Transactions of the Institute of Marine Engineers,* vol. 98, paper 11.

Tuthill A H 1987 Guidelines for the use of copper alloys in seawater, *Materials Performance* **26**(9):12–22.

White R A and Ehmke E F 1991 *Materials selection for refineries and associated facilities,* NACE International, Houston TX.

16 CATHODIC AND ANODIC PROTECTION

It is better to wear out than to rust out.
(Bishop Richard Cumberland, 1631–1718)

To someone with merely a passing interest, corrosion control by electrical techniques may seem as unbelievable as building one of the world's largest steel structures then immersing it unpainted in the sea. Perhaps they would be shocked to learn that both of them are true. Indeed, only by virtue of the first is it possible to do the second.

The principles and applications described in most of this chapter relate to **cathodic protection**, a name which is almost self-explanatory. This important and widely used method of corrosion control involves two techniques: the **sacrificial anode** method, which uses the principle of bimetallic corrosion; and the **impressed current** method, which is an electrically controlled process. The latter is also known as **impressed current cathodic protection (ICCP)**.

Another electrical technique which involves an impressed current is **anodic protection** and is used for some metal/electrolyte combinations. This will be described in Section 16.6.

16.1 THE THEORETICAL BASIS

By far the most important application of electrical techniques to control corrosion is for steel. Steel hulls of ships, offshore drilling platforms and oil and gas undersea pipelines are all protected against attack in one of the most aggressive natural environments by methods which will be discussed in this chapter. Additionally, the steel reinforcement within concrete structures, such as bridges and the containers which hold potentially corrosive chemicals, have also been protected by such techniques.

Except in Section 16.6, which is devoted to anodic protection, this chapter will deal only with the protection of carbon steel. Let us begin by considering the E/pH diagram for iron in water (Fig. 16.1) which shows the regions of thermodynamic stability under different environmental conditions.

The free corrosion potential, E_{corr}, of iron in aerated water is in the range

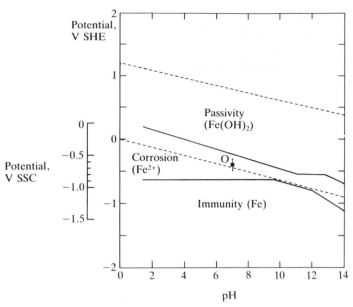

Fig. 16.1 E/pH diagram for iron in water.

−600 to −700 mV SSC at a pH of 7. It is represented by point O in the figure. In sea-water, the only significant change to the figure is a lack of passivation below pH 5. This is largely irrelevant in natural sea-water, which has a pH of 8.2–8.5 approximately. Point O is thus shifted to the appropriate pH value, but the range of values of E_{corr} remains about the same. Whether it is in a chloride-containing environment or not, point O is situated within a zone of corrosion and the iron will rust rapidly if a supply of oxygen is maintained.

Let us consider four different ways of affecting the condition of the system at O.

Decrease the pH: the solution is made more acidic. Inspection of Fig. 16.1 shows that the specimen remains in the corrosion zone at all values of pH < 7, i.e. the zone in which the soluble iron ions are the most stable species. The *rate* of corrosion actually increases as pH falls, but such *kinetic* information is not obtainable from Fig. 16.1.

Increase the pH: a small increase moves the iron into a region of passivity where the most stable species is an insoluble iron hydroxide (or hydrated oxide). If it is allowed to coat the iron we should expect a reduction in the corrosion rate because the film will tend to separate the iron from its corrosive environment. However, it cannot be assumed that an insoluble film will behave in a protective manner. If the film is not uniform, is conducting and permeable to ions, or if it is damaged by flow of electrolyte across the surface or by mechanical means, then corrosion is likely to continue, probably by pitting. Indeed, the local rate may well be accelerated because the exposed anode area is smaller.

Apply a more negative potential: the metal condition moves into a zone of immunity. Remember that the distinction between a corrosion zone and an immunity zone is merely one of definition, i.e. a concentration of iron ions at equilibrium of 10^{-6} M determines the boundary between the two zones. From Fig. 16.1, this occurs at potentials less than -800 mV SSC at pH 1–9. Although the metal is in the immunity region according to our definition, a corrosion reaction may still occur. Consult Fig. 4.9(a) and remind yourself how the anodic and cathodic reactions vary with potential. The more negative the potential, the smaller the anodic reaction but the greater the cathodic reaction; the metal is more cathodic. This is the principle of cathodic protection of metals.

Make the potential more positive: the metal is once again brought into the passive region in which there is a chance that the rate of corrosion may be reduced by the formation of a barrier between metal and electrolyte. Although it was said that iron was difficult to passivate, this principle has been used effectively for certain steel/electrolyte combinations, as well as for other metals and electrolytes. The principle is known as **anodic protection** (see Section 16.6).

The example above suggests three ways of protecting against corrosion, all except decreasing the pH. With regard to increasing the pH, the corrosion engineer very often has no control over the electrolyte. The offshore industry engineer cannot change the pH of the sea, neither can the process engineer alter the composition of the product being manufactured, product which corrodes containers and pipework. Sometimes, inhibitors can be added to the electrolyte, and examples where pH increases were tried are given in Chapter 13. Very often, however, engineers must consider other methods of safeguarding their expensive metal structures. Altering the potential forms the basis of corrosion control by cathodic and anodic protection.

Figure 16.2 is based upon the work of LaQue [1] and shows, in schematic form, the variation of cathodic potential against current density for carbon steel in sea-water. It also shows the effect of potential and current changes upon the rate of corrosion, measured in terms of weight losses. LaQue found a good correlation of a corrosion rate/current density curve with a potential/current density curve. Let us now consider how this information can be used to advantage.

Figure 16.2 shows that less of the metal corrodes as potential is made more negative. This is hardly surprising, for a metal which is the cathode in a wet corrosion cell does not generally corrode. It is possible for small local anodes to exist on the metal surface, and it is not until the anode reaction has been completely suppressed that all corrosion will cease. A designer begins by stating a maximum acceptable corrosion rate, r_p, and uses a graph such as Fig. 16.2 to obtain a value of current density, i_p, which will lead to the desired corrosion rate. This, in turn, yields a protection potential, E_p. The actual protection potential used in cathodic protection systems is dependent upon the applications; more discussion of this will occur in following sections. From Fig. 16.1, we have seen that it is probably more negative than -800 mV SSC,

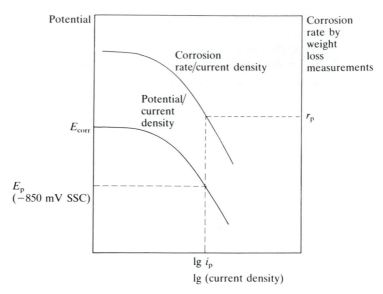

Fig. 16.2 Schematic diagram showing the variation of cathodic potential with current density for steel in sea-water, and the correlation with corrosion rate measured by weight-loss data.

and many designers choose −850 mV SSC. There have been many instances where the choice of potential required for adequate protection has been arrived at by trial and error, and uncertainty remains to some degree. As a general rule, the protection range is −800 to −900 mV SSC.

The following example shows how the expected corrosion rate is calculated for different applied cathodic potentials.

The corrosion rate, i_p, at a given polarisation cathodic to E_{corr} (at which the corrosion rate is i_{corr}) is expressed by

$$i_p = i_{corr} \exp\left(\frac{\alpha \eta Z F}{RT}\right) \tag{16.1}$$

where we assume that $\alpha = 0.5$ (see Section 4.7), $z = 2$ (for iron), $F = 96,494$ C mol^{-1}, $R = 8.314$ J mol^{-1} K^{-1} and $T = 283$ K (a sea-water temperature of 10°C). For a polarisation of −200 mV, we get

$$i_p = i_{corr} \exp\left(\frac{0.5 \times (-0.2) \times 2 \times 96,494}{8.314 \times 283}\right)$$

$$i_p = i_{corr} \exp(-8.20)$$

$$i_p = 0.00027 i_{corr}$$

Thus, if we assume the free corrosion potential for steel in sea-water to be −650 mV SSC, then the corrosion rate at −850 mV SSC will be reduced to 0.03% of the rate for unprotected steel. At −750 mV SSC, the corrosion rate is reduced to 2% of the rate at the free corrosion potential.

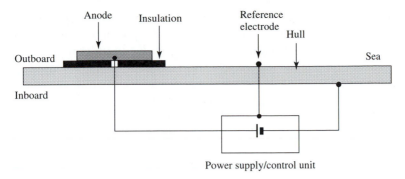

Fig. 16.3 The principle of impressed current cathodic protection using a potentiostat.

In order to carry out such protection, we use an electrical system, similar to that shown schematically in Fig. 16.3. The components shown have already been described in detail in Section 4.8 which concerned the three-electrode cell. The control system is essentially a potentiostat, but with an automatic compensation device; the role of the control unit will become more apparent later.

The control unit is set to impress a current, I_p, so that, for a given wetted surface area, A, the current density is $I_p/A = i_p$, and the chosen protection potential, say -850 mV SSC, is measured at the metal surface. Assuming there are no further changes in the system, the metal surface is protected against corrosion.

Question: *Potentials more negative than -850 mV SSC show even less metal loss. Why not use -1000 mV SSC, or lower?*

There are two reasons why it is not a good idea to use more negative potentials. The evolution of hydrogen, which occurs at more negative potentials, is often found to cause damage in other ways, notably hydrogen embrittlement (see Chapter 10). In addition, the large currents associated with more negative potentials produce high local concentrations of hydroxyl ion, which may cause excessive chalking (see below) or damage any barrier coatings such as paint. The potential of -850 mV is a convenient value which gives efficient protection, while reducing the likelihood of these other forms of damage.

With a suitable control unit, the system can be made self-regulating. If the surface were protected by a barrier coating or paint film, the currents observed would be much lower, but the current density for protection of the bare metal areas would still be the same. Let us now consider what happens if the protective coating is damaged, causing an increase of effective surface area. If the control unit continued to impress current I_p, then the current density would fall. Figure 16.2 shows that this would result in a change of potential back towards the free corrosion potential, together with a corresponding increase in corrosion rate. In an ICCP system, the instrument has an automatic compensation circuit which senses the onset of a potential change and counters it by maintaining the potential at the set level, E_p, In order to achieve this, the control unit must maintain current density i_p, and must therefore impress a greater current.

The opposite effect is equally important. If there were a reduction in the effective surface area, there would be a tendency for the current density to increase, and the potential to become more negative. This is also undesirable, as was mentioned above. By continuously sensing the potential of the metal surface, the control unit is able to maintain the constant potential required for protection and impress a smaller current. Instruments which maintain constant applied potentials are known as **potentiostats**. It is important to appreciate that the magnitude of the current varies considerably with time, and the circuitry must be able to cope with such changes. Thus, a significant input to the design is knowledge of the maximum **current demand** of the system.

Remember also (with the aid of Fig. 16.2) that in order to maintain the corrosion rate below the maximum acceptable value, r_p, the same current density, i_p, is required to protect steel, whether painted or bare metal. The problem is that the area is not known accurately. In Chapter 14 we saw how paint films are imperfect and usually contain defects or holidays which allow restricted access of an electrolyte. When immersed in the sea, bare metal becomes covered in marine growth and is coated with films called **calcareous deposits** when cathodically protected. These effects mean that current density cannot be determined precisely. There is a great advantage in applying a given potential using a system similar to Fig. 16.3. The current is varied until the reference electrode senses that the structure is at the required potential. In this way, the necessary current density for protection is achieved without the input of an accurate area value and irrespective of the presence or absence of films on the surface. When current densities are quoted in the discussion which follows, remember the limitations introduced by not knowing the precise area of metal exposed to the electrolyte.

16.2   SACRIFICIAL ANODE DESIGN

Perhaps the simplest way to describe the workings of sacrificial anode cathodic protection is to recall the basic wet corrosion cell (Fig. 4.2) and the general rule established in Section 4.2. In any cell it is the anode which corrodes and the cathode which does not. An engineer, faced with the problems of preventing a metal structure from corroding, apparently has a ready solution: make the structure the cathode in a corrosion cell. Sometimes, cathodes may be damaged if too great a current density occurs. Usually, this results in damage to paint coatings, but hydrogen embrittlement of the metal is also a possibility if the potential is sufficiently negative (See section 10.3).

Question: *If the metal structure is the cathode, what is the anode?*

Using the galvanic series (Fig. 5.6) the engineer selects a material which becomes an anode when coupled to the metal requiring protection. Because engineers are most frequently concerned with the protection of iron and steel, a glance at the galvanic series shows that any of the metals at more active potentials than iron will, in theory, be suitable. But in practice, it does not make sense to try to protect iron by coupling it to sodium because sodium reacts explosively with water. Sodium will protect iron, but for such a short time as to be useless. It is therefore necessary to find another anodic metal which corrodes more slowly than sodium. Metals such as zinc, magnesium and aluminium are suitable and are much used. If the engineer attaches a piece of zinc to an iron structure, the zinc will corrode preferentially and leave the iron unaffected.

Question: *Is it really that simple?*

Almost. It is obvious that a 1 kg lump of zinc on an offshore drilling platform will not have a significant effect in reducing corrosion on the platform. Whichever method is used, protection is obtained by electrons flowing from anode to cathode. In a very large cathode, the current density effect means that a very large number of electrons will be required. And 1 kg of zinc will provide only 2,000 faradays of charge before dissolving completely. Thus, we need to take into account the relative areas of anode and cathode and the rate at which the zinc corrodes in order to predict how much anode material is required initially and when a replacement will be necessary.

Figure 4.11(a) is an Evans diagram which shows the effect of coupling metals such as zinc and copper. The potential on the copper cathode becomes more negative as current is drawn, whereas the potential of the zinc becomes more positive. The potentials of both metals converge to the corrosion potential, E_{corr}, at which there is a negligible resistance in the cell and the limiting current is passed. The difficulty with these schematic diagrams is that they do no more than *indicate* answers to problems; in this case, the actual potential of the couple is dependent upon the two metals used and their relative areas, but more important for the protection of large structures is the *distribution of potential* over the surface. If there is a large cathode/anode ratio, protection is unlikely to be satisfactory because most of the cathode has not been sufficiently polarised away from its free corrosion potential. Remember the anodic reaction still occurs at potentials more negative than the free corrosion potential, so it is usual to define a **protection potential**, E_p — the least negative potential necessary to achieve a satisfactory level of protection. For steel in aerated sea-water, this is considered to be about -800 mV SSC. The correct amount and type of anode material, when distributed over the surface of a structure, will result in its cathodic polarisation to a potential more negative than this.

Anodes connected to structures for corrosion protection in this way are termed **sacrificial anodes**. Because they rely upon the galvanic effect, it is essential that they make a good electrical contact with the structure being protected. Anodes are usually welded to special lugs which are integrated into the structure at predetermined points.

Question: *What anode materials are actually used in practice?*

Zinc, magnesium and, more recently, aluminium. The traditional sacrificial anode material for steel in sea-water is zinc. In 1824, Sir Humphrey Davy [2] reported the successful use of zinc and cast iron anodes to reduce corrosion of the copper sheathing on warships which had been put there to protect the hulls from marine life (Chapter 1). Within the year they were duly installed on HMS *Samarang*. It is amusing to realise that although the sacrificial protection provided by the anodes successfully reduced the anodic dissolution of the copper, it allowed the growth of marine organisms once more and the fouling returned. For this reason, it is said that cathodic protection was not seriously used until a century later in the United States to protect underground pipelines.

Although sacrificial anodes were used before World War I to prevent condenser tube corrosion in warships, it was not until the 1950s that the technique was used on a significant scale. Since then C-Sentry,* a zinc alloy containing 0.1–0.5% aluminium and 0.025–0.15% cadmium, has been used extensively throughout the marine industry.

The protective action of zinc would be excellent if it dissolved at a reasonably constant rate. Unfortunately, this is not usually the case. Normal commercial purity zinc corrodes in sea-water with the formation of an impermeable skin which severely limits its current output. Of the principal impurities — iron, copper and lead — the most detrimental to anode performance is iron. Its solubility in zinc is so low (<0.0014%) that any excess is present as discrete particles. These in turn yield local galvanic cells which produce a coating of insoluble non-conductive zinc hydroxide/zinc carbonate, which eventually renders the anode ineffective. Addition of aluminium is beneficial because a less noble aluminium/iron intermetallic is formed which reduces the effect of the local corrosion cells. The addition of cadmium acts in a similar way to reduce the adverse effect of lead impurity. Figure 16.4 shows two zinc sacrificial anodes. The top anode is unused and measures 360 mm length, 60 mm thickness and about 4 kg weight. The holes allow it to be mounted over studs and bolted on to a structure. The lower anode is partly consumed and shows how the optimisation of the alloying components ensures uniform dissolution over the surface.

Aluminium normally undergoes pitting corrosion in sea-water because of the cathodic oxide layer which always exists on the metal in air. The unpredictable nature of this form of corrosion makes the pure metal most unreliable for use as a sacrificial anode, so alloying additions are made to prevent the formation of a continuous adherent protective oxide film, and permit continued and regular galvanic activity. As a result, aluminium alloys containing zinc and mercury or

---

*Trade mark of Impalloy Ltd.

Fig. 16.4 Two sacrificial zinc anodes. The top anode is unused and measures 360 mm length, 60 mm thickness and 4 kg weight. The holes allow it to be mounted over studs and bolted onto a structure. The lower anode is partly consumed.

zinc and indium have been developed. These have much greater electrical power/weight ratios than zinc alloys.

The highly negative free corrosion potential of magnesium leads to a rather too vigorous corrosion rate in sea-water. Its use is thus restricted to the protection of pipelines in soil, or structures in estuarine waters where the resistivity is high enough to limit the effectiveness of zinc or aluminium alloys. The protection of storage tanks containing fresh or brackish water is another suitable application for magnesium anodes. Magnesium represents a significant fire hazard which precludes its use in certain applications. But surprising applications sometimes emerge.

Case 16.1: The Living Seas exhibit at EPCOT Center, Florida, is a 22,000 m^3 artificial sea-water aquarium constructed from good quality reinforced concrete to a depth of 75 mm over the reinforcing steel. However, rust stains and corrosion were noticed only two years after completion. Potential gradients in sea-water generate abnormal behaviour in marine creatures, while release of zinc or aluminium ions from sacrificial anodes or chlorine from impressed anodes is detrimental to marine life. Magnesium ions are abundant in natural sea-water and represent no hazard. Indeed, in normal use it was necessary to replenish them from time to time.

The rest potential of the unprotected steel was found to be between -0.07 and -0.1 V SSC. Four groups of six Galvomag anodes were used, with rheostat controls to alter the resistance in the circuit and enable the system to be tuned for optimum effect. The system, though designed to achieve -0.9 V SSC, was operated only to -0.6 V, at which potential all the anodic sites just disappeared. The system was simple and easy to install, economic to run at only 350 kg of anode material per annum, and had no detectable effects on the fish. [3]

Although the theoretical free corrosion potential of magnesium is -2.12 V SCE, it is found in practice to be about -1.7 V SCE. This is reflected in its low (50–60%) efficiency. If we assume that 1 mol of magnesium produces 2 F (every atom yielding two electrons) we calculate that each kilogram of material should yield about 2,200 A h. In reality we rarely get more than 1,200 A h kg^{-1}. The efficiency of the anode material is thus 1,200/2,200 $\times$ 100%. This compares unfavourably with zinc or aluminium alloys, which have efficiencies greater than 90%. On the other hand, magnesium alloys are non-toxic, as Case 16.1 showed.

There are two commonly available magnesium alloys; Galvomag is a high purity alloy with 1% Mn; the other contains 6% Al, 3% Zn, 0.2% Mn. They are the result of attempts to increase the efficiency by alloying additions, albeit with limited success. The reasons for their inefficiency are too complex to be described here, but are thought to be concerned with changes in anion and cation concentrations close to the metal surface, as well as the evolution of hydrogen at local cathodes in the metal.

A complete review of sacrificial anodes and their properties has been published by Schreiber. [4] Table 16.1 lists three typical sacrificial anode materials and their properties.

Table 16.1 Sacrificial anode materials and their properties

Property	Zinc alloy* (C-sentry)[†]	Aluminium alloy (Galvalum I)[‡]	Magnesium alloy (Galvomag)[‡]
Percent composition	Al: 0.4–0.6 Cd: 0.075–0.125 Cu: <0.005 Fe: <0.0014 Pb: <0.15 Si: <0.125 Zn: remainder	Al: remainder Cu: <0.006 Fe: <0.1 Hg: 0.02–0.05 Si: 0.11–0.21 Zn: 0.3–0.5 Others: each <0.02	Al: <0.01 Cu: 0.02 Fe: <0.03 Mg: remainder Mn: 0.5–1.3 Ni: 0.001 Pb: <0.01 Sn: <0.01 Zn: <0.01
Density/kg m^{-3}	7,060	2,695	1,765
Capacity/A h kg^{-1}	780	2,640	1,232
Wastage by weight/kg A y^{-1}	10.7	3.2	4.1
Wastage by volume/ml A y^{-1}	1,518	1,180	2,296
Output/A m^{-2}	6.5	6.5	10.8
E_{corr}/mV SSC	$-1,050$	$-1,050$	$-1,700$

*US Dept of Defense specification for zinc sacrificial anode material requires stricter control of impurity levels than in this alloy.
[†]Trade mark of Impalloy.
[‡]Trade mark of Dow Chemical Company.

Question: *How does the possible use of a coating affect cathodic protection?*

It has been shown that a combination of cathodic protection and protective coating is the most economical means of protecting a steel structure. No paint coating is perfect; there will always be a number of defects in the coating which leave areas of bare metal exposed to the environment. When extra care is taken to reduce the number of such defects, the cost of painting the structure rises. If attempts are made to eliminate the last few defects, the cost rises dramatically. Conversely, the cost of cathodic protection falls as the surface is coated because less anode material is needed. Logically, there exists a point at which the cathodic protection, together with a good (but not perfect) coating, is the most economic option.

Question: *What are the general principles of design of a sacrificial anode system?*

There are two general approaches in designing a cathodic protection system. In the first, a protection potential, E_p, is defined and a system is designed to achieve this potential over the whole structure. In the second, assumptions are made about the current density required to protect the structure and the area of steel which will be exposed to the sea. A paint film may be present, in which the number of defects will increase as the structure ages. Even bare steel surfaces may become coated with scales or other barriers which will alter the area needing protection. Having made assumptions about the current density, the designer uses the output of the proposed anode material to calculate the weight required.

Various standards have been laid down for the current density which is necessary to protect steel surfaces. A surface which has been given a fresh coal tar epoxy paint coating is deemed to be well protected if the current density is $20-30$ mA m^{-2} to accommodate holidays in the coating; a bare steel surface is reported to require in excess of 100 mA m^{-2}. There is much debate about optimum current densities and the figures quoted are subject to considerable variation, depending upon the severity of the environment. The reasons for the uncertainty stem from the problems in knowing the area of (coated) metal exposed to the electrolyte. This was explained at the end of section 16.1.

Question: *How do designers calculate how much material to use?*

Sacrificial anodes are evaluated partly by means of their **capacity**, the number of ampere-hours which can be supplied by each kilogram of material. Other useful parameters are the current output per unit of exposed surface area; so-called **wastage rates** which express the rate of loss of metal by volume or by mass; and **throwing power**, a general description of how the corrosion protection level is affected by the distance of the anode from the metal it is protecting. If the distance is great, part of the potential is used up in overcoming the resistivity of the electrolyte.

In the design process, the total weight of anode material required to protect a structure for its projected life is calculated using

$$W = \frac{i_{av} \times A \times L \times 8,760}{C} \qquad (16.2)$$

where W = the total mass of anode material (kg), A = area of structure (m^2), i_{av} = mean current density demand of the structure (A m^{-2}), L = design life (y) and C = anode capacity (A h kg^{-1}). Thus, the total weight of anode material used must equal or exceed W. Each individual anode must be able to supply current for the design lifetime, and be able to meet the varying demands at the beginning and end of the system life. The complicating factor is that the current output from an anode will vary according to its shape, because there is a different surface area/weight ratio. The anode output, I, is therefore dependent on the anode shape, the resistivity of the environment, the protection potential of the structure, E_p (usually -0.8 V SSC), and the anode operating potential, E_a, that is,

$$I = \frac{(E_p - E_a)}{R_a} \qquad (16.3)$$

R_a in eqn (16.3) is the anode resistance, which incorporates the effect of shape, and values are calculated from formulae according to Table 16.2. Anode life, L, is then given by

$$L = \frac{M_a U}{I w_w} \qquad (16.4)$$

Table 16.2   Resistance formulae for anodes of varying geometries

Anode type	Resistance formula
Slender anodes mounted at least 300 mm from the surface of the structure, $L > 4r$	$R_a = \dfrac{\rho}{2\pi l_a}\left\|\dfrac{4l_a}{r_a} - 1\right\|$
Slender anodes mounted at least 300 mm from the surface of the structure $L < 4r$	$R_a = \dfrac{\rho}{2\pi l_a}\left(\ln\left\|\dfrac{2l_a}{r_a}\left\{1 + \left[1 + \left(\dfrac{r_a}{2l_a}\right)^2\right]^{1/2}\right\}\right\|\right)$ $+ \dfrac{\rho}{2\pi l_a}\left\{\dfrac{r_a}{2l_a} - \left[1 + \left(\dfrac{r_a}{2l_a}\right)^2\right]^{1/2}\right\}$
Flat plate anodes	$R_a = \dfrac{\rho}{2S}$
Other shapes and bracelet anodes	$R_a = \dfrac{0.315\rho}{A^{1/2}}$

For non-cylindrical anodes, $r_a = c_a/2\pi$ where c_a = circumference.
R_a = anode resistance $\qquad\qquad\qquad$ ρ = electrolyte resistivity
l_a = anode length $\qquad\qquad\qquad\qquad$ r_a = anode radius
w_a = anode width $\qquad\qquad\qquad\qquad$ $S = (l_a + w_a)/2$

where M_a is the mass of a single anode, I is the anode current from eqn 16.3), and w_w is the waste rate by weight, measured in kg $(Ay)^{-1}$ (see Table 16.1). U is a utilisation factor which depends upon anode geometry and is the fraction of material consumed when the remaining material is no longer able to provide the required current.

Question: *How are the anodes distributed over the structure?*

It may not be until you have tried some of the relevant problems in Chapter 18 that you appreciate the true difficulties of designing cathodic protection systems. There is no single correct answer to these questions and, as often, engineers must compromise. Obviously, good protection will be achieved the more anodes are used, but this will become more expensive and will add too much weight to the structure. On the other hand, too few anodes will either provide insufficient protection over the complete structure, or will produce an inadequate lifetime. The precise methods used to determine the positioning of anodes on a steel structure cannot be described here. Most designers rely upon experience gained in the field and it may involve more art than science. Calculations involving potential field gradients are possible when the structures are simple. They give some scientific basis for the location of anodes where the aim is to maintain a uniform potential field over the whole structure.

Particular importance is attached to maintaining the correct protection potential at the points of the structure considered to be the most vulnerable to stress; these are the welded joints of an offshore platform, the **nodes**. But remember the dangers associated with overprotection and hydrogen embrittlement (Section 10.3). Other practical problems concern the anode-to-cathode distance. If the distance is too great it will lead to inefficiency because the anode must overcome the resistance between itself and the structure. If the distance is too small the anode output will be non-uniform over its exposed surface and the build-up of anode dissolution products can be a problem. This design problem lends itself to solution by computer systems and is a rapidly developing area for new knowledge-based systems (Section 11.4).

16.3 SACRIFICIAL ANODE APPLICATIONS

The attachment of sacrificial anodes to structures in both marine and soil environments has been common practice for decades and continues to be a very important means of corrosion protection. Sacrificial anodes are relatively inexpensive, easy to install, and in contrast to impressed current techniques, can be used where there is no power supply. The method has the added advantage that there is no expensive electrical equipment to buy and current cannot be supplied in the wrong direction (see Case 16.3). Sacrificial anodes are very suitable in small-scale applications, though they are also used extensively and with equal effect on large-scale structures. They do, however, need frequent replacement and, if large amounts are necessary, extra stress may be placed upon the structure.

Ships

The use of sacrificial anodes to protect ship hulls has become less favoured than impressed current techniques, but it is still found on smaller vessels, where the impressed current method is uneconomic. Zinc is the commonest anode material, preferably used in conjunction with a paint coating (Section 14.3). Anodes are welded or bolted to fixtures on the hull (Fig. 16.4), often in arrays located in the stern area where protection is most likely to be needed. Here, the extreme turbulence caused by the propellers damages protective coatings and results in impingement corrosion. Cavitation corrosion is also common in this vicinity. Furthermore, fittings such as propellers are often made from non-ferrous (copper-based) alloys and provide such excellent cathodes that protection, in addition to paint, is essential at the stern of a ship. Anodes may be found further forward on bilge keels, if they are present, and are also extensively used inside the numerous sea-water intakes for the machinery systems (even when the main part of the hull is protected with an impressed current system).

Regular dry-docking of ships is essential so that consumed anodes may be replaced. Naval ships which require both good fuel economy and high performance need to minimise corrosion and fouling. Docking occurs every 18–24 months, and after removing the water from the dock, it is usual to find that most of the anodes have been consumed. Complete replacement of all sacrificial anodes is thus a routine maintenance procedure. Anodes are sometimes painted by overzealous or ill-informed painters during the application of the various hull coatings, as on the stern of the commercial ship shown in Fig. 16.5. This, of course, renders them useless.

Fig. 16.5 Sacrificial anodes correctly located on the stern of a ship, but painted. Anodes must never be painted.

The need for frequent docking means that sacrificial systems are unpopular with shipping operators who maximise fuel economy by keeping their ships at sea for as long as possible. Impressed current systems can use non-consumable anodes and therefore tend to be favoured.

Offshore platforms

The oil and gas industries have probably been responsible for the greatest applications of cathodic protection, using it on marine and natural water platforms, on undersea and underground pipelines and on tanks. The first use of platform technology began in the natural waters of Lake Maracaibo, Venezuela, but soon moved into the more aggressive environments of the Gulf of Mexico. The technology progressed gradually until alternative fossil fuel supplies had to be found during the period of instability in the Middle East in the mid 1970s. This led to the urgent exploitation of the North Sea resources, and new technology had to be developed so it could take place. Corrosion protection of the vastly expensive structures had to be based upon cathodic protection systems, for which the available scientific data were sparse. Designs were often based more upon inspired guesswork than the application of science, particularly for impressed current systems. Designers preferred to use copious quantities of inexpensive 'zincs' in the belief that overprotection was safer than the risk of underprotection. The complexities of the environment have made it very difficult to determine reliable quantitative data. Section 16.5 highlights some typical problems which have been experienced with impressed current systems.

In the North Sea, the tapping of the vast reservoir of natural resources has been possible only by giant feats of engineering achievement. Some of the world's largest steel structures have now been operating in extremely aggressive environments over 20 years and the offshore technology in the relatively tranquil Gulf of Mexico has come of age in one of the world's most hostile marine environments. In 1994, as many as 250 platforms operate in the North Sea in depths ranging from 25 m to over 150 m. Of these, about 90% are of bare steel, with large numbers of sacrificial anodes welded all over the submerged parts of the platforms. Meanwhile, the technology has been returned to the Gulfs of Mexico and California where, now able to operate in much deeper waters than before, oil exploration and exploitation has gained new momentum.

Many systems in current use were designed and installed after the basic design of the platform had been formulated. (The use of the word 'retrofit' is common.) The extra surface area of the anodes, coupled with areas created by the growth of marine organisms, results in an increase in the impact energy of the waves. Stresses on the platforms may sometimes be greater than the designers anticipated.

A favourable natural phenomenon has also (somewhat unexpectedly) assisted the designer of marine structures. Calcium, magnesium and other metal ions are present in significant quantities in sea-water. The negative potentials of cathodic steel surfaces generate hydroxyl ions (eqn (4.22)) which react with dissolved ions

389

and cause precipitation of insoluble calcium and magnesium salts, known as calcareous deposits. A strongly adherent film is formed which reduces the current required for protection and protects extensive surfaces against corrosion at any local anodes which may be present. As films form, they block cracks and prevent the opening and closing which can lead to propagation. [5]

Buried pipelines

Pipelines, too, need protection. Steel pipelines buried in soil have been successfully protected by sacrificial anodes for many years. Anodes are buried at fixed intervals along the length of the pipe, and at a constant distance from it. The method relies upon the existence of a conducting path through the soil to the pipe. Electrical connection is made from the anode to the pipe, the anode dissolves and a current flux is created which polarises the pipe to potentials shown diagrammatically in Fig. 16.6. The importance of the correct anode spacing is obvious from the figure. If the distance is too great, the polarisation at points furthest from the anodes will not be sufficient to give protection. The potential profile is extremely soil-dependent and reliable performance requires care, both at the design stage and in subsequent monitoring. A defect in the coating can also cause a localised potential which is outside the protection criterion. Such a situation may lead to perforation of the pipe and render the cathodic protection useless. The reaction which occurs at the steel surface is dependent upon the type of soil. In acid soils, the preferred cathode reaction is the reduction of hydrogen ions to hydrogen gas, eqns (4.20) and (4.21). In well-aerated non-acid soil, the cathode reaction reduces oxygen to hydroxyl ions, eqn (4.22).

In practice, the monitoring of pipeline potential is fraught with difficulties.

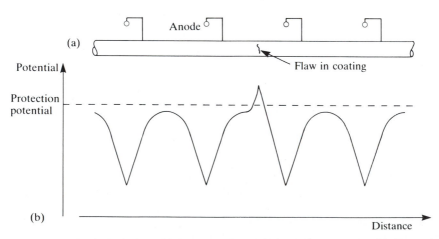

Fig. 16.6 Protection of buried steel pipelines using sacrificial anode cathodic protection: (a) schematic plan view of buried pipeline; and (b) distribution of potential along pipeline with a coating defect.

For many years, serious errors were consistently made because of lack of consideration of IR drops between the protected surface and the reference electrode. Buried pipelines, in particular, suffer from this problem because of their inaccessibility. Methods in which a stabbing electrode takes readings as close to the metal surface as possible have been most successful for subsea pipelines (see below), though this, too, is difficult if the pipelines are buried or cased in concrete. In the past, overdesign of underground pipelines probably assisted in the generally acceptable performance levels, but current trends towards higher operating pressures, higher stress levels and thinner wall thickness have increased the importance of better monitoring of pipeline corrosion resistance. Numerous other methods are currently being developed for underground pipelines and a complete review is published elsewhere. [6]

Marine pipelines

Since the 1970s the installation of 4,000 km of subsea steel pipelines in the North Sea has been achieved almost universally with zinc alloy sacrificial anodes combined with an insulating coating. Until 1985, aluminium alloy had been used only rarely. Subsea pipelines usually employ a different anode arrangement from their land-based counterparts. Anodes, mostly in the form of 'bracelets' weighing 300–400 kg, are fixed around the circumference of the pipes at intervals of about 150 m. A pipeline may stretch along the seabed, either exposed or buried, for hundreds of kilometres and link the platforms with the mainland.

Fear of pollution leaking from a corroded pipe prompted the United Kingdom and Norway to require oil companies to safeguard against such an event. This has led to the formation of teams whose task is to traverse and survey the length of the line, sometimes as much as 250 km and often in appalling weather conditions. A survey has several requirements.

The most important parameter to measure is the local cathodic protection potential of the pipe steel along its entire length. Steel which is at a potential of −850 mV SSC is adequately protected and it is of paramount importance that accurate potential measurements be made. However, this is not as simple as it seems. Local variations are found at anodes, flanges, joints and coating defects. They can only be measured accurately at points close to the pipe. Superimposed on the local variations are long distance changes which result from a number of sources. A platform has considerable influence on potential when connected to a pipeline. At platform approaches, when no isolation joints are fitted, the average pipe potential may change by 100–200 mV in a few kilometres. Variations will also be caused by a poorly coated pipe connected to a well-coated pipe or a pipeline with a poor sacrificial system connected to one with a good system. Many methods used to determine pipeline potentials involve errors which greatly undermine the value of the survey.

Measurements of the output currents and potentials of bracelet anodes are of secondary importance but can be used to predict anode life.

Determination of the field gradients along a pipeline can allow the detection of coating defects. At first sight, it might appear that a coating defect is serious, but it should be remembered that the sacrificial anodes have been used for that very purpose. A defect will cause the adjacent anodes to be consumed faster than normal, but the steel should be protected.

Inspection of the performance of the corrosion protection of a pipeline is achieved using the equipment illustrated in Fig. 16.7(a). A multi-electrode probe is attached to a remotely operated vehicle (ROV), itself attached to a survey ship by an umbilical. An electrode which is remote from the probe is attached to the umbilical so that field gradient measurements can be made. The probe travels underwater along the length of the pipe, periodically stabbing the coating on the pipe to take a local measurement of the potential field gradient. An analogue-to-digital converter connected to a transmitter on the ROV then

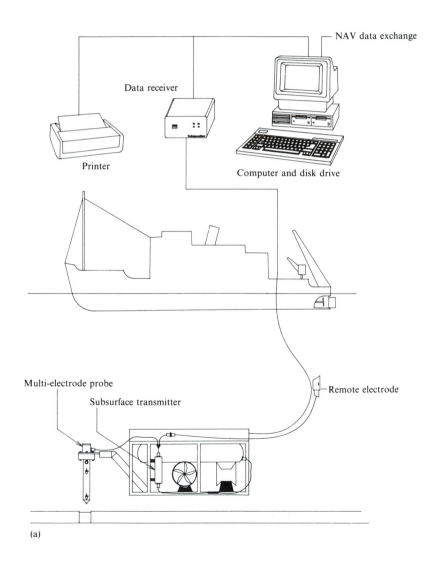

(a)

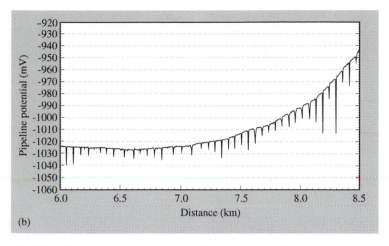

(b)

Fig. 16.7 Measurement of potential profiles for subsea pipelines. (a) The equipment used in the survey. (b) The result of a typical potential survey on a subsea pipeline. Spikes on the profile are the positions of the anodes. The drift to less negative potentials on the right is caused by the polarising effect of the platform. (Reproduced by kind permission of Subspection Ltd)

sends the data to an on-board computer which is coupled to the ship's navigational computer. Thus it is possible to produce an accurate chart of the potential of each of the positions chosen for stabbing. Figure 16.7(b) shows a chart for a pipeline approaching a platform. The polarising effect of the platform on the pipeline is clearly visible at the right of the chart. The positions of each of the anodes are visible as negative potential spikes.

16.4 IMPRESSED CURRENT PROTECTION

Although the principles underlying impressed current cathodic protection described in Section 16.1 are very similar to those of the sacrificial anode method, there are some important differences:

Insulation of anodes: in the sacrificial anode method, the anode material and the structure must be in electrical contact. In the impressed current method, the anode, when mounted on the structure it protects, is separated from it and surrounded by an insulating shield. The shield protects the adjacent metal from excessive current densities in the vicinity of the anode. Anodes for the few ICCP systems on offshore platforms may, however, be sited on the seabed, about 100 m away from the structure. This is thought to give better potential distribution, though some say no better than having anodes mounted on the structure.

Non-consumable electrodes: a major advantage of the impressed current method

393

is that it can use anodes which are virtually non-consumable. The anode reaction is the oxidation of water (Section 4.11) rather than metal dissolution:

$$2H_2O \rightarrow O_2 + 4H^+ + 4e^- \tag{4.80}$$

The use of permanent anodes obviates renewal of large amounts of anode material at regular intervals and is cheaper to run.

In sea-water, it is usually found that oxidation of chloride ions to chlorine gas is the favoured anode reaction:

$$2Cl^- \rightarrow Cl_2 + 2e^- \tag{16.5}$$

although in the diluted waters of estuaries, the oxidation of water, eqn (4.80), may predominate.

Control electronics: the use of electronics enables the system to be self-regulating, as described in Section 16.1. This is a significant advantage over the sacrificial method, where there is no control of the current supply once the anodes have been installed.

The power supply of the control unit is a transformer/rectifier which converts a locally available alternating current supply into direct current of the required voltage. Such sources are usually custom-built for each application; there may be separate control units for each anode or group of anodes, known as a **zone**, depending on the size of the system to be employed.

Anode materials

Early anode materials consisted of large pieces of scrap iron or steel which were slowly consumed by the normal process of anode dissolution. Today, the use of consumable anodes is usually restricted to sites in buried mud or seabed sand, where the dispersal of the gaseous anode reaction products from the non-consumable anodes is difficult. Most ICCP applications use anode materials such as lead–silver alloy, platinised titanium and platinised niobium. Table 16.3 lists some materials and their properties. [7] Although they are capable of supplying large current densities (platinised tantalum and niobium are rated at $2{,}000 \ A \ m^{-2}$ in sea-water), the anodes are usually distributed at regular intervals over the whole of a structure, rather than in small numbers which must protect large areas. There are two main reasons for this:

A large current density in the immediate vicinity of an anode is damaging to many types of paint film. The use of more anodes reduces the current output of each anode and lessens the damage to the protective coating.

In the complex geometrical arrangements of offshore platforms, it is difficult to predict the distribution of potentials. Consequently, it is safer to employ more anodes to protect smaller areas. If any doubt exists about the ability of an anode to protect a particular part of the structure, sacrificial anodes may be used in conjunction with the impressed current system.

Table 16.3   Impressed current anode materials and their properties

Material	Consumption/kg Ay^{-1}	Recommended uses
Platinised tantalum and niobium	8×10^{-6}	Marine environments, potable water, carbonaceous backfill and high purity liquids
Platinised titanium	8×10^{-6}	Marine environments and potable water
High silicon iron	0.25–1.0	Potable waters and soil or carbonaceous backfill
Steel	6.8–9.1	Marine environments and carbonaceous backfill
Iron	Approx 9.5	Marine environments and carbonaceous backfill
Cast iron	4.5–6.8	Marine environments and carbonaceous backfill
Lead–platinum	0.09	Marine environments
Lead–silver	0.09	Marine environments
Graphite	0.1–1.0	Marine environments, potable waters and carbonaceous backfill

Reference electrodes

When used in ICCP systems, reference electrodes are either of the zinc, silver/silver chloride (SSC), or copper/copper sulphate (CSE) types. Copper/copper sulphate is favoured for applications involving reinforced concrete. Reference electrodes (REs), sometimes also called control electrodes, are vital components which determine the current to be provided by the power supply. Malfunction must always be guarded against; physical damage to the anode system or reference electrodes by any one of the hundreds of daily operations around an offshore platform is always a possibility. And so is damage to an electrode caused by the numerous pollutants in estuarine waters.

16.5   IMPRESSED CURRENT APPLICATIONS

The first recorded application of ICCP is uncertain, although the protection of buried pipelines was probably the first. One of the earliest applications which was actually demonstrated to be immediately successful occurred in the power industry.

Case 16.2: In the late 1920s, pitting of condenser tubes was experienced at a UK power station. It was decided to apply impressed current cathodic protection using cast iron anodes. These were fed from a dc generator which supplied power into the water-box via a cable through a wooden stuffing gland. The problem was completely solved. [8]

395

ICCP for ships

ICCP techniques have been used to protect ship hulls since the early 1950s, though not always with such a successful outcome.

Case 16.3: One experimental installation in the Royal Navy vessel, HMS *Blackwood*, was destined to become an ignominious disaster. Two control systems were used, one to protect each side of the ship. Unfortunately, during installation, the connections to one of them were transposed and the starboard side of the ship was actively dissolved away, while corrosion on the piece of old iron used as the nominal anode was controlled instead. It was only when serious leaks occurred that the problem was realised.

In today's vessels, optimum designs usually place anodes in symmetrical dispositions, but in bulk carriers, there is a need for internal access and for cable-runs to be away from anodes and REs. This usually precludes electrodes from being sited external to storage tanks. Instead electronics are placed either well forward or well aft, where the adjacent machinery spaces provide convenient access to the various pieces of equipment (Fig. 16.8).

Impressed current cathodic protection of ships is always used in conjunction with protective coatings. The coatings are intended as the primary protection, and the cathodic protection is a backup in those areas where coating defects may be present. In the period immediately subsequent to the application of the coating, there is very little demand on the impressed current system. During the operational life of the ship, the coating deteriorates and the demand on the cathodic protection increases, in a manner described in Section 16.1. Eventually, the demands placed upon the system may exceed the capabilities of the design, with high anode currents causing even more damage to the coating, particularly in areas adjacent to the anodes, where the concentrations of hydroxyl ions are high. In addition, the high local currents around the anodes

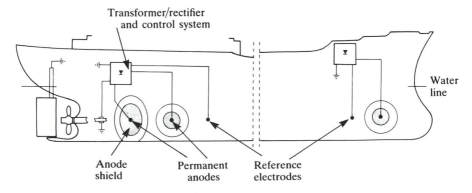

Fig. 16.8 Typical impressed current cathodic protection system for a commercial bulk carrier.

may reduce the protection supplied to the rest of the structure. On a supertanker, the initial current of 10 A may rise to over 1,000 A during the course of its operational life. Although it is possible to extend the period between dockings by use of impressed current rather than sacrificial anode cathodic protection, hull maintenance periods are still a part of any ship's programme in order to maintain the protective coatings.

Cathodic protection of ships is commonly thought to be well understood, but there is good reason to doubt the effectiveness of many systems. There are several reasons why potentiostatic systems frequently fail to provide adequate protection. [9]

The galvanic effect of unpainted non-ferrous propellers and their rotation

Varying sea-water flow from underway and tidal/current conditions

Increasing exposure of bare steel as a result of paint damage

REs monitor electrodes only in their own vicinity

Poor positioning of zones, REs and anodes

The underwater area of a ship is a large complex cathode with at least three components: painted steel, bare steel and bronze. These have different current density requirements for achieving the desired polarisation and respond differently to changes in operational conditions, particularly sea-water flow. For example, it was found [9] that a protection current density of 30 mA m^{-2} in still water rises to over 110 mA m^{-2} at 15 knots and above. Ship anodes are few in number and very small in relation to the cathodic surface area, thus the distribution of potential from them is inevitably uneven. In the design process, it is the difficulty of knowing the expected potential distribution over the structure that leads to reliance on current density measurements as a means of assessment. Yet many ship ICCP designs utilise no more than two or three REs, and since REs measure the potential only of metal immediately adjacent, this increases the uncertainty about their ability to protect a large area under very variable conditions. The siting of these REs for inputs to the control system is vital to the correct functioning of the system.

Until recently, there was very little understanding of the performance of impressed current systems under dynamic conditions. Much of the published research reports data obtained in static conditions, apparently because of the difficulties of obtaining consistent results in flowing environments. One trial for an Egyptian patrol boat indicated that, although a current of about 4 A was sufficient to protect the hull while alongside a jetty, the requirement was about 35 A at 45 knots. Current demand on steel in the wave-affected zone of an offshore platform can increase by up to 25%, with concomitant change of potential to less negative values.

The most significant part of a ship in determining the potential distribution over the hull is the stern area, where non-ferrous alloys are used for the propeller. Modelling techniques have produced accurate simulations of data measured on a real warship under static conditions. In conditions which simulated cruising speed, the protection system was found unable to provide the required potential at the stern (Fig. 16.9(a)). By adding an additional control electrode and repositioning the others, good protection levels were obtained over the whole of the model, under both static and dynamic conditions. This demonstrates the critical nature of the reference position to the effective operation of the system (Fig. 16.9(b)).

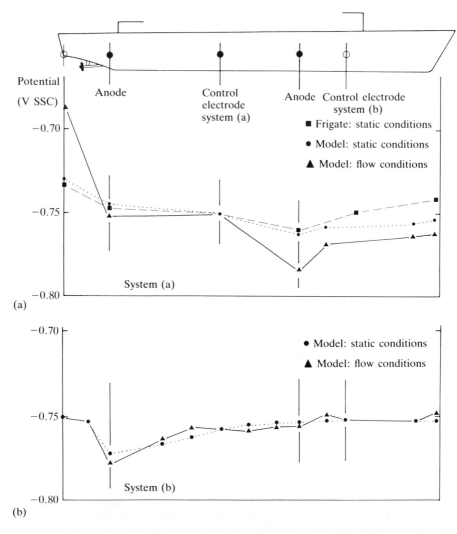

Fig. 16.9   Effect of reference electrode location on potential distribution on a ship hull in both static and flow conditions: (a) with one reference electrode amidships; and (b) with an extra reference electrode aft and the original electrode repositioned ahead of the forward anode.

ICCP for offshore platforms

The large degree of uncertainty which, until recently, arose in the design of impressed current systems may account for the reluctance of designers to use such systems on offshore platforms. Very few of the enormously expensive structures in the North Sea are protected by impressed current installations. In many early impressed current systems, anodes were removed by wave action and the reliability over a 25 year lifetime was thought to be greater with the sacrificial anode method. Conoco's Murchison platform (Fig. 16.10) is set in one of the most exposed parts of the northern North Sea. Its 80,000 m^2 surface area is protected by 100 anodes and 50 REs.

Case 16.4: The Piper and Claymore platforms are located about 120 miles north-east of Aberdeen. When the Piper drilling platform was installed in 1975, the current density for protection was 86 mA m^{-2}, but during seven years of operation, current density was gradually increased to 130 mA m^{-2} on the Piper platform, and to 160 mA m^{-2} on the modified Claymore system. Both systems utilised platinum–iridium anodes: 42 with 6 REs on Piper, and 55 with 12 REs on Claymore. Both systems

Fig. 16.10 Conoco's Murchison platform. The field was ready for production by September 1980, only five years after a wildcat well first struck oil on the location. (Photograph courtesy of Conoco (UK) Ltd)

suffered severely from the rigours of natural and artificial environments. On many occasions, cable runs to anodes were damaged, and anodes broken off by falling scaffolding. Many anodes were rendered inoperative because the pulsed current system was causing platinum consumption 30 times greater than expected. Surprisingly, despite this catalogue of problems, the platforms seemed to be resisting corrosion satisfactorily. [10]

If systems are found to be inadequate it is not necessarily the result of bad design; it may reflect the complexity of the task. An ICCP designer should always work alongside the structural engineer to decide the areas most critical to structural integrity and ensure they are adequately monitored. Even so, most users regularly inspect both structures and systems to ensure correct functioning of the equipment and adequate protection levels.

ICCP for buried pipelines

On land, ICCP of buried pipelines is currently the preferred method of protection. As with subsea pipelines, regular monitoring is an essential requirement; the following examples provide illustrations.

Case 16.5: After a survey on a length of pipe it was decided to investigate six possible defects associated with inadequate potentials. In each case, extensive corrosion was revealed, with average pitting of 2.8 mm in just six years. Pipeline potential was found to vary between $-1,820$ mV CSE and -560 mV CSE, depending on the resistivity of the soil.

Case 16.6: The results of surveying 3,560 miles of a gas pipeline over a period of five years pinpointed 1,567 possible problem areas; 625 were deemed to be serious enough to require investigation and repair. Without the survey there is no doubt that many leaks would have occurred, but more seriously, four defects were so serious that there was a real possibility of total rupture and a catastrophic accident.

The ICCP of buried pipelines is not without its difficulties. The method relies upon buried localised arrays of anodes known as ground beds, which distribute current along the length of the line. The material in which the anodes are buried is not soil, but a specially formulated backfill which reduces the soil/anode resistance, allows the escape of anode gases and increases the current capacity. Two common backfill materials are coal coke breeze and petroleum coke breeze, both 95% carbonaceous. Despite this, the great variation in water content and acidity of soils along the length of the pipeline can cause problems during the operation of the system, highlighting the need for regular monitoring. Anode materials vary from graphite, magnetite and silicon iron to the more recently developed lead–platinum and platinised titanium. A comprehensive review of the subject has been published by Shreir and Hayfield. [11]

ICCP for concrete structures

Impressed current methods have been used to protect steel in concrete in such applications as buried prestressed pipelines and tanks, bridge decks and marine structures. When embedded in a few centimetres of sound concrete, steel is passivated and corrosion resistant because of the highly alkaline conditions which exist (pH 12.5–13.5). Corrosion problems occur when the iron is depassivated by penetration of reactive species. The most common reaction occurs when carbon dioxide (acidic) penetrates the concrete and neutralises the alkaline constituents; pH falls and corrosion occurs. A much more dangerous reaction occurs when chloride ions, either from a marine environment or from de-icing salt, are able to reach the iron surface. Chloride ions depassivate iron, even at high pH, and corrosion problems occur similar to those suffered by Pelham Bridge (Case 1.8). Not only is the concrete weakened, but the rusting steel sets up severe tensile stresses within the concrete and large pieces may become detached from the structure, a process known as spalling (see Fig. 14.6).

Case 16.7: In Ontario, Canada, 32 bridges were given impressed current systems after a successful trial on Kingston's Division Street Bridge. Action became necessary after masonry failing on to Toronto's Gardiner Expressway necessitated the use of safety nets. In North America, in 1985, approximately 100 bridges had ICCP systems, [12] while in 1988 a survey by the Strategic Highway Research Program found 840,000 m^2 of concrete surface under cathodic protection on the US and Canadian interstate highway system. [13]

When oxygen concentration is low, iron is not passivated but seems to corrode at an insignificant rate. This is particularly true for concrete structures which are immersed in sea-water. It is thought that the steel reinforcing bars are polarised to as low as -800 to $-1,100$ mV CSE in low oxygen, water-saturated conditions; steel in this environment is considered safe. The splash zone, where oxygen access is considerable, is the most vulnerable area of an offshore steel-reinforced concrete structure. There is good reason, therefore, to attempt to protect such concrete structures by impressed current techniques. [14]

Figure 16.11 summarises the natural and impressed potentials for steel in concrete. Potentials more negative than -850 mV CSE should be used to provide adequate protection, but -600 mV has been used with a measure of success under known operating parameters (see Case 16.1). The possibility of hydrogen evolution and consequent embrittlement of high strength steels has been considered a danger at potentials more negative than $-1,100$ mV CSE, and overprotection at such potentials should be avoided.

One advantage of ICCP of steel in concrete is that current densities are lower than in many other applications, principally because the conductivity of the 'electrolyte' is much lower. In a bridge deck, values of 20 mA m^{-2} are typical, while concrete structures in soil require 1–4 mA m^{-2}, depending upon the quality of the concrete. Values as low as <0.15 mA m^{-2} have been reported for steel in high quality concrete in sea-water, but this is rare.

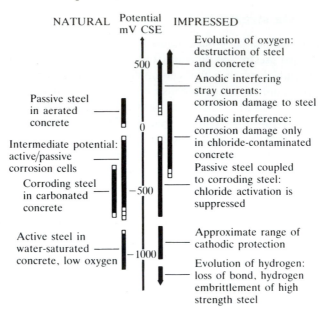

NATURAL   Potential   IMPRESSED
mV CSE

Evolution of oxygen:
destruction of steel
and concrete

Anodic interfering
stray currents:
corrosion damage to steel

Passive steel
in aerated
concrete

Anodic interference:
corrosion damage only
in chloride-contaminated
concrete

Intermediate potential:
active/passive
corrosion cells

Passive steel coupled
to corroding steel:
chloride activation is
suppressed

Corroding steel
in carbonated
concrete

Active steel in
water-saturated
concrete, low oxygen

Approximate range of
cathodic protection

Evolution of hydrogen:
loss of bond, hydrogen
embrittlement of high
strength steel

500

0

−500

−1000

Fig. 16.11   Natural and impressed potentials of steel in concrete.

Such applications of ICCP are not without their difficulties. These can be

Establishing effective electrical bonding of all the steelwork

Distribution of the protective current through the concrete, a relatively high resistivity electrolyte

Monitoring potentials and corrosion

Probably the most serious difficulty is the monitoring of potential. Doubts have been expressed [14] about the use of reference probes, and for optimum performance a system should be carefully designed and installed before the concrete is cast. There have been numerous attempts to reduce corrosion by retrofit techniques.

Case 16.8: The reinforced concrete substructure of the Tay road bridge at Perth in Scotland was found to be suffering from corrosion of the reinforcement resulting from chloride attack centred on the splash zone. The bridge is 2250 m long with 42 spans and consists of a composite structure of twin steel box girders and concrete deck slab supported on a reinforced concrete marine superstructure of twin columns on river piers. It was found that 80 of the columns required treatment and, after an extensive period of trials, the affected columns underwent a programme of cathodic protection involving an overlay method. This consisted of a titanium mesh anode covering the relevant area, overlaid with 50 mm thick shrinkage-compensated flowable concrete with further reinforcement. Two silver/silver chloride electrodes were used in each anode zone, together with a 12V, 1A transformer rectifier. Current densities were 15–20 mA m^{-2}. [15]

16.6 ANODIC PROTECTION

In Section 16.1 it was established that, in principle, the application of a potential to steel, such that the metal was anodically polarised from its free corrosion potential, could lead to the formation of a passive film and afford protection against corrosion. The criterion for protection must be a coherent insulating film which is sufficiently robust to withstand mechanical damage.

In Figs 7.6 and 7.7, we studied the polarisation behaviour of passivating materials. When potential is made more positive than E_{corr}, if stable passivity is possible, a point is reached at which current density falls to an extremely low value, i_{pass}, and the corrosion rate is comparatively low. This is the principle of anodic protection. A dangerous aspect is that if, for any reason, the metal potential wanders even slightly outside this range, corrosion may be worse than if anodic protection had not been used. Careful monitoring of the protection potentials and currents is therefore essential.

It is important to appreciate that such behaviour is not specific to the material alone, but to material/electrolyte combinations. For example, mild steel can be passivated in pure nitric acid, but not in acid which has been diluted with water, or in any concentration of hydrochloric acid (see Expts 3.12 and 3.13). Notice that two current densities are relevant: one to initially passivate the metal, i_{max}, and one to maintain the film when it has been formed, i_{pass}. In many cases, they may differ by factors of 100 to 1,000. Thus, high current densities are needed in the first instance, but once passivated, large areas of metal can be efficiently protected by very small current densities.

The metals frequently protected by this means are iron, nickel, aluminium, titanium, molybdenum, zirconium, hafnium and niobium, together with alloys containing them in major amounts. Electrolytes can vary from very acid to very alkaline, and the method has a major advantage over comparable cathodic protection, which is rarely applicable in such severe environments. Thus, anodically protected carbon steel can be used in the chemical and fertiliser industries to store a variety of oxidising acids, as well as caustic solutions. Similarly, anodically protected stainless steel is best employed for oxidising acid environments, but not for alkalis. Storage of non-oxidising acids such as hydrochloric and hydrofluoric acids is best done in protected chromium or titanium vessels. Like its cathodic counterpart, anodic protection is only suitable in applications which involve continuous immersion.

The aggressive nature of the environments in which anodic protection is employed means that special reference electrodes are often necessary. All solid references made from the noble metals are usually suitable, for example, silver/ silver chloride, or platinum/platinum oxide. Noble metals such as platinum are often used for the cathode, too.

There are numerous advantages of using anodic protection. Industrial applications benefit from low operating costs and controlled predictable conditions. Furthermore, it is often possible to replace an expensive alloy in unprotected plant with a cheaper material which is anodically protected, and the slow corrosion rates involved lead to less contamination of the products being stored or processed.

The disadvantages of anodic protection are that failure of the electrical supply may be hazardous because of depassivation. In addition, the requirement for electrical current makes it useless for protection in organic liquid environments, or for components which are not continuously immersed. A comprehensive review of anodic protection has been published. [16]

16.7 REFERENCES

1. LaQue F L 1975 *Marine corrosion: causes and prevention,* John Wiley, New York, p. 203.
2. Davy H 1824 *Philosophical Transactions of the Royal Society* **114**(1):197–204.
3. Swain G W, Muller E R and Polly D R 1994 The design and installation of a cathodic protection system for the Living Seas, EPCOT Center. Paper 495 in *Corrosion 94, Baltimore MD*, NACE International, Houston TX.
4. Schreiber C F 1986 Sacrificial anodes. In *Cathodic protection theory and practice,* edited by V Ashworth and C J L Booker, Ellis Horwood, Chichester UK, pp. 78–93.
5. Scott P M 1983 Design and inspection related applications of corrosion fatigue data, *Mémoires et Etudes Scientifiques Revue de Metallurgie,* **Novembre 1983,** 651–60.
6. Martin B A 1986 Potential measurements on buried pipelines. In *Cathodic protection theory and practice,* edited by V Ashworth and C J L Booker, Ellis Horwood, Chichester UK, pp. 276–92.
7. Brand J W L F and Lydon P 1994 Impressed current anodes. In *Corrosion,* edited by L L Shreir, R A Jarman and G T Burstein, Butterworth-Heinemann, Oxford, pp. 10:56–92.
8. Burbridge E 1986 Process plant and cooling water systems. In *Cathodic protection theory and practice,* edited by V Ashworth and C J L Booker, Ellis Horwood, Chichester UK, pp. 191–213.
9. Tighe-Ford D J 1994 A systematic approach to the design of warship impressed current cathodic protection systems. In *Modelling aqueous corrosion,* edited by K R Trethewey and P R Roberge, Kluwer Academic, Dordrecht, The Netherlands, pp. 381–398.
10. Tischuk J L 1984 Operation and maintenance of impressed current cathodic protection systems. In *Corrosion and marine growth on offshore structures,* edited by J R Lewis and A D Mercer, Ellis Horwood, Chichester UK, pp. 61–8.
11. Shreir L L and Hayfield P C S 1986 Impressed current anodes. In *Cathodic protection theory and practice,* edited by V Ashworth and C J L Booker, Ellis Horwood, Chichester UK, pp. 94–127.

12. Anon 1985 *New Scientist*, no. 1448, p. 18.
13. Broomfield J P 1994 Five years research on corrosion of steel in concrete: a summary of the Strategic Highway Research Program structures research. Paper 318 in *Corrosion 93, New Orleans LA*, NACE International, Houston TX.
14. Wilkins N J M 1986 Cathodic protection of concrete structures. In *Cathodic protection theory and practice,* edited by V Ashworth and C J L Booker, Ellis Horwood, Chichester UK, pp. 172–82.
15. Watters A 1991 Tay Bridge road experience, *Industrial Corrosion* **9**(3):4–9.
16. Walker R and Ward A 1969 The theory and practice of anodic protection, *Metals Review* **137**:143–51.

16.8 BIBLIOGRAPHY

Anon 1983 *Solving Rebar Corrosion Problems in Concrete,* NACE International, Houston TX.

Anon 1985 *Corrosion protection of reinforced concrete bridge decks,* NACE International, Houston TX.

Anon 1988 *Corrosion of metals in concrete,* NACE International, Houston TX.

Anon 1989 *Cathodic protection criteria — a literature survey,* NACE International, Houston TX.

Anon 1989 *Collected papers on cathodic protection current distribution,* NACE International, Houston TX.

Benedict R L 1986 *Anode resistance fundamentals and applications — classic papers and reviews,* NACE International, Houston TX.

Bogaerts W F and Agema K S 1992 *Active library on corrosion,* CD-ROM for DOS-PC, Elsevier, Amsterdam.

Lehmann J A 1993 Fifty years of cathodic protection on underground storage tanks. Paper 7 in *Corrosion 93, New Orleans LA*, NACE International, Houston TX.

Lye R E 1990 A corrosion protection system for a North Sea jacket, *Materials Performance* **29**(5):13–18.

Morgan J H 1987 *Cathodic Protection,* 2nd edn. NACE International, Houston TX.

Peabody A W 1967 *Control of pipeline corrosion,* NACE International, Houston TX.

Riordan M A 1990 A modern model for cathodic protection technology, *Materials Performance* **29**(3):23–33.

Tighe-Ford D J 1994 A systematic approach to the design of warship impressed current cathodic protection systems. In *Modelling aqueous corrosion*, edited by K R Trethewey and P R Roberge, Kluwer Academic, Dordrecht, The Netherlands, pp. 381–98.

17 CORROSION AT ELEVATED TEMPERATURES

> Let all the greatest minds in the world be fused into one mind and let this great mind strain nerve beyond its power; let it seek diligently on the earth and in the heavens; let it search every nook and every cranny of nature; it will only find the cause of the increased weight of the calcined metal in air.
>
> (Jean Rey, 1630)

The previous chapters have defined corrosion as the degradation of a metal by an electrochemical reaction with its environment, but the environments considered have been almost entirely aqueous. Corrosion on a metal surface can occur even though a liquid electrolyte is not present, and not surprisingly, the process is often referred to as dry corrosion. The definition of corrosion which we have used so far is unchanged in dry corrosion processes, as also is the description of a corrosion process by eqn (2.7).

Perhaps the most obvious dry corrosion process is the reaction of a metal with the oxygen of the air. Although nitrogen is the major constituent of air, its role is unimportant when metals are heated in air because of the dominating influence of oxygen. At high temperatures, nitrogen does react with chromium, aluminium, titanium, molybdenum and tungsten, but reactions such as these are outside the scope of this text. Though the reaction with oxygen is, in principle, very simple, early scientists experienced great difficulty in understanding the weight changes which accompanied the calcination (oxidation) of a metal in air. Even today, studies of the oxidation and other high temperature reactions of modern alloys have shown that the processes involved are very complex. Therefore, the treatment in this text is limited to an overview of this large and expanding subject.

Oxygen reacts readily with most metals, though the thermal energy required to produce an oxidation rate of engineering significance may vary considerably for different metals at the same temperature. At ambient temperatures, most engineering materials are either already oxidised such that the oxide layer screens the underlying metal from further reaction, or they react sufficiently slowly in dry air for oxidation not to be a problem. At elevated temperatures, however, the rate of oxidation of metals increases. Thus, if an engineering component is exposed for a prolonged period in a high temperature environment it may be rendered useless. For example, in pure dry air at temperatures below about 480°C, a thin protective oxide film forms on the surface of polished mild steel, but at a rate which is considered to be negligible for engineering purposes; a threshold rate has been defined as 10^{-3} kg m^{-2} h^{-1}.

However, during hot rolling and forging of mild steel (processes which take place at about 900°C) the rate of oxidation is sufficiently great to produce a layer of oxide called mill scale, which is non-protective. We have already seen (Section 15.1) that mill scale may have an important effect upon the corrosion rate of mild steel in aqueous environments. On the other hand, the usefulness of metals such as aluminium and titanium depends upon their ability to form protective oxide films at room temperature (see Sections 15.4 and 15.6).

In Chapter 1 we saw that not all corrosion processes are undesirable. The controlled oxidation of iron and steel in the manufacture of arms and armour was a well-established craft, intended to make the materials both decorative and long-lasting. Decoration was achieved by the creation of colours on the metal surface. Titanium is oxidised electrochemically to produce beautifully coloured jewellery and works of art. In both cases, the effects are caused by oxide films. You may be more familiar with these effects if you have seen the spectral colours at the engine end of stainless steel motor cycle exhausts.

Experiment 17.1
Take a piece of clean, brightly polished steel strip, approximately 300 mm × 20 mm × 3 mm. Heat one end of the strip in a Bunsen flame so that a steep temperature gradient is formed along its length (Fig. 17.1). After a short time a series of coloured bands will be seen along the length of the strip.

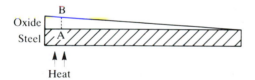

Fig. 17.1 Development of interference colours on a steel strip heated at one end. In a transparent oxide layer, the colour of the reflected light depends upon the thickness, AB, of the oxide film at that point. (AB is greatly exaggerated for clarity.) Movement of the colours reveals the gradual thickening of the oxide as heat travels along the strip.

The colours are produced in transparent oxide films in the same way as in oil films on water. Interference takes place between the light reflected from a point such as A in the metal/oxide interface, and that reflected from point B, eliminating certain wavelengths from the spectrum. Hence the actual colour of the reflection will depend upon the thickness of the layer AB, and changes as AB increases. The temperature gradient causes the film to grow in the shape of a wedge, gradually thickening as heat flows along the strip. This, in turn, causes the movement of the coloured bands with time. Only thin oxide films are transparent and show interference tints. Eventually the oxide becomes sufficiently thick to be opaque and the colours disappear; then the film is normally renamed a scale.

Before the days of sophisticated temperature control in heat treatment processes, the temperature of steel strip and bars was often judged by the colours developed on the metal surface during tempering heat treatments. The method is surprisingly accurate: for each step of 10°C between 230°C and 280°C, the colour changes through the sequence, pale straw, dark straw, brown, brownish purple, purple and dark purple. The metal appears blue at 300°C.

Until the development of the gas turbine for modern aircraft began with the Whittle engine in 1937, the uses of metals and alloys for engineering in high temperature environments were rarely severe enough to cause materials selection problems. Although the steam turbine had been developed in the late 1800s and used by Parsons in 1897 for marine propulsion, the operating temperatures required were low enough for existing materials to be used. The development of gas turbine aero-engines in World War II changed this situation dramatically.

The operating conditions were severe; materials were required which were capable of withstanding temperatures of 800–1,000°C, combined with large stress levels created by high speeds of rotation. This led to the development of a large class of alloys known as the superalloys. These are mostly nickel-based alloys, though there are also groups of iron- and cobalt-based materials. Today, superalloys are used in marine, aircraft, industrial and vehicular gas turbines, as well as in spacecraft, rocket engines, nuclear reactors, steam power plant, petrochemical plant and many other applications.

Steels still represent the major material for use in gas turbines, though their percentage share has declined in favour of the superalloys and titanium alloys. The contribution made by aluminium alloys to gas turbine development is small, but as we shall see, aluminium as an alloying addition is very important.

This chapter is a brief study of the behaviour of metals in elevated temperature, non-aqueous environments. The discussion tends to concentrate upon oxide formation, although some other types of film will be introduced.

17.1 METAL OXIDES

Metal oxides (together with other compounds such as sulphides and halides) can be divided into two classes: those which are stable over the range of temperatures likely to be met in engineering structures, and those which are not. We will consider the unstable oxides first.

When an unstable metal oxide is heated, it decomposes to release the metal, which may be deposited on to the substrate surface. For example, silver oxide decomposes above 100°C, mercury (II) oxide above 500°C and cadmium oxide in the temperature range 900–1,000°C. Today, unstable oxides are of little use to the engineer, but they were of paramount importance to scientists in determining the fundamental mechanism of oxidation.

Early chemists, notably Stahl, had postulated the erroneous theory that a metal lost a substance which they called phlogiston when it was heated to form a calx (metal oxide):

$$\text{metal} - \text{phlogiston} \rightarrow \text{calx} \qquad (17.1)$$

Stahl [1] said

> Phlogiston is lighter than air, and, in combining with substances, strives to lift them, and so decrease their weight; consequently a substance which has lost phlogiston must be heavier than before.

In the 1780s, Lavoisier used the decomposition of mercury oxide to demonstrate that the phlogiston theory of oxidation was untenable. He heated mercury at just below its boiling point (357°C) in a sealed apparatus and showed that approximately 20% of the air was absorbed by the mercury. By collecting the red mercury oxide and heating it at about 500°C he dissociated the unstable oxide to obtain a volume of gas equal to that lost in the first part of the experiment. He demonstrated that the recovered gas could support combustion, while that remaining in the apparatus after the first stage of the experiment could not. He further showed that the weights of mercury and recovered gas (oxygen) obtained by heating mercury calx, exactly equalled the weight of calx. Also, the increase in weight of the mercury in the formation of the calx is equal to the weight of the oxygen taken from the air. By this method it was established beyond doubt that the mechanism of oxidation was

$$\text{metal} + \text{oxygen} \rightarrow \text{metal oxide} \qquad (17.2)$$

While this seems so obvious today, the problem had been vexing scientists for many years, as the quotation at the start of the chapter shows. The solution was a great advance for scientific thinking in the late 1700s.

The much broader class of stable oxides can again be divided into two groups: a minor group which volatilise at relatively low temperatures, and a major group which normally remain on the metal surface, unless they are physically or chemically removed.

Volatile oxides form on the metal surface, but immediately change to the gaseous state, leaving a fresh, reactive surface to continue the oxidation process until the metal is completely lost. The rate of reaction is constant and normally increases as the temperature rises. Molybdenum is the classic example, oxidising at significant rates in air above 300°C. Two oxide layers develop on the metal surface: an inner layer of MoO_2 and an outer layer of MoO_3. Above 500°C, MoO_3 begins to volatilise, and at about 770°C the rate of volatilisation equals the rate of oxidation. Further increases in temperature lead to extremely rapid metal loss. The effect becomes catastrophic when the MoO_3 begins to form a molten phase at temperatures above 815°C.

Case 17.1: Molybdenum plates of thickness 3 mm were employed as heat reflecting shields around a furnace used to determine the thermal conductivity of ceramics at temperatures up to 1,400°C. The whole unit was placed in a vacuum chamber and kept at a very low pressure during elevated temperature runs to prevent damage to the ceramics and the heat shields. One evening, while the chamber was operating at 900°C, a small leak developed and air bled into the system overnight. In the morning, holes extended over 70% of the shields and the metal which remained was severely thinned.

Stable, non-volatile oxides would be expected to remain on a metal surface and it may be thought that all such oxides would protect the substrate metal. Such is not the case. The rate of continued oxidation depends upon several factors, three of which are

The rate of diffusion of the reactants through the oxide film.

The rate of supply of oxygen to the outer surface of the oxide.

The molar volume ratio of oxide to metal.

The slowest process at any given temperature controls the rate of corrosion. In general, the rate will fall as the oxide thickens.

The molar ratio of the volume of oxide formed to the volume of metal consumed in producing the oxide is a most important factor in determining the corrosion rate over an extended period of time. If M is the molecular mass of an oxide of density D, the volume occupied by 1 mol will be M/D. If m is the mass of metal in the mass M of oxide, and its density is d, a volume of metal m/d will have been converted into oxide. Table 17.1 lists the ratios (M/D) and (m/d) for a selection of metals. When the volume of oxide is smaller than that of the metal, i.e. $Md/mD < 1$, as in lithium, calcium and magnesium, the oxide is stretched over the metal surface to produce a porous non-protective oxide. The oxidation process proceeds at a linear rate with time.

Table 17.1 Values for the ratio volume of oxide produced/volume of metal consumed in producing the oxide

Metal	Md/mD
Li	0.57
Ca	0.64
Mg	0.81
Al	1.28
Ni	1.52
Zr	1.56
Cu	1.68
Ti	1.77
Fe	1.77
U	1.94
Cr	1.99
Mo	3.24
W	3.35

If the oxide volume is larger than the metal from which it formed, i.e. *Md/mD* > 1, then we would predict that the oxide is continuous and protective. This is true for aluminium, but other complications may arise. Often compressive internal stresses are developed in the oxide as it thickens. Should a small stress develop it will tend to force together the sides of any cracks or defects, slowing down the rate of oxidation. Larger stresses tend to disrupt the bond between the oxide and the metal; the oxide then blisters and cracks. The disruption occurs because the fracturing of the interface between the metal and the oxide relieves the compressive internal stress. The magnitude of the stress in the oxide will increase as it thickens. The oxidation rate may be small for long periods of time, as there is sufficient compressive stress in a thin oxide film to maintain a tightly adherent and compact barrier layer. As the film very slowly thickens, a stress level is generated where spontaneous rupture of the interface occurs and the oxidation rate suddenly increases. This is one type of **breakaway corrosion** and is dealt with in the next section.

The oxidation of metals that form stable non-volatile oxides occurs with an increase in weight of the sample which is relatively simple to measure in the laboratory. Indeed, much of the current knowledge about oxidation mechanisms has been obtained from studies of time-dependent weight increases. The rates of thickening have been found to fall mostly into three categories, examples of which are shown in Fig. 17.2. Also shown is the weight loss with

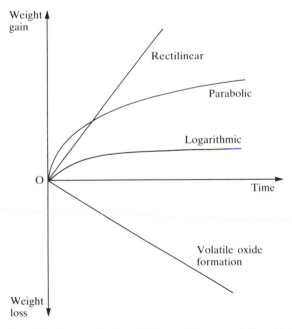

Fig. 17.2 Four oxidation rate laws: stable non-volatile oxide formation leads to weight gain with linear, parabolic or logarithmic kinetics; formation of volatile oxides leads to linear weight loss with time.

411

time which occurs when volatile oxides are formed. In the equations which follow, y = oxide thickness, t = time, and c_1 to c_5 are constants.

Parabolic growth

When the oxide film remains intact on the metal surface and offers a uniform barrier to the diffusion of metal or oxide ions through the film (Section 17.3), the rate of growth of the oxide is inversely proportional to its instantaneous thickness:

$$dy/dt = c_1/y \qquad (17.3)$$

The integration of eqn (17.3) yields a parabolic rate law:

$$y^2 = c_1 t \qquad (17.4)$$

When $t = 0$, $y = 0$, hence no constant of integration is necessary. Metals which oxidise according to the parabolic rate law are common and associated with thick coherent oxides. Examples are cobalt, nickel, copper and tungsten, though these, as with the other examples, may follow a different rate law, depending upon the experimental conditions.

Rectilinear growth

The rate of oxidation is constant with time:

$$dy/dt = c_2 \qquad (17.5)$$

which, when integrated, gives a linear relationship:

$$y = c_2 t \qquad (17.6)$$

Rectilinear growth occurs whenever the oxide is unable to hinder the access of oxygen to the metal surface, as occurs when the oxide formed from a given volume of metal is too small to cover the metal surface completely. If the oxide cracks or spalls owing to large internal stresses, a series of short parabolic-type weight increases will be observed which will appear linear overall. Such behaviour is termed **paralinear**. It may occur when the temperature cycles sufficiently for differential contraction and expansion between the metal and the oxide to spall the oxide from the metal.

Rectilinear growth is typically a high temperature process for the metal involved; two examples are iron above 1,000°C and magnesium above 500°C.

Logarithmic growth

At low temperatures, a thin oxide film covers the surface. The rate of diffusion through the film is very low and after an initial period of rapid growth, the rate of thickening becomes virtually zero. The rate law is:

$$y = c_3 \lg(c_4 t + c_5) \tag{17.7}$$

Examples of metals which oxidise in such a manner are magnesium below 200°C and aluminium below 50°C.

17.2 BREAKAWAY CORROSION

We have already mentioned breakaway corrosion in connection with the compressive stresses developed in oxide scales (see Section 17.1). The mechanism of breakaway can be very complicated and involves the interaction of many factors, including temperature, gas composition (particularly minor constituents), gas pressure, metal composition, component geometry and surface finish. It is an insidious form of attack and is frequently catastrophic in its outcome.

Two typical oxidation curves showing breakaway behaviour appear in Fig. 17.3. For long periods of time the oxidation rate appears to fall; low rates of weight increase are observed which may represent an acceptable and predictable rate of metal loss. Suddenly, the rate of oxidation increases. Then it either mimics the parabolic kinetics at the start of the oxidation, as shown in curve A, or else it continues at a high linear rate of loss to failure, as in curve B.

There have been many cases of breakaway corrosion. Zirconium undergoes breakaway in the conditions found in pressurised water atmospheres. Before

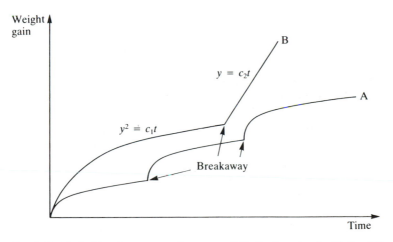

Fig. 17.3 Typical breakaway corrosion curves. Initially, the rate of oxidation falls with time and obeys a parabolic growth law. At the breakaway point, the existing oxide no longer protects the metal and linear growth ensues. In A, the new oxide is similar to the initial scale and the growth curve again follows a parabolic law until breakaway is repeated. In B, the new oxide no longer protects the substrate, which corrodes according to the rectilinear growth law.

413

breakaway occurs, zirconium oxide is a shiny black, tightly adherent film; after the transition, it is a loose powdery white oxide. Thus, zirconium is used in pressurised water reactors in the form of Zircaloy 2 for cladding fuel rods. This alloy contains tin to reduce the likelihood of breakaway corrosion.

A classic case of breakaway corrosion was found in 1969. [2] It illustrates not only the complex nature of the interaction of many variables in causing the phenomenon, but also the large financial losses which may be incurred when breakaway occurs. We shall discuss it in some detail.

Case 17.2: The first two generations of nuclear power station used in the United Kingdom, the magnox and the advanced gas-cooled reactor (AGR), used carbon dioxide as the cooling medium in the reactor. In 1983, the operating conditions were 370°C and 28 atmospheres for the magnox, and 650°C and 41 atmospheres for the AGR. Maintenance and replacement of the components in a reactor are very difficult and many components are designed for a life of 30 years. Of necessity, an oxidation allowance is designed into the structure to accommodate the effects of metal wastage and deformation of bolted and welded components as a result of interfacial oxide growth. These allowances were derived by pretesting the materials in simulated environments, many of the tests lasting for over a year. The materials selected formed protective oxides and showed an oxidation rate which decreased with time. These tests demonstrated that mild steel would be satisfactory up to 400°C, meeting the requirements of the magnox station. The AGR requirements could be met by using mild steel up to 400°C, a steel containing 9% chromium and 1% molybdenum for components operating between 400°C and 550°C, and an austenitic stainless steel for those above this temperature.

In 1969, during a biennial inspection of a magnox station, some fractured mild steel bolts were found. Subsequent examination showed they had failed as a result of bolt straining caused by excessive oxidation of the interfaces between the bolts, washers and nuts. Such a rate of oxide growth was not predicted by extrapolation of the laboratory test data: breakaway oxidation had occurred. The porous oxide produced by the breakaway occupied twice the volume of the metal consumed in producing it, and had the ability to continue to form even under the large compressive stresses generated at the interfaces by its formation. Immediately, the maximum operating temperature of all the magnox stations had to be reduced with a consequent loss of generating capacity.

As a result of the problems with the mild steel, all the other steels were re-examined and it was shown in accelerated corrosion tests that the 9% chromium steel had also suffered breakaway oxidation, although over a much longer time. Even so, on the basis of the then available data, it was predicted that there would be an unacceptable number of failures in boiler tubes. Only after further work, which resulted in a more complete understanding of oxidation mechanisms and an enlargement of the database, could operating conditions be defined to achieve design life. (Courtesy CEGB)

The mechanism of the breakaway oxidation in the nuclear reactors was found to be very complex. The protective oxide on a ferritic steel is in two layers, both of which are porous to the carbon dioxide coolant gas. The inner layer consists of small crystallites containing chromium and silicon if these elements are present in the steel. The outer layer has a columnar structure and consists of magnetite, Fe_3O_4. An equilibrium is established between the ingress of carbon dioxide and the outward, rate-controlling solid-state diffusion of the iron. The carbon dioxide oxidises the iron, a reaction which produces carbon monoxide:

$$3Fe + 4CO_2 \rightarrow Fe_3O_4 + 4CO \tag{17.8}$$

followed by deposition of carbon:

$$2CO \rightarrow CO_2 + C \tag{17.9}$$

This carbon dissolves partly in the metal and partly in the oxide. When the carbon level of the inner oxide reaches about 10% by weight, the individual crystallites become separated from each other by a grain boundary carbon film. This makes the oxide porous and it loses its ability to protect the substrate. Breakaway oxidation follows. Owing to the low solubility ($<0.01\%$) of carbon in ferrite, most of the carbon enters the oxide on mild steel components and breakaway occurs after about 1–5 years of exposure. In the case of the 9% chromium steel, carbon entering the metal is precipitated as chromium carbide. This allows a much greater proportion of the carbon to be absorbed by the metal and it takes a longer time to reach the critical carbon concentration for breakaway in the oxide film. The oxidation of mild steel follows curve B in Fig. 17.3, but there is evidence to show that 9% chromium steel exhibits an oxidation pattern similar to curve A in some cases.

Many other factors influence the time at which breakaway occurs; three of them are temperature, the amount of water vapour in the carbon dioxide and the amount of silicon in the mild steel (up to 0.2% silicon is very beneficial). Breakaway occurs preferentially at sharp corners on the 9% chromium steel, where three oxidising surfaces contribute carbon to a small volume element of metal, compared with two at an edge and one at a plane surface.

In the magnox stations, the chance of component failure by steel oxidation has been limited by careful control of the gas composition and temperature to such an extent that all stations can be safely operated up to and beyond their original planned life.

However, carbon dioxide environments which are beneficial to steel oxidation (i.e. low moisture and hydrogen levels) can be harmful to the graphite used to moderate the reactor. A balance has to be drawn between the requirements for long steel life and long graphite life. The situation in the AGRs is more complicated and operating ranges have been defined for gas composition and temperature such that the needs of the graphite moderator, carbon deposition, coolant performance and 9% chromium steel are optimised.

The above example clearly shows the complex interaction between a large number of variables and the different requirements of various parts of a plant. Such problems often confront engineers running large industrial processing and production units. It also emphasises once more the care needed when extrapolating short-term tests in simulated environments to real engineering structures.

17.3 GROWTH OF OXIDE FILMS

We must now consider the mechanism by which oxygen and the metal are brought together through the film so that oxidation can continue. If this diffusion through the film did not occur, the oxidation would cease once a monomolecular layer of oxide had formed all over the surface.

At one time it was assumed that oxidation always involved the movement of oxygen ions inwards through the film towards the metal. However, Pfeil was able to demonstrate that this was not the mechanism for oxide growth on iron. He took a clean abraded piece of iron on which he coated a layer of chromium oxide. After the iron had been heated in air for some time, it was covered by a thick layer of iron oxide, but the chromium oxide was either on or just above the iron surface and and below the iron oxide. The position of the chromium oxide on the iron surface clearly indicated that the iron ions had diffused outwards to form the oxide, passing through the chromium oxide layer. It did not indicate that oxide ions had diffused inwards. Further experiments have shown that copper ions also diffuse outwards to form an oxide film, whereas in zirconium and titanium, oxide ions move inwards to react at the metal/oxide interface.

Metal oxides are largely ionic compounds in which the metal and oxide ions are arranged in regular arrays in their respective crystal lattices. Some oxides contain excess metal ions at interstitial positions (Section 2.6); they are called *n*-type or negative carrier type materials. Others are deficient in metal ions, so vacant sites exist in the cation (metal ion) lattice; these are called *p*-type or positive carrier type materials. Typical *n*-type oxides are ZnO, CdO and Al_2O_3, while some *p*-type oxides are Cu_2O, NiO, FeO and Cr_2O_3.

Let us consider how the diffusion of different species can occur through a layer such as Cu_2O, which might be expected to behave as an insulator. Figure 17.4(a) shows how the diffusion of copper ions occurs. Careful chemical analysis of the copper (I) oxide shows there is a slight deficiency of copper compared to the chemical formula, Cu_2O. Such an oxide is said to be **non-stoichiometric**. Vacancies exist in the singly charged cuprous sublattice of the crystal structure, but to retain overall electrical neutrality there are sufficient doubly charged cupric ions to compensate. In particular, the number of vacancies is greater at the air/oxide interface than at the metal/oxide interface. The existence of a concentration gradient of vacancies causes the cuprous ions to migrate

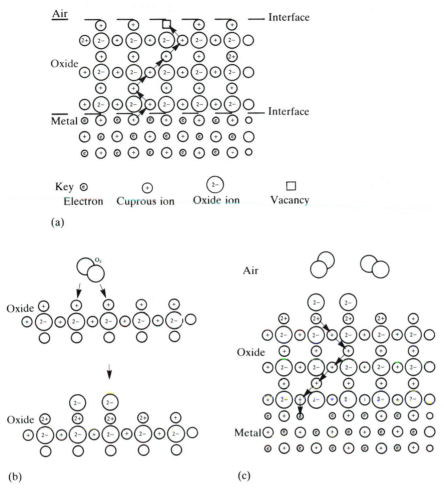

Fig. 17.4 Schematic diagram of the mechanism of oxidation of copper. (a) Diffusion of Cu$^+$ ions from the metal to the air/oxide interface via cation vacancies. The presence of cation vacancies causes an equivalent number of copper ions to be in the divalent oxidation state. (b) Reaction of oxygen molecules with Cu$^+$ ions at the air/oxide interface. Reaction of one molecule leads to the attachment to the lattice of two O^{2-} ions and the oxidation of four Cu$^+$ ions to Cu^{2+} ions. (c) Diffusion of positive charge inwards (electron outwards) to neutralise the excess of electrons in the metal.

outwards to the air/oxide interface by a stepwise movement shown in Fig. 17.4(a). Conversely, the vacancy diffuses inwards to the metal/oxide interface, where a spare electron becomes available.

If this were the only process, it would lead to a surplus of electrons in the metal. However, oxygen atoms attach themselves to the surface layer, as shown in Fig. 17.4(b), where they become oxide ions:

$$O_2 + 4e^- \rightarrow 2O^{2-} \tag{17.10}$$

The reaction in eqn (17.10) is achieved by oxidising positive Cu^+ ions at the surface to Cu^{2+} ions:

$$Cu^+ \rightarrow Cu^{2+} + e^- \tag{17.11}$$

The electrons left behind when the Cu^+ ions diffused outwards can now diffuse to restore the Cu^{2+} ions to the Cu^+ state. This is a sequential process. An electron from an adjacent cuprous ion diffuses into the cupric ion and restores it to the cuprous state:

$$Cu^{2+} + e^- \rightarrow Cu^+ \tag{17.12}$$

leaving the adjacent cuprous ion in the cupric state. This process continues between adjacent ions until the electron left in the metal crosses the metal/oxide interface. This is equivalent to a flow of positive charges in the opposite direction, as shown in Fig. 17.4(c). The mechanism for the growth of the film is complete.

This type of mechanism can only occur for oxides of metals with variable valency, such as copper and iron. Aluminium has only one valency and an oxide of fixed stoichiometry. In this case, oxidation would be expected to be very slow, as is indeed the case.

In *n*-type oxides, such as ZnO, where excess metal ions are located at interstitial positions, there must be an excess of electrons to maintain electrical neutrality. The simplest picture of oxide growth is the simultaneous diffusion of positive zinc ions and negative electrons in opposite directions. The zinc ions move through the defect sites to the outer surface, where they react with oxygen and build additional layers of oxide.

The situation occurring in oxidation processes is analogous to the basic wet corrosion cell. The oxide acts like all four components:

An electrode for the oxidation of a metal (an anode in the wet corrosion cell)

An electrode for the reduction of oxygen (a cathode)

An ionic conductor (the electrolyte)

An electron conductor (the external circuit)

The outward diffusion of metal ions can produce some peculiar effects. If iron wire is heated at about 800°C, an oxide film forms on the outer surface. Iron ions diffuse outwards through this film, while vacancies diffuse inwards. Gradually, a tube of oxide is formed as the iron migrates to the outer surface to react with oxygen. Frequently, the oxide spalls during the process and a misshapen but still hollow section may be produced. The same process occurs with nickel heated in air at 1,250°C. But cobalt oxide at this temperature is more plastic than nickel oxide and slowly collapses into the void, giving a solid oxide with only a few small holes at its centre.

17.4 OXIDATION OF ALLOYS

The influence of lattice defects on diffusion through an oxide film led both Hauffe and Wagner to propose a series of rules for the effects of metallic alloying additions on the oxidation rate of alloys. These rules, summarised in Table 17.2, are not inviolable, but they give general guidance in cases where the alloying metal is present in the oxide film of the parent metal. In some cases they predict unexpected, but nevertheless observable, effects.

For example, the addition of 0.1% aluminium to zinc, which forms an *n*-type oxide, causes a reduction in the oxidation rate by a factor of about 100. Only two Al^{3+} ions instead of three Zn^{2+} ions are associated with three O^{2-} ions. This leaves an extra hole in the metal lattice which is occupied by one of the interstitial Zn^{2+} ions. This ion is trapped in the hole and is restricted in the part it can play in the diffusion of metal ions through the lattice; the oxidation rate is consequently reduced.

On the other hand, adding small quantities of chromium (which has a higher valency than the nickel ion) to nickel oxide (a p-type oxide) also increases the number of vacancies. But as the oxide is already deficient in metal ions, the additional holes make it easier for the nickel to diffuse; the oxidation rate rises.

If lithium, which forms a monovalent ion, is added to nickel oxide, two Li^{+}

Table 17.2 The effect of alloying upon the rate of oxidation

Oxide type	Valency of alloying element compared to parent metal	Effect	Diffusion controlled oxidation rate
p-type Cu_2O NiO FeO Cr_2O_3 CoO	Higher valency	Increases number of vacancies Decreases number of parent metal ions with a higher valency	Increases
Ag_2O MnO SnO	Lower valency	Decreases number of vacancies Increases number of parent metal ions with a higher valency	Decreases
n-type ZnO CdO Al_2O_3 TiO_2	Higher valency	Decreases concentration of interstitial metal ions Increases number of free electrons	Decreases
V_2O_5	Lower valency	Increases concentration of interstitial metal ions Decreases number of free electrons	Increases

ions are required to replace one Ni^{2+} ion. The number of vacant sites in the lattice is reduced in maintaining electrical neutrality; the nickel diffusion is hindered and the diffusion rate falls.

These examples illustrate two rather peculiar situations. Lithium, an active metal with a very low oxidation resistance, reduces the oxidation rate of nickel in oxygen, while chromium, an alloying addition noted for its oxidation resistance, increases the oxidation rate of nickel. The latter effect occurs only for concentrations of below 5%. Above 5% chromium, the solubility of Cr^{3+} ions in the NiO lattice is exceeded; a separate protective layer of Cr_2O_3 forms on the metal surface and keeps the oxidation rate at very low levels. For example, nichrome electric fire elements last much longer than nickel wire.

The most effective way to control the oxidation of iron and steel is to form a stable surface layer of protective oxide from an alloying element. This hinders the diffusion of iron ions and electrons and so reduces the rate. A complex triple-layer scale forms on the surface of iron heated in air at temperatures above about 500°C. The inner layer, FeO, is by far the thickest. The next layer is Fe_3O_4 and the outermost layer is Fe_2O_3. Chromium and aluminium are the most effective alloying elements added to form stable scales on steels. Additional protection is conferred by adding nickel, silicon and some rare earth elements such as yttrium to iron–chromium alloys.

An alloy of Fe–9Al compares favourably in oxidation resistance to Ni–20Cr but the iron–aluminium alloys have poor mechanical properties and are unsuitable for use as general engineering materials. This can be overcome to some extent by forming aluminium-rich outer layers on iron and steel in a process called calorising. Small components are heated in a mixture of aluminium powder, aluminium oxide and ammonium chloride. The aluminium-rich outer layer resists oxidation, and the inner core of aluminium-free metal provides the desired mechanical properties of the component.

Chromium additions confer good oxidation resistance upon iron and steel, and on many other types of alloy. The chromium enriches the innermost layers of the iron oxide coating and often generates a layer of chromium oxide on the metal surface below the iron oxides. These layers are more resistant to ion or electron diffusion than iron oxides alone, and the oxidation rate is decreased. Iron–chromium alloys containing 4–9% chromium are used as oxidation-resistant metals in many areas, including oil processing plant. The 12% chromium alloys make good blade materials for steam turbines, and alloys with up to 30% chromium are used in the chemical industry and for heat treatment plant and burners. Further additions of silicon, nickel and yttrium make the high chromium alloys a suitable choice for petrol engine valves and other components which operate at high temperatures in aggressive environments. When 5% aluminium is also added, excellent oxidation resistance is obtained, but the alloy has poor mechanical properties and suffers embrittlement at room temperature. Iron–aluminium alloys are used as windings in electrical furnaces, where they give a long life provided they are protected from mechanical shock.

Table 17.3 lists the nominal compositions of some high temperature alloys for industrial environments, and Table 17.4 summarises their performance in high

Table 17.3 Nominal compositions of high temperature alloys for industrial environments

| Material | Percent nominal weight | | | | | | | | | | |
	Ni	Co	Fe	Cr	Mo	W	Mn	Si	C	La	Others
Haynes 188	22	39	<3	22	–	14	<1.25	0.35	0.10	0.03	–
214	75	–	3	16	–	–	<0.5	<0.2	0.05	–	4.5 Al, 0.01 Y, <0.1 Zr
230	57	<5	<3	22	2	14	0.5	0.4	0.10	0.02	0.3 Al
242	65	<2.5	<2	8	25	–	<0.8	<0.8	<0.03	–	<0.5 Al, <0.006 B, <0.5 Cu
556	20	18	31	22	3	2.5	1	0.4	0.10	0.02	0.6 Ta, 0.2 Al, 0.2 N, 0.02 Zr
HR-120	37	<3	33	25	<2.5	<2.5	0.7	0.6	0.05	–	0.7 Cb, 0.2 N, 0.1 Al, 0.004 B
HR-160	37	30	<3.5	28	<1.0	<1.0	0.5	2.75	0.05	–	<1.0 Cb
Hastelloy N	71	<0.2	<5	7	16	<0.5	<0.8	<1	<0.08	–	<0.35 Cu, <0.5 (Al + Ti)
Hastelloy S	67	<2	<3	16	15	<1	0.5	0.4	<0.02	0.05	0.25 Al, <0.015 B
Hastelloy X	47	1.5	18	22	9	0.6	<1	<1	0.1	–	<0.008 B

421

Table 17.4 Alloy performance in high temperature industrial environments

Alloy*	Environment							
	Oxidising	Sulphur-bearing	Carburising	Nitriding	Molten chloride salts	Oxidising chlorine-bearing	Reducing chlorine-bearing	Reducing fluorine-bearing
214	Excellent to 1,250°C	–	Excellent	Excellent	Good	Excellent	Very good	–
230	Excellent to 1,150°C	Good	Very good	Excellent	–	–	–	Good
HR-160	Very good to 1,100°C Good to 1200°C	Excellent	Very good	Very good	Very good	Excellent	Very good	Good
Hastelloy X	Very good to 1,100°C	Good	Good	Good	Good	Very good	Good	–
Hastelloy S	Very good to 1,100°C	–	Good	Very good	Very good	–	Excellent	Very good
Hastelloy N	–	–	–	–	Excellent	–	–	Very good
Haynes 188	Very good to 1,100°C	Very good	Good	Very good	Very good	–	–	–
556	Very good to 1,100°C	Very good	Very good	–	Very good	Excellent	Good	–
HR-120	Very good to 975°C Good to 1,100°C	Very good	Very good	–	–	Good	Good	–
242	–	–	–	Excellent	Excellent	–	Excellent	Excellent

*Haynes, Hastelloy, 214, 230, 242, 556, HR-120 and HR-160 are trade marks of Haynes International Inc.

temperature environments. Hastelloy X was one of the most widely used alloys during the early days of jet engines and is still much used. In the 1960s, Haynes 188 was developed to give a 150°C advantage over any other available alloy and was used in the Pratt & Whitney F100 which powered the F15 fighter. Today, the Haynes 230 and 242 alloys offer even better performance. Haynes 230 is used for heat treatment plant operating at 850°C.

Materials for Incinerators

Waste incineration is a growing industry, whether for municipal, industrial, hospital or radioactive waste disposal. Temperatures range from 600–1,100°C in atmospheres containing sulphur and chlorine compounds, as well as a variety of other chemical species and metals. Alloys 556, HR-120 and HR-160 are valuable in these applications because of superior resistance to oxidation, sulphidation and molten salt attack.

Alloy 556 has a broad spectrum of resistance to high temperature corrosion environments. Besides incinerators, it is used for heat treating, calcining, chemical processing, refinery, boiler and gas turbine components of various types. It has excellent fabricability and makes a good filler metal for welding nickel or cobalt alloys to ferrous alloys.

HR-120 is a nitrogen-strengthened Fe–Ni–Cr alloy with better oxidation resistance and hot workability than 800 and 330 series alloys. The 25% chromium content provides good oxidation and environmental resistance and stabilises nitrogen, while the balance of nitrogen, carbon and niobium leads to excellent mechanical properties.

HR-160 is a Ni–Co–Cr–Si alloy developed for outstanding resistance to severely sulphidizing environments or those containing chlorine and chlorides. Used for numerous incinerator components, ranging from combustion liners to superheater tube shields where it is exposed to high temperature corrosion and fly-ash corrosion, it typically performs 10–20 times better than stainless steels in the same applications.

17.5 HOT CORROSION

Despite the great developments in gas turbines, their principles of operation have remained the same. In brief, a gas turbine ingests air from the atmosphere, mixes it with fuel then compresses and ignites the mixture. This produces gases at temperatures in the region of 730–1,370°C. A fraction of the hot gas drives the turbine, which keeps the compressor running. The remainder of the hot gas is used to provide thrust, in the case of a turbojet engine, or shaft horsepower in a turboshaft engine.

Although the operating principles of the gas turbine have not changed, the improvements in performance have been remarkable. Specific fuel consumption has been reduced to one-third of the early jet engines, while thrust/weight ratios have been tripled and time between overhauls increased a hundredfold. This has been made possible only by advances in materials technology, which have resulted in great weight reductions and allowed far higher operating temperatures to be used. In 1995, the good safety record of the aircraft industry is beyond doubt, an achievement made possible by meticulous inspection and maintenance procedures. There have been very few major disasters directly attributable to high temperature corrosion failure within aero-engines. Hence, the significant effect of high temperature corrosion is upon the life of an engine and the periods between major overhauls.

At present, there remains one particular problem associated with the operation of gas turbines in marine environments. This problem is known as hot corrosion. It should not be confused with general high temperature oxidation. Hot corrosion is a combination of oxidation and reactions with sulphur, sodium, vanadium and other contaminants which are present, either in the inlet air or in the fuel. It produces a non-protective oxide on the blade surface in place of the normal protective oxide of chromium or aluminium. Hot corrosion can severely reduce the life of turbine blades. It could also lead to engine failure, although careful and regular inspection procedures can usually reduce the likelihood considerably.

Case 17.3: The nozzle guide segments of one engine corroded so badly in 800 operating hours that the engine failed completely.

Case 17.4: After only a few months of operation carrying out low-level flights among islands in the Caribbean the turbine sections of Royal Air Force Harriers suffered a drastic loss of efficiency. The turbines of Royal Navy Sea Harriers in similar conditions were found to perform satisfactorily. The problem was caused because the turbine materials used in the RAF aircraft were not resistant to hot corrosion, in contrast to the Royal Navy aircraft which had been 'marinised' by employing different materials and coatings for increased corrosion protection.

Civil aviation gas turbines operating at high altitude or inland airports do not normally suffer from hot corrosion, and are designed primarily for good mechanical/creep strength and oxidation resistance. This is best achieved using alloys of low chromium content but high aluminium content; the aluminium forms an oxide which provides an efficient barrier to further oxidation.

However, if the inlet air is laden with sea-salt as well as having a high moisture content, conditions for hot corrosion are ideal and the protection offered by the aluminium oxide is much reduced.

Hot corrosion presents the design engineer with a materials selection problem. To obtain high creep strength (a prerequisite for rapidly rotating turbine blades and discs) a high nickel content is required, whereas good resistance to hot corrosion is best conferred when large amounts of chromium are present in the alloy. Thus guide vanes, which are stationary and therefore

Fig. 17.5 Hot corrosion of a marine gas turbine blade.

operate at lower stress levels, can be made from high chromium alloys such as the cobalt-based X40. Rotating blades are made from high nickel alloys typified by the Nimonic series.

Case 17.5: Figure 17.5 shows a rotor blade from a ship gas turbine engine. The blade is suffering from hot corrosion characterised by large green/black blisters with accompanying loss of dimensions, especially on concave surfaces and trailing edges.

The mechanism of attack by hot corrosion is complex and not fully understood. This is only a cursory discussion.

In marine environments, sodium chloride will either be ingested into the engine from the atmosphere, or be present as a contaminant in the fuel. In the hottest part of the engine, the sodium chloride reacts with the sulphur and other components of the gas stream to produce sodium sulphate:

$$2NaCl + S + \frac{3}{2}O_2 + H_2O \rightarrow Na_2SO_4 + 2HCl \tag{17.13}$$

Sodium sulphate and sodium chloride react together to form a slag which melts on the surface of the component. The slag melts at approximately 620°C, fluxes the layer of chromium and aluminium oxides, which would otherwise protect the metal, and leaves it vulnerable to attack by the aggressive atmosphere in the engine.

Sulphur from the slag then diffuses into the alloy and reacts with the chromium to form internal sulphides. The substrate becomes depleted in chromium and the metal oxidises to yield a mixed nickel/chromium oxide with the spinel crystal structure. This oxide is much less protective than the chromium oxide alone. The chromium sulphide subsequently oxidises and releases sulphur. The sulphur diffuses further into the metal, where it reacts with more chromium to allow the oxidation process to continue in the newly

425

chromium-depleted matrix metal. Metals which have a high chromium content are therefore better able to resist hot corrosion as they are capable of maintaining the protective chromium oxide for longer periods.

Research is still in progress to determine the best method to reduce the incidence of hot corrosion. Improving the quality of the fuel/air mixture ingested by the engine will obviously be beneficial; reducing the sulphur content of the fuel or filtering out the sodium chloride will minimise the formation of low melting point slag. Alloy coatings have also been applied to blades to provide a barrier between the metal and the aggressive atmosphere. Typical of these is Cocraly, an alloy of cobalt, chromium, aluminium and yttrium, although zirconium is now being substituted for yttrium.

17.6 REFERENCES

1. Mellor J W 1946 *A comprehensive treatise on inorganic and theoretical chemistry,* Longmans, Green and Co, London.
2. Rowlands P C, Garrett J C and Whittaker A 1983 The corrosion of ferritic steels in gas-cooled nuclear reactors, *CEGB Research* **Nov**:3–12.

17.7 BIBLIOGRAPHY

Birks N and Meier G H 1983 *Introduction to high temperature oxidation of metals,* Edward Arnold, London.

Graver D L 1985 *Corrosion data survey — metals section,* 6th edn. NACE International, Houston TX.

Kofstad P 1988 *High temperature corrosion,* 2nd edn. Elsevier, New York.

Lai G Y 1990 *High temperature corrosion of engineering alloys,* ASM International, Ohio.

Levy A V 1982 *Corrosion-erosion wear of materials in emerging fossil energy systems,* NACE International, Houston TX.

Levy A V 1991 *Corrosion-erosion wear at elevated temperatures,* NACE International, Houston TX.

Michels H T and Friend W Z 1980 Nickel-base superalloys. In *The corrosion of nickel and nickel-base alloys*, edited by W Z Friend, John Wiley, New York.

Nicholls J R and Hancock P 1987 Advanced high temperature coatings for gas turbines, *Industrial Corrosion* **5**(4):8–17.

Rapp R A 1983 *High temperature corrosion,* NACE International, Houston TX.

Sahm P R and Speidel M O 1974 *High temperature materials in gas turbines,* Elsevier, Amsterdam.

Sims C T and Hagel W C 1972 *The superalloys,* John Wiley, New York.

Viswanathan R 1989 *Damage mechanisms and life assessment of high temperature components,* ASM International, Ohio.

18 WORKED EXAMPLES AND PROBLEMS

Examinations are formidable even to the best prepared, for the greatest fool may ask more than the wisest man can answer.

(Charles Caleb Colton, 1780–1832)

18.1 WORKED EXAMPLES

E18.1 Express the standard electrode potential, $E°$, of a metal in terms of the standard Gibbs free energy change, $\Delta G°$. Hence calculate the value of $\Delta G°$ at standard temperature and pressure for the corrosion of iron, assuming a divalent reaction.

$$\Delta G° = -zE°F$$
$$= -2(+0.44) \times 96,494$$
$$= -84.9 \text{ kJ mol}^{-1}$$

Notes
1. Negative $\Delta G°$ indicates that corrosion occurs spontaneously.
2. The reduction potential (Table 4.1) is -0.44 V, so for the oxidation process, we use $+0.44$ V. See Section 4.5 for details of the sign convention.

E18.2 Iron is connected to copper and then immersed in a solution containing both Fe^{2+} and Cu^{2+} ions.
(a) Which metal corrodes?
(b) Write equations to describe the reactions which occur at each electrode, assuming each metal has a valency of 2.
(c) Calculate the maximum possible potential of the resulting corrosion cell.

The metal with the most negative reduction potential will be the anode. From Table 4.1, $E°$ for iron is -0.44 V, while for copper it is $+0.34$ V. The iron will thus be the anode and will corrode. Convention requires that we write the cell as

$$Fe|Fe^{2+}||Cu^{2+}|Cu$$

and the cell potential as the reduction potential of the electrode on the right minus the reduction potential of the electrode on the left. Thus

$$E_{(cell)} = E_{(Cu\ redn)} - E_{(Fe\ redn)}$$
$$= (+0.34) - (-0.44)$$
$$= +0.78 \text{ V}$$

The electrode on the left is the anode, and thus the electrode reactions are

$$Fe \rightarrow Fe^{2+} + 2e^-$$

$$Cu^{2+} + 2e^- \rightarrow Cu$$

Note: A positive cell potential leads to a negative $\Delta G°$ and tells us that the cell, as described, will corrode spontaneously.

E18.3 Calculate the rest potential, versus the saturated calomel electrode, of a piece of copper in equilibrium with a solution containing 10^{-6} M copper (II) ions.

Using the Nernst equation:

$$E = E° + (0.059/2) \lg (10^{-6})$$
$$= +0.34 + (-0.177)$$
$$= +0.163 \text{ V SHE}$$

To convert from SHE to SCE we subtract 0.242 V, thus

$$E = 0.163 - 0.242$$
$$= -0.079 \text{ V SHE}$$

E18.4 A metal, M, of valency, z, atomic mass, W, and density, D kg m^{-3}, is corroding uniformly over its exposed surface area with a current density of i_{corr} A m^{-2}. Derive an expression for the number of millimetres of metal which will be lost during one year, assuming that the build-up of corrosion product does not stop the corrosion reaction.

Over 1 m^2 of exposed metal, the number of coulombs passed in one year will be

$$i_{corr} \times 60 \times 60 \times 24 \times 365 = 3.154 \times 10^7 \times i_{corr}$$

1 mol of metal of valency z converted into ions gives

$$z \times 96,494 \text{ coulombs}$$

(Remember that a mole is expressed in grams.)

Thus the number of moles per square metre lost in a year

$$= (3.154 \times 10^7 \times i_{\text{corr}})/(z \times 96,494)$$

$$= (326.8 \times i_{\text{corr}})/z$$

Converting moles into kilograms, the number of kilograms lost per square metre per year

$$= (326.8 \ W \times i_{\text{corr}})/(1,000 \times z)$$

The metal is of density D $\text{kg}\,\text{m}^{-3}$, thus if the mass lost were D $\text{kg}\,\text{m}^{-2}$ then the depth of penetration of the corrosion would be 1 metre (1,000 mm). The actual penetration (in mm) is therefore

$$= (326.8 \ W \times 1,000 \times i_{\text{corr}})/(1,000 \times z \times D)$$

$$= (326.8 \ W \times i_{\text{corr}})/(z \times D)$$

Note: The expression derived above is commonly used as a measure of corrosion rate. It is known as the wastage rate, is measured in $\text{mm}\,\text{y}^{-1}$ (sometimes written mmpy), and is popular because it gives engineers a better 'feel' for the rate of corrosion. In the case of copper, a corrosion current density of 0.01 $\text{A}\,\text{m}^{-2}$ is commonly observed. Thus, with $z = 2$, $W = 63.5$, and $D = 8,960$ $\text{kg}\,\text{m}^{-3}$, we calculate a wastage rate of

$$= (326.8 \times 63.5 \times 0.01)/(2 \times 8,960)$$

Thus we see that the wastage rate for copper is almost the same as the corrosion current density of copper in $\text{A}\,\text{m}^{-2}$. The units of i_{corr} are important. It is both surprising and fortuitous that for many metals the quantity, $W/(zD)$, is approximately constant. Hence the observation for copper is applicable to many other metals too:

$$i_{\text{corr}} \ (\text{Am}^{-2}) \simeq \text{mmy}^{-1}$$

E18.5 A univalent metal, M, has $\beta_a = +0.2$ V, $\beta_c = -0.2$ V, and $i_0 = 20$ mA m^{-2}. Calculate values for i_a and i_c when the metal is anodically polarised to $+0.20$ V. Hence determine the experimentally measured current density for the same polarisation.

Figure E18.5 plots the anodic and cathodic current densities for M. At $+0.20$ V:

$$i_a \quad = 200 \ \text{mA}\,\text{m}^{-2}$$
$$i_c \quad = 2 \ \text{mA}\,\text{m}^{-2}$$
and
$$i_{\text{meas}} = i_a - i_c$$
$$= 198 \ \text{mA}\,\text{m}^{-2}$$

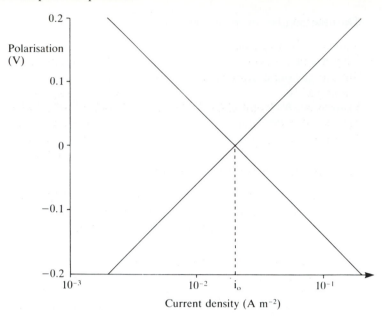

Fig. E18.5

Alternatively, we can use the Tafel equation:

$$\eta = \lg(i/i_0)$$

Taking antilogs we get for the anodic reaction:

$$10^{\eta/\beta} = i_a/i_0$$

But $\quad \eta\beta = 1$

Thus $\quad i_a = 10i_0$

$$= 200 \text{ mA m}^{-2}.$$

Similarly for the cathodic current,

$$\eta/\beta = -1$$

Thus $\quad i_c = 0.1i_0$

$$= 2 \text{ mA m}^{-2}$$

Therefore $i_{\text{meas}} = 198 \text{ mA m}^{-2}$, as before.

Note: The slope of the Tafel plot — the beta constant (+0.2 or –0.2 in the example) — is obtained by measuring the voltage change per logarithmic decade of current, e.g. the voltage change between 10^{-1} and 10^{-2} A m^{-2}.

E18.6 (a) Calculate the theoretical capacity of zinc sacrificial anode material.
(b) The measured capacity is 780 Ah kg^{-1}. Calculate the efficiency of

the material.

Atomic weight of zinc $\qquad$ = 65.4
1 kg of zinc contains $\qquad$ 1,000/65.4 = 15.29 mol
But 1 mol of zinc converted into divalent zinc
  ions yields $\qquad$ 2 x 96,494 = 192,988 C
Therefore, 15.29 mol yield $\qquad$ 15.29 × 192,988 = 2.95 × 10^6 C
But 1 coulomb $\qquad$ = 1 ampere-second
Therefore charge created $\qquad$ = 2.95 × 10^6 As
$\qquad$ = 819 Ah

Thus, the maximum capacity of zinc is 819 Ah kg^{-1}.

Allowing for the presence of impurities and other irregularities in the dissolution process, the practical quoted capacity of 780 Ah kg^{-1} leads to a dissolution efficiency of 780/819 × 100% = 95%.

E18.7 Two gas pipelines, each 75 km long and 1 m diameter, are laid in different parts of a Middle Eastern oil state at the same time. Pipeline A receives a very poorly applied coal tar epoxy coating, while pipeline B is given a high quality protective coat. Each line is further protected by a standard ICCP installation which uses a 100 V, 100 A (10 kW) rectifier/ground-bed system to cover the whole length of the line. After several years it is found that pipeline A requires a current density of 1.72 mA m^{-2} to achieve the protection potential, while line B requires 86 μA. The recommended protection current density is 130 μA m^{-2}. For each pipeline:
(a) Comment upon any excess or shortfall in the capacity of the protection system.
(b) Describe the effect of the corrosion protection system.

Total area of each pipeline: $\pi \times 1 \times 75,000$ $\qquad$ = 239,000 m^2
Total current:
  line A = 1.72 × 10^{-3} × 239,000 $\qquad$ = 411 A
  line B = 86 × 10^{-6} × 239,000 $\qquad$ = 21 A

Line A requires more current than the equipment is able to provide. Much of the line is thus unprotected by the ICCP and with its poor coating is therefore liable to leak or rupture. (This question is based upon a real case history. In fact only 7 km of line A was protected by the ICCP.) Line B requires only 21 A of the 100 A available and is seriously overdesigned. (Only 120 W of the 10 kW capacity was required.) In the case of line A which required such a high current, the best solution is to use more current drain points, each with a lower current.

E18.8 It is planned to place an uncoated steel drilling platform in the sea. The continuously immersed parts of the structure will be protected with a sacrificial anode cathodic protection system which must have a life of 10 years. The anodes available are semicylindrical, 100 mm diameter, 400 mm long, with a steel core suitable for welding the plane side to the structure. (The ends and backs may be ignored for calculation purposes.)

(a) Select a suitable anode material.

(b) Calculate the minimum number of anodes necessary to meet the requirement at the start of the period.

(c) Show by calculation whether 2,510 anodes will be sufficient to provide protection throughout the whole of the 10 years.

Data

Wetted surface area of legs = 2,000 m^2

Wetted surface area of
   cross-members   = 500 m^2

Current requirement for
   uncoated steel   = 110 $mA\,m^{-2}$

(a) The alloy should be either zinc or aluminium, not magnesium. (Magnesium is for freshwater applications only.)

(b) Assume all anodes corrode at a constant rate. Output of both zinc and aluminium alloy (Table 16.1) is 6.5 $A\,m^{-2}$.

Total current requirement   = 2,500 m^2 × 110 $mA\,m^{-2}$
                                      = 275 A

Anode area required to
   supply 275 A   = 275/6.5
                               = 42.3 m^2

Approximate area of one
   anode   = 3.142 × 0.4 × 0.05
                               = 0.063 m^2

Minimum number of
   anodes   = 42.3/0.063
                               = 671

(c) 275 A are required for
   10 years   = 2,750 Ay

This output is required
   from 2,510 anodes.

Output required from
   each anode   = 2,750/2,510
                               = 1.10 Ay

Approximate volume of
   one anode   = 3.142 × 0.05 × 0.05 × 0.4/2
                               = $1.57 × 10^{-3}\ m^3$

From Table 16.1, volume
   wastage rate for zinc   = $1.518 × 10^{-3}\ m^3\,(Ay)^{-1}$

Thus total wastage by
volume of one zinc
anode = $1.518 \times 10^{-3} \times 1.1$
= 1.67×10^{-3} m^3

Thus all zinc anodes will be consumed after a 10 year period.
Similarly, for anodes of aluminium, wastage by volume (Table 16.1)
= $1.18 \times 10^{-3} \times 1.1$ m^3

Volume left after 10 years = $(1.57 \times 10^{-3}) - (1.3 \times 10^{-3})$
= 0.27×10^{-3} m^3

Assuming regular dissolution and using $V = \pi r^2 l/2$ for a semicylinder, the radius of anode remaining is given by the following expression:
$$r^2 = (2 \times 0.27 \times 10^{-3})/(3.142 \times 0.4)$$
$$r = 21 \text{ mm}$$

Area left after 10 years = $3.142 \times 0.021 \times 0.4 \times 2500$
= 66.2 m^2.

Since 42.3 m^2 is required to supply the current, then aluminium anodes are still effective after 10 years.

18.2 PROBLEMS

P18.1 For each of the following give one possible reason why:
(a) So few metals exist in the uncombined state in nature.
(b) Gold exists in the uncombined state in nature.
(c) Well-preserved iron artefacts have been uncovered after centuries of immersion in peat bogs.

P18.2 Briefly define each one of the following:
(a) Corrosion.
(b) Standard electrochemical potential.
(c) Exchange current.
(d) Polarisation.
(e) Double layer.

P18.3 Consider the following statements:
(a) All natural corrosion reactions are spontaneous.
(b) Measurements of electrical potential form an important part of corrosion monitoring.
Show how it is possible to relate them.

P18.4 Draw and label a diagram to describe each of the following:
(a) The use of a polarisation curve to calculate a cathodic Tafel constant.
(b) An anodic polarisation curve exhibiting passivation.

P18.5 Illustrate each of the following with a simple labelled diagram:
(a) The measurement of a standard electrochemical potential.
(b) The measurement of a free corrosion potential in 3.5% sodium chloride solution.

P18.6 For the electrochemical cell having nickel and cadmium electrodes in equilibrium with solutions of their ions:
(a) Write two equations to describe the reactions which occur.
(b) Determine the greatest potential which may be obtained from such a cell under standard conditions.
(c) State one other condition necessary to achieve this potential.

P18.7 Calculate the rest potential, versus the saturated calomel electrode, of a piece of nickel in equilibrium with a solution containing 10^{-6} M nickel ions.

P18.8 (a) Give two useful functions of E/pH diagrams.
(b) Give two limitations to their use.

P18.9 (a) Explain what is meant by *passivity* in the context of corrosion.
(b) Give two examples of natural passivity of metals.

P18.10 The following couples with equal areas are immersed in fresh water: Fe/Cd, Fe/Ti, Fe/Zn and Fe/Cu.
(a) In which one of the four will the iron corrode the fastest?
(b) Which combination offers the best protection to the iron?

P18.11 (a) What is mill scale and how is it formed?
(b) Why is its formation important in determining the subsequent corrosion resistance of the steel?
(c) With the aid of diagrams, explain the standard mechanism for the formation of rust scabs on a piece of horizontal unpainted mild steel exposed to rain.

P18.12 (a) List four ways in which the environment may influence an aqueous corrosion process.
(b) Briefly summarise the effect of each parameter you have listed.
(c) List four properties which you would expect from a metallic coating in order for it to be suitable for use in the corrosion protection of a metal.
(d) Name three metals which can be used successfully for protective coatings on steel and give three different methods by which they can be applied.

P18.13 *Stray current corrosion* is often used to describe a particular type of corrosion attack.
(a) Explain the meaning of the term and describe the mechanism of the corrosion process.
(b) Give an example and indicate how the problem may be reduced.

P18.14 (a) Give the general name for the form of corrosion which can occur when turbulent water flows through brass pipes.
(b) Give two likely causes of this turbulence.

P18.15 (a) What precipitates are formed when weld decay occurs in an unstabilised stainless steel?
(b) Where are the precipitates formed?
(c) Explain the mechanism of weld decay.
(d) What elements are added to stainless steels to reduce weld decay?
(e) State two other measures which may be taken to alleviate the problem.

P18.16 (a) Why is the combination of pitting corrosion and oscillating stress a serious problem in a metal component?
(b) Suggest two ways of assessing the severity of pitting corrosion.

P18.17 Explain how the addition of an anodic inhibitor can aggravate a corrosion problem.

P18.18 Outline the role of sulphur in the hot corrosion of an Ni–Cr alloy turbine blade.

P18.19 (a) Sketch two curves which illustrate the weight gained by materials exhibiting rectilinear and parabolic oxidation kinetics.
(b) What is the role played by the oxide in each case?

P18.20 In the oxidation of metals, explain the importance of the ratio:

$$\frac{\text{molar volume of oxide produced}}{\text{molar volume of metal oxidised}}$$

P18.21 (a) Explain clearly the meaning of the following terms when used in the context of E/pH diagrams: immunity, corrosion, passivation, pH.
(b) Use Fig. 4.18 to obtain a value for the free corrosion potential of zinc in water at pH 8.
(c) Calculate the same value by means of the Nernst equation and thus show how the equation is used in constructing part of the E/pH diagram.
(d) Describe, in principle, how it is possible to complete the remainder of the E/pH diagram.

(e) Explain how inspection of the $E/$pH diagram suggests two possible methods for controlling the corrosion of zinc in water.

(f) What factors, if any, might make these suggestions impractical?

(g) What other factors limit the application of $E/$pH diagrams to real situations?

(h) State the effect of an increase in temperature upon the potential you found in parts (b) and (c). Give a reason for your answer.

P18.22 (a) Using the following data construct the $E/$pH diagram for the beryllium/water system at 25°C for pH values between 0 and 10:

$$\text{Be}^{2+} + 2e^- \rightleftharpoons \text{Be} \qquad\qquad E^\circ = -1.849 \text{ V}$$
$$\text{BeO} + 2\text{H}^+ \rightleftharpoons \text{Be}^{2+} + \text{H}_2\text{O} \qquad\qquad \log K = 2$$
$$\text{BeO} + 2\text{H}^+ + 2e^- \rightleftharpoons \text{Be} + \text{H}_2\text{O} \qquad\qquad E^\circ = -1.79 \text{ V}$$

(b) A beryllium component has been stored in a damp atmosphere (relative humidity >70%). Using the information given below, calculate the corrosion rate in $\text{g cm}^{-2}\,\text{day}^{-1}$. Assume the concentration of Be^{2+} in the condensed moisture film to be $10^{-6} \text{ mol dm}^{-3}$.

Data

$2.303\ RT/F = 0.0591 \text{ V}$

$\text{Be}^{2+} + 2e^- \rightleftharpoons \text{Be} \qquad E^\circ = -1.849 \text{ V}$

i_0 for the process $\text{O}_2 + 2\text{H}_2\text{O} + 4e^- \rightarrow 4\text{OH}^- = 10^{-1} \text{ A m}^{-2}$

Assume $E_{\text{equilibrium}}$ for the cathodic reaction $= -0.05 \text{ V}$

Quasi-cathodic Tafel constant $= -0.4 \text{ V}$

Anodic Tafel constant $= 0.05 \text{ V}$

i_0 for the process $\text{Be} \rightarrow \text{Be}^{2+} = 10^{-3} \text{ A m}^{-2}$

The relative molecular mass (rmm) of Be $= 9.01$

Faraday constant (F) $= 96,500 \text{ C mol}^{-1}$

$i_{\text{corr}} = zFM$, where M is the number of moles of material lost per second.

P18.23 (a) Sketch an energy profile for Gibbs free energy versus reaction coordinate for the corrosion reaction:

$$\text{M} \rightarrow \text{M}^{z+} + ze^- \qquad\qquad (1)$$

(b) On your profile show the overall Gibbs free energy change and the Gibbs free energy of activation.

(c) Briefly describe the feature of your energy profile which is directly related to the rate of the corrosion reaction.

Corrosion rate is an important parameter in the determination of the lifetime of components and structures. Two electrochemical methods of measuring corrosion rates involve (i) potentiodynamic scans and (ii) polarisation resistance measurements. Figure P18.23(a) is a potentiodynamic scan for mild steel in a chemical reactor environment, pH 5,

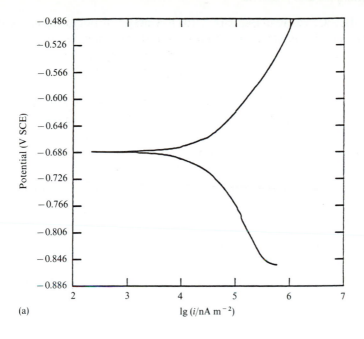

(a)

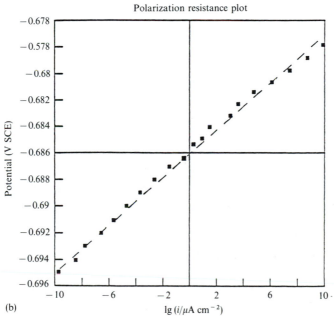

(b)

Fig. P18.23

and 25°C, whereas Fig. P18.23(b) is a graph of polarisation resistance taken from the same set of data, but plotted differently. It can be shown that

$$\frac{\Delta E}{\Delta i} = \frac{\beta_a \beta_c}{2.303 \, i_{(corr)} (\beta_a + \beta_c)} \tag{2}$$

where E = potential, i = current density, β_a and β_c are the slopes of the anodic and cathodic Tafel lines obtained from the potentiodynamic scan, and i_{corr} is the corrosion rate.

(d) Calculate the rate of corrosion from the potentiodynamic scan and then compare it with the value calculated by means of eqn (2)

(e) Which of the two is likely to be most accurate and why?

P18.24 Imagine that you are asked to investigate the rate of corrosion of an Inconel 625 reactor in the particular environment of a chemical process.

(a) Describe the basic equipment you would need to carry out a laboratory investigation.

(b) Explain the role of each piece of equipment mentioned in (a).

(c) Describe one experiment which you would perform, clearly stating the conditions for the experiment.

(d) Explain how you would interpret data obtained from your experiment.

(e) List the other parameters which might usefully be varied and explain how further information relating to the corrosion performance in the reactor might be obtained as a result.

(f) Discuss the limitations of the experiment you have described in the context of other forms of corrosion to which the material may be susceptible.

P18.25 (a) Explain why pitting corrosion is called an autocatalytic process.

(b) Give one reason why stainless steels are particularly susceptible to this form of attack.

(c) Name one elemental addition to stainless steels which improves their resistance to pitting corrosion.

In a programme of experiments to determine the resistance of a series of alloys to crevice corrosion, suitable specimens were subjected to two different tests which consisted of: (i) potentiodynamic scanning measurements in sea-water at 25°C and (ii) sea-water exposure trials for four years, followed by weight loss measurements. A typical plot of potential versus current density obtained from potentiodynamic scans is shown in Fig. P18.25. Values of E_c - E_p obtained from the scans for the range of alloys, together with the results of the exposure trials are summarised in the table.

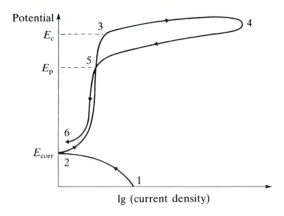

Potential

E_c

E_p

E_{corr}

lg (current density)

Material type	% composition			$E_c - E_p$ (mV)	Weight loss (mg ml⁻¹)
	Cr	Ni	Mo		
316	18	8	3	80	9.2
304	18	8	—	110	10.8
410	12	—	—	220	26.7
430	18	—	—	160	18.9
446	25	—	—	140	17.4
Hastelloy C	16	59	16	0	0.2

Fig. P18.25

Material type	Percent composition			$E_c - E_p/mV$	Weight loss/mg ml⁻¹
	Cr	Ni	Mo		
316	18	8	3	80	9.2
304	18	8	–	110	10.8
410	12	–	–	220	26.7
430	18	–	–	160	18.9
446	25	–	–	140	17.4
Hastelloy C	16	59	16	0	0.2

(d) Describe the sequence of events taking place at the metal/electrolyte interface at each of the stages represented by 1 to 6 in the figure, and interpret the significance of the parameters denoted by E_{corr}, E_c and E_p.

(e) By means of a suitable graph compare the laboratory-derived data with those from the exposure trials.

(f) Comment on the usefulness of the experiments in determining the resistance to crevice corrosion and the relative performance of the alloys listed.

(g) Show how the data can be used to make a useful prediction of the crevice corrosion resistance of an alloy containing 15% chromium and 4% nickel.

P18.26 A number of notched specimens were used in an experimental programme in which they were subjected to constant tensile load equal to 0.5 times the yield stress, while immersed in an electrolyte of pH 6 at 25°C. A different potential was applied to each specimen such that anodic and cathodic current densities, i, were obtained, and the times to failure, t_f, recorded. The data obtained from the experiments are summarised in the table.

$i/mA\ mm^{-2}$	t_f/min	$i/mA\ mm^{-2}$	t_f/min
+8.0	12	−0.33	415
+5.1	13	−2.4	92
+3.0	15.5	−4.1	39
+1.2	31	−7.5	21
0	90	−8.8	19

(a) Plot a graph of i versus $\lg(t_f)$.
(b) Interpret the significance of the graph in relation to possible corrosion failure modes.
(c) Describe the mechanisms of the corrosion processes involved.

P18.27 (a) Discuss the contribution which the *stress-corrosion spectrum*, as devised by Parkins, makes to the understanding of stress-corrosion cracking.
(b) Corrosion fatigue behaviour has traditionally been examined using endurance testing and the creation of *S–N* curves. Explain how this would be done for a structural steel in a moist atmosphere.
(c) Draw a diagram of the sort of results you would expect to obtain, compared to the performance of the same material in dry air.
(d) Highlight the significant features of the diagram you have drawn.
(e) Before constructing an offshore platform, corrosion fatigue tests were carried out on the structural steel which was to be used. Testing methods used the principles of linear elastic fracture mechanics to quantify susceptibility to corrosion fatigue. Figure P18.27 shows the experimental data obtained from the tests, which used steel of composition: Fe–0.17C–0.35Si–1.35Mn, in sea-water. Specimens were of the single edge-notched type. Explain the significance of (i) the axes, (ii) the values of R, (iii) the frequency and (iv) the applied potential.

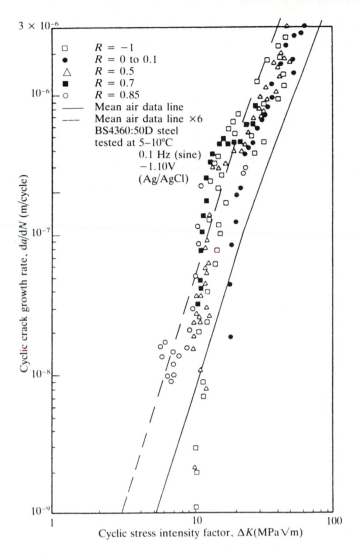

Fig. P18.27

P18.28 Decohesion by adsorption is a factor which may influence the mechanism of environment-sensitive cracking.

(a) Explain the meaning of (i) environment-sensitive cracking and (ii) decohesion by adsorption.

(b) Distinguish, as far as possible, between stress-corrosion cracking, corrosion fatigue and hydrogen embrittlement, indicating how decohesion by adsorption can be implicated in each form of corrosion.

It is planned to use a type of steel known as 4340 for wire ropes in a marine environment. The wire rope diameter is 127 mm and is made up

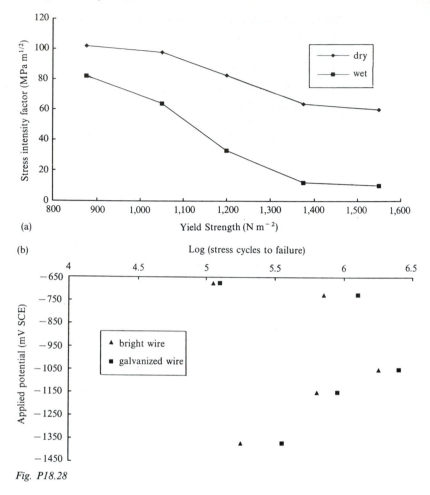

Fig. P18.28

of closely packed strands, each of 3.44 mm diameter galvanised 4340 steel. Figure P18.28(a) illustrates the variation of yield strength for the steel, varied by changes in the tempering temperature, versus the threshold stress intensity factor, K_{1SCC}, in flowing sea-water and air. Figure P18.28(b) summarises experimental data of a series of corrosion fatigue tests. The graph plots applied potential versus the number of cycles to failure for type 4340 steel wires in sea-water. The tests were carried out at 0.5 Hz with a load pattern of $R = 0.5$ and $\Delta\sigma = 500$ MPa. Specimens of both galvanised and bright steel were used.

(c) Define K_{1SCC} and explain how the data provides useful information about the proposed use of the material.

(d) Interpret the graph in Fig. P18.28(b) in terms of the corrosion fatigue performance and its implications for the use of the material for marine wire rope.

(e) The largest crack which can be tolerated by a strand is 1 mm at 500 MPa and $R = 0.5$. Using

$$\Delta K_1 = \Delta\sigma(\pi a)^{1/2}$$

find the maximum yield stress of the material which will not lead to stress-corrosion cracking.

(f) Assuming the conditions used in the laboratory tests accurately reflect service conditions, how many days of continuous service would you expect the galvanised rope to last before suffering a strand failure?

P18.29 In a programme of experiments to investigate the corrosion fatigue behaviour of a ferrous alloy in a nuclear reactor environment, a series of 17 compact tension test specimens were manufactured to conform with the specifications required for a valid plane strain corrosion fatigue test. The test pieces, of width $W = 50$ mm and thickness $B = 25$ mm were precracked such that the ratio of crack length to width, $a/W = 0.50$. Next they were exposed to the following environments: (i) specimens 1 to 8, dry air at 373 K and (ii) specimens 9 to 17, simulated reactor environment at 373 K. Cyclic loading patterns were applied such that in every case, frequency = 0.1 Hz and the ratio of minimum to maximum load, $R = P_{min}/P_{max}, = 0$. The table lists for each specimen the maximum load and the calculated crack growth rate in the appropriate environment. If the plane strain intensity factor $K_1 = PY_2/BW^{1/2}$ and Y_2, the stress intensity factor coefficient = 9.60 for $a/W = 0.50$,

(a) Plot a graph which can be used to predict the susceptibility of the alloy to corrosion fatigue in the reactor.

(b) Identify three important points on your graph and discuss the significance of each in relation to the mechanism of crack growth under the conditions used for the test.

(c) Given that such a material/environment combination is unavoidable, discuss the operating conditions which you consider would lead to acceptable corrosion fatigue resistance.

number	P_{max}/kN	da/dN (mm cycle^{-1})	number	P_{max}/kN	da/dN (mm cycle^{-1})
1	2.10	1.2×10^{-7}	9	1.43	1.3×10^{-7}
2	2.39	5.5×10^{-7}	10	1.63	5.0×10^{-7}
3	3.49	1.8×10^{-6}	11	2.10	1.3×10^{-6}
4	5.82	7.25×10^{-6}	12	3.49	4.8×10^{-6}
5	12.52	6.5×10^{-5}	13	5.82	1.7×10^{-5}
6	21.00	2.7×10^{-4}	14	12.52	1.1×10^{-4}
7	45.10	2.5×10^{-3}	15	21.00	4.0×10^{-4}
8	58.23	1.0×10^{-2}	16	45.10	2.6×10^{-3}
			17	58.23	1.0×10^{-2}

P18.30 (a) List three factors which control the rate of atmospheric corrosion on steel, and give brief details of the effect each has on the rate.

A number of precision components made from mild steel are to be transported by sea. They will travel and be stored together in a single closed wooden case which has a hinged lid. On arrival at the destination the case will be placed in an unheated warehouse. Individual components will be removed from the case at approximately 3 week intervals over a period of 2 years. It has been suggested that perforated containers of vapour phase inhibitors (VPIs) placed in the case will give adequate protection to the components during transit and storage.

(b) What type(s) of VPI would you recommend for this purpose? Mention any special characteristics which make them suitable for the task.

(c) What additional measures would be necessary to make the VPIs effective?

(d) How would the decision to use VPIs be affected if the components were cadmium plated?

(e) Explain the roles of the cation and anion in a VPI.

(f) Describe the mechanism of the particular corrosion hazard which wood presents to such components.

P18.31 Figure P18.31 shows details (not to scale) of the cooling system for an engine in an open, wooden-hulled launch operating in estuarine waters which are polluted by ammonium and sulphide ions. The pipe system is made from 70/30 brass tube with cast 60/40 brass elbows and T-joints. There are two flexible stainless steel bellows adjacent to the engine, which has a grey cast iron cylinder block. The engine is bolted to wooden bearers using mild steel fasteners. The pump has a cast iron body, a stainless steel shaft and a nickel aluminium bronze impeller. The valves are made with leaded bronze bodies and brass trims. A copper clip is used to connect the earth from the battery to the outlet pipe just before the pipe passes through the hull. If the engine block must be made from cast iron, identify four areas in which corrosion is

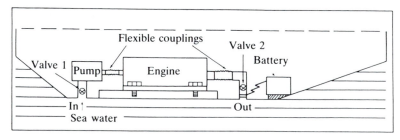

Fig. P18.31

likely to occur and suggest remedial actions which will minimise corrosion damage.

P18.32 Figure P18.32 shows a section (not to scale) from the machinery space in a bulk carrier. A cold sea-water pipe passes through the bulkhead.
(a) Identify the corrosion problems which are likely to occur.
(b) Suggest modifications which will minimise corrosion damage.

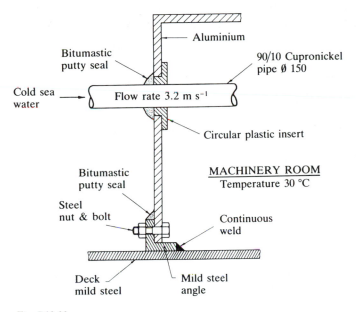

Fig. P18.32

P18.33 Figure P18.33 shows the arrangement (not to scale) of the stern fittings of a seagoing wooden-hulled vessel. The electrically driven pump fills the water storage tank at 2 day intervals, working for about 2 hours on each occasion. The system has a projected life of 10 years. The table lists some components and their materials.

(a) Identify the corrosion problems which are likely to occur.
(b) Comment on the protection measures already taken.
(c) Suggest modifications which will minimise corrosion.

P18.34 Figure P18.34 shows the layout (not to scale) of a subsidiary heat exchanger unit for a large land-based power station sited on an estuary. The unit has a projected life of 20 years and is closed for major overhaul at 2 year intervals. The pipe sizes, flow rates and the order of the valves and pumps were formulated in a computer-aided design study; these cannot be changed. The vapour side of the condenser is

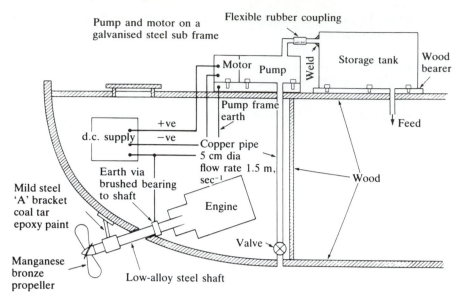

Fig. P18.33

Component		Material
Pump	Body	Cast iron
	Impeller	Nickel aluminium bronze
	Shaft	Mild steel
	Motor case	Aluminium
Valve	Body	Nickel aluminium bronze
	Trim	Cupronickel (alloy 400)
Storage tank		Mild steel, galvanised inside and out to a thickness of 130 μm
Fasteners	Pump and tank	Cadmium-plated mild steel

compatible with all conceivable tube/tube-plate materials. The table lists some of the components and their materials. The system is in continuous use.

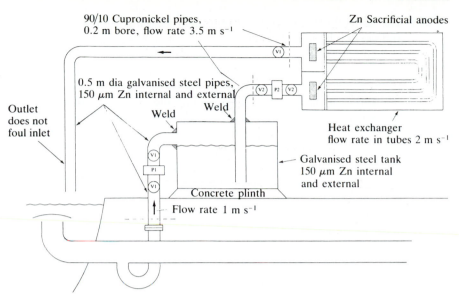

90/10 Cupronickel pipes,
0.2 m bore, flow rate 3.5 m s⁻¹

Zn Sacrificial anodes

0.5 m dia galvanised steel pipes,
150 μm Zn internal and external

Weld

Weld

Outlet
does not
foul inlet

Heat exchanger
flow rate in tubes 2 m s⁻¹

Galvanised steel tank
150 μm Zn internal
and external

Concrete plinth

Flow rate 1 m s⁻¹

Fig. P18.34

Component		Material
Estuarine water		$[O_2]$ high; pH = 8; NH_4^+ and S^{2-} present as pollutants; inlet temp. 4–10°C; outlet temp. 40°C
Valve V1	Body	Cast iron
	Trim	Brass
Valve V2	Body	Leaded bronze (Cu–6Sn–4Zn–2Pb)
	Trim	Alloy 400
Pump P1	Body	Cast iron
	Impeller	Nickel aluminium bronze
	Shaft	316 stainless steel
Pump P2	Body	Nickel aluminium bronze
	Impeller	Alloy 400
	Shaft	Alloy 400
Heat exchanger	Water-box	Low alloy steel with Zn sacrificial anodes
	Tube-plate	Naval brass
	Tubes	Aluminium brass

(a) Identify the corrosion problems which are likely to occur.
(b) Comment on the protection measures already taken and suggest modifications to minimise corrosion.

P18.35 The alloy, 90/10 Cu/Ni, has been much used for water piping systems
and heat exchangers. Although it has provided reasonable performance
in many sections of these systems, some areas have been plagued with
continued and often accelerated failures. A number of cases of Cu/Ni
sea-water service piping and electronic cooler tube failures occurred in
less than one year of service. These failures were diagnosed as mostly
erosion corrosion of Cu/Ni, induced in zones of highly turbulent water.
(a) With the aid of diagrams, describe three ways in which such
corrosion typically occurs.
(b) An assessment of the available remedial measures suggested
replacement with one of the following materials: aluminium, stainless
steels, high nickel alloys, GRP, or titanium. Of these, it was concluded
that titanium represented the best option. List reasons why the other
materials might have been considered unsuitable and why titanium
should be considered the best.
(c) Figure P18.35(c) shows one way in which titanium performed
worse than 90/10 Cu/Ni. It also shows the solution to the problem.
Explain.

A water cooling system is fed by a pump which supplies $0.5 \text{ m}^3 \text{min}^{-1}$ of
water at 1.4 MPa. The pipework is of 90/10 Cu/Ni, of internal diameter
58 mm and of tensile strength 262 MPa. In the design process, the wall
thickness of the pipe was obtained using

$$T = \frac{PD}{2(S + 0.4P)} + A$$

where T = minimum thickness, P = system design pressure, D = pipe
outer diameter, S = maximum allowable stress for material at design
temperature, and A = additional thickness to allow for corrosion

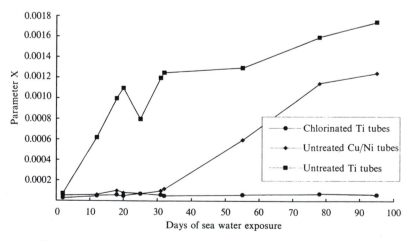

Fig. P18.35

during the service lifetime. The appropriate design code requires the maximum stress to be 0.25 × tensile strength.

(d) Calculate the value of A if the minimum wall thickness used for the Cu/Ni pipe is 2 mm.

(e) What value of A would you use to calculate the thickness for pipework of titanium?

(f) The yield stress of the replacement titanium is 345 MPa and the internal diameter must remain the same as for the Cu/Ni. Calculate the minimum thickness of titanium pipe suitable to replace the Cu/Ni pipework.

(g) What benefits other than those you mentioned in part (b) would accrue from the conversion of the 90/10 Cu/Ni system to one of titanium?

P18.36 It is proposed to use unstabilised austenitic stainless steel to construct a temporary staging in a tidal basin. Tubes will be used for the general platform, held together by clamps which are galvanically compatible with the steel in this environment. The deck plates will be welded to the tubes. It is expected that the staging will be in place for 9–12 months.

(a) Describe the development of two major corrosion phenomena you would expect to find in the structure.

(b) Give details of the methods normally used to control the corrosion you have described.

(c) What materials and corrosion control methods would you recommend for such a structure?

P18.37 Fig. P18.37 shows part of the superstructure (not to scale) of an offshore platform and a sea-water cooling system.

(a) List the locations in which you would expect to find corrosion.

(b) Describe the measures you would take to control the corrosion you have identified in (a).

P18.38 (a) Describe the mechanisms whereby the semipermeable nature of paints may give rise to breakdown of paint films on mild steel plates in a marine environment.

(b) The corrosion problems caused by diffusion through paint films on mild steel plates may be alleviated by the incorporation of sacrificial or inhibitive pigments in the primer. Describe the mechanisms of the corrosion protection afforded by these two classes of pigments, giving examples of each.

(c) The design for a low alloy steel shaft called for a microcracked chromium coating in preference to conventional chromium plating. Use diagrams to compare and contrast the corrosion behaviour of each coating, stating clearly the advantages the designer hoped to gain.

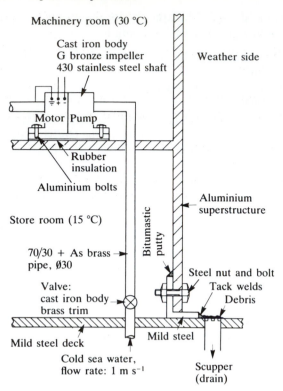

Machinery room (30 °C)

Cast iron body
G bronze impeller
430 stainless steel shaft

Weather side

Motor | Pump

Rubber
insulation

Aluminium bolts

Store room (15 °C)

Aluminium
superstructure

70/30 + As brass
pipe, Ø30

Bitumastic
putty

Valve:
cast iron body
brass trim

Steel nut and bolt
Tack welds
Debris

Mild steel deck

Mild steel

Cold sea water,
flow rate: 1 m s^{-1}

Scupper
(drain)

Fig. P18.37

P18.39 One method of protecting the hull of a ship is to use impressed current cathodic protection. For a system which uses lead/silver anodes rated at 20 A.

(a) Describe the nature, function and location of all the components.

(b) Outline the theoretical and practical principles of the design.

(c) Using the data given below, calculate the minimum number of anodes necessary to provide the protective current.

(d) The lead anodes do not dissolve. Suggest what the electrode reactions might be.

Data

Wetted surface area
    hull and appendages   = 1,390 m^2
    propellers   = 18 m^2
Current requirement for protection
    hull and appendages   = 32 mA m^{-2}
    propellers   = 540 mA m^{-2}.

P18.40 It is required to protect the hull and appendages of a ship with aluminium alloy sacrificial anodes in longitudinal arrays. The designer has specified the use of 85 sacrificial anodes, each semicylindrical of diameter 100 mm and length 400 mm. If the designer's specifications are carried out:

(a) Estimate the maximum length of time which may elapse before the spent anodes must be replaced, assuming that current requirements are just satisfied.

(b) Calculate the current output of the anodes and hence comment on the designer's specifications.

(c) Give details of any approximations or assumptions implicit in your method of calculation. Discuss the importance of any errors which may have arisen from them.

(d) Briefly describe where you would expect the sacrificial anodes to be located on the hull.

Assume that all anodes corrode at the same rate.

Data
Total wetted surface area
  hull and appendages $= 972 \text{ m}^2$
  propellers $= 14 \text{ m}^2$
Current requirement for protection
  hull and appendages $= 32 \text{ mA m}^{-2}$
  propellers $= 540 \text{ mA m}^{-2}$.
Other useful data may be found in Table 16.1.

P18.41 (a) Sketch a plan of a modern ship on which is shown the layout of an impressed current cathodic protection system. Label all the components and on the circuit diagram indicate the direction of the electron flow.

(b) Describe in detail the nature and role of each component.

(c) Over a period of nine months during the operation of the ship, problems were experienced in obtaining stability of the protection potential. After several months of high current readings it was decided to switch the system off. Upon subsequent docking and inspection, damage was found in local areas on the hull. Describe where you would expect to find this damage and suggest reasons why it might have occurred.

P18.42 Two methods of predicting the current requirement for the ICCP of a coated buried pipeline have been proposed. Method 1 assumes that the contractor is likely to apply the coating with an efficiency of 95% and that a current density of 12 mA m^{-2} is required for protection. Method 2 uses the assumption that the coating will completely cover the pipe and that a current density of less than 130 μA m^{-2} will be sufficient for a

451

pipe of diameter >760 mm. Assume that you must design a protection system for a pipeline similar to those in E18.7.

(a) Calculate the protection current density for a line of 75 km length and 1 m diameter.

(b) Bearing in mind the case history of E18.7, comment upon the suitability of each method for use in the design.

P18.43 Imagine that you have been tasked to decide upon a cathodic protection system for a new type of small ship. The choice is between (i) a commercially supplied impressed current system costing £20,000 per ship, all inclusive; (ii) a sacrificial system, fitted by labour in your own dockyard for £10,000 per ship, excluding the cost of the anodes. Rectangular zinc anodes measuring 400 mm × 100mm × 50 mm can be purchased for £45 each. They have an output of 6.5 $A m^{-2}$ and a consumption rate of 1.518×10^{-3} $m^3 (Ay)^{-1}$. Current density requirements for protection are 32 $mA m^{-2}$ for hull and fittings, and 540 $mA m^{-2}$ for propellers. The wetted surface areas are respectively 1,000 m^2 and 15 m^2. Anodes are fixed with the largest face in contact with the hull. This area can be ignored in calculations.

(a) Calculate the current needed to protect the ship.

(b) Calculate the minimum number of anodes necessary to protect the ship. Hence compare the cost of the two cathodic protection systems.

(c) How long will the zinc anodes last before replacement is necessary?

(d) Discuss the other factors which would have to be taken into consideration before you would be able to state which system should be adopted.

P18.44 Figure P18.44 shows the proposed design (not to scale) of the jacket of an offshore platform of tubular steel. The four corner legs are each of diameter 1 m, while the cross-members are of diameter 0.3 m. The bare structure is protected by twelve slender sacrificial anodes mounted on brackets at a distance 0.3 m from the outside of the legs in the locations shown in the figure. The anodes are cylindrical, 3 m in length and 0.2 m diameter. The mean current density required for the protection of steel in this application is 100 $mA m^{-2}$.

(a) Calculate the lifetime of the anodes.

(b) Discuss the selected locations of the anodes and whether you would expect them to provide effective corrosion protection.

(c) Discuss the environmental parameters which are likely to influence the performance of the cathodic protection system on this platform.

P18.45 A power plant geared its corrosion control efforts toward a policy of applying available technology to prevent corrosion, as opposed to the traditional 'chip and paint' method of corrosion control. One of the most time-consuming corrosion control tasks was the maintenance of ferrous steam valves and lagged steam piping systems. These components reach temperatures of over 500°C when the steam plant is in operation and whether the steam plant is operating or shut down, the environment is always damp with the relative humidity often at 100%. Testing showed that sprayed aluminium gave excellent high temperature corrosion protection to lagged steam valves and piping sections.

(a) Explain, giving the relevant theory, why ferrous components may be unacceptable for use at high temperatures.

(b) The Pilling–Bedworth ratio is 2.16 for iron and 1.24 for aluminium.

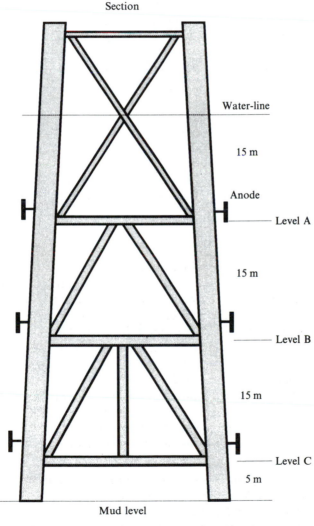

Section

Water-line

15 m

Anode

Level A

15 m

Level B

15 m

Level C

5 m

Mud level

Fig. P18.44 *continues*

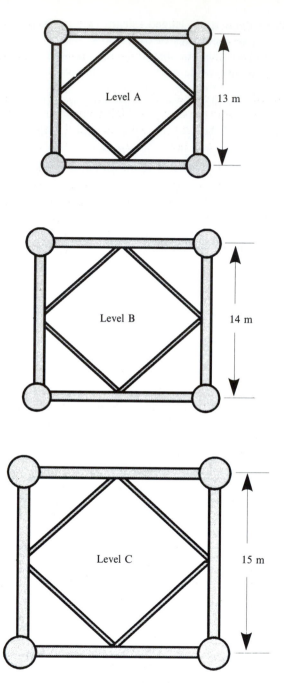

Fig. P18.44 (continued)

Define the Pilling–Bedworth ratio and explain why aluminium is more suited to high temperature use.

(c) What extra corrosion protection does the aluminium coating give the ferrous components?

You are required to investigate the performance of several metals intended for use as a spray coating for high temperature protection of steel. Describe a suitable laboratory experiment and sketch the possible results, indicating the kinetic laws that apply in each case.

(d) It was decided to replace a T-section of mild steel pipe with a stainless steel alloy containing 6% molybdenum. The section was annealed in an air furnace. After cooling in rapidly flowing air, an unusual oxidation was noticed on the surface. On cleaning away the scale, the surface was found to be extensively pitted. Explain what has happened to the component and suggest ways to avoid the problem in future.

P18.46 As design engineer in charge of a project to build an offshore platform, you decide to use HY100 steel, instead of the normal HY80. HY100 is a weldable high strength alloy with superior notch toughness and ductility. Before the design is finalised, you commission a series of fatigue crack propagation experiments using compact tension specimens in three environments: (i) dry air, (ii) sea-water using freely corroding conditions and (iii) sea-water using cathodic protection provided by zinc sacrificial anodes. Figure P18.46(a) shows the results obtained. Similar experiments were carried out using sea-water and sacrificial anode protection for weldments made by the shielded metal arc weld process (SMAW). These were compared with the results for unwelded plates; the data are shown in Fig. P18.46(b).

(a) What particular problems are associated with the experimental assessment of the susceptibility of materials to corrosion fatigue?

(b) What steps have been taken in these experiments to try to alleviate these problems?

(c) What information about the crack growth rates is yielded from the experimental data?

(d) Analyse the information you have reported in (c) and comment on the performance of both plate and weldments in the given environments.

(e) The Paris Law is often used in the analysis of such experimental data. Evaluate the constants C and m in the Paris Law, in SI units, for the set of data for the plate, freely corroding in sea-water.

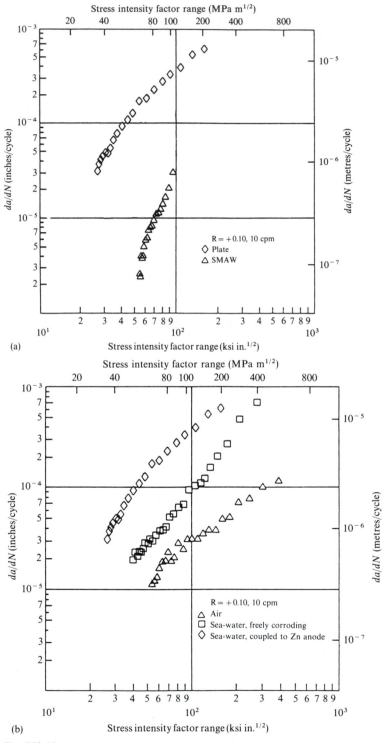

Fig. P18.46

INDEX

457